God's World Science Series

God's World Science Series

God's Wonderful World

Grade 5

with special reference
to the
Book of Psalms

Teacher's Manual

Rod and Staff Publishers, Inc.
P.O. Box 3, Hwy. 172
Crockett, Kentucky 41413
Telephone: (606) 522-4348

In Appreciation

For the greatness of God and His wonderful world, we give thanks and praise. From the biggest galaxy to the smallest atom, we see the power and wisdom of God. To inspire our children with these wonders, that they might love and fear God, is a sacred privilege and responsibility. We are thankful for the freedom, ability, and resources to have Christian schools. We are thankful for the vision to publish textbooks that exalt God and are based on the truth of His Word. We are thankful for the church to whom God has given gifts to produce this science textbook, *God's Wonderful World.*

We are grateful to God for enabling Brother Steven Mast to do the original writing. Many were involved in reviewing, classroom testing, and revising. Brother Melvin Burkholder, Seth Rudolph, and Marvin Eicher were the editors. As each did his part in response to the Lord of the church, there was a blending of efforts, for which we are thankful.

The publishing of this text does not yet meet its objective. These pages have not served their purpose until they become a tool for increasing knowledge about God's created world and, with that knowledge, inspiring the rising generation to worship and serve their Creator. As God has blessed the efforts to produce this textbook, may He further bless the teachers and students who use it.

—The Publishers

Cover photo: Bryce Canyon

Rod and Staff Publishers, Inc.

P.O. Box 3, Hwy. 172

Crockett, Kentucky 41413

Printed in U.S.A.

ISBN 978-07399-0616-3

Catalog no. 14591

5 6 7 8 9 — 25 24 23 22 21 20 19 18 17 16

Illustration Credits

Aaron Martin: 27, 96 (top right), 117, 121 (bottom right), 124, 152, 161 (both right), 164, 165 (top), 169 (top), 176 (top), 226 (top left).
Arlin Weaver: 57 (right).
Architect of the Capitol: 226 (top right).
Bureau of Reclamation, 1996 photo by Andrew Pernick: 79 (right).
Cartesia: 65 (map), 75 (right),* 84, 85 (bottom),* 96 (bottom),* 97.*
C. A. Lang, Stanford University: 156.
Comstock.com: 103 (all photos).
Corel Corporation, © 2004: front cover, 10, 13 (left), 14 (left), 18 (right, all except atom),* 32 (bottom, all except ant and butterfly), 33 (all except spider crab), 38 (both left), 39, 40 (bottom left and top right), 44 (top left and both bottom), 48, 51, 52 (all bottom), 53 (all bottom), 56 (all except alfalfa weevil and cockroach), 59 (all photos except Japanese beetle, boll weevil, and butterfly), 64, 65 (all photos), 68 (bottom left), 70, 80, 81 (all photos), 83 (left), 120 (both cars), 135,* 143 (right),* 154, 160 (all except battery),* 163 (various components),* 165 (bottom), 176 (bottom), 186 (bottom), 188 (all except whole spine), 191 (right), 193 (right),* 196 (both top), 196 (bottom, heart),* 197 (both right),* 200,* 201, 202 (right), 220 (front larynx),* 244, 245 (left), 247, 249, 261 (bottom left), 263.
Digital Vision: 89 (right), 90, 91, 92 (left).
Dover Publications, 2004 *Snow Crystals,* Wilson A. Bentley and W.J. Humphreys: 19.
Edith Burkholder: 232 (right).
ImageClub: 18 (atom), 160 (battery),* 163 (various components).*
Jennifer Balholm: 52 (top, larva and pupa), 53 (top).
Jim Brickman: 76.
Lester Miller: 11, 22, 33 (spider crab), 34, 36, 40 (all drawings), 41 (all drawings), 44 (fairy fly), 46, 47, 58, 68 (top), 69, 73, 74 (top), 78, 79 (left), 81 (Angel Falls), 85 (top), 86, 89 (left), 94, 95, 96 (top left), 101, 113 (bottom left), 115 (left), 125, 126, 128 (right), 132 (top), 133 (top right), 134, 138 (right), 139 (top right and both bottom), 140, 144, 145, 146, 147, 148 (straws), 158, 186 (drawing), 189, 191 (left), 193 (left), 197 (left), 198, 204, 205, 211 (right), 220 (all except front larynx), 221 (top), 225 (left), 226 (bottom), 231 (classroom), 236, 238, 239, 241, 253, 256 (drawing), 261 (top drawing), T–148, T–278, T–285.
Lester Showalter: 103 (top left), 104, 138 (bottom photo), 141, 159.
LifeART, © 2004 Lippincott Williams & Wilkins: 187,* 188 (whole spine), 190, 213 (top)* 220.
Marian Baltozer: 16, 130, 131 (top right), 132 (bottom), 184, 202 (smiling), 206, 209, 213 (bottom), 218, 219, 231 (right, all except classroom).
Nova Development Corporation: 45 (bottom), 52 (top, butterfly), 178 (right).
Peter Balholm: 74 (bottom).
PhotoDisc, Inc., © 2004: 38 (both right), 248 (green).
PhotoSpin, Inc., © 2004: 32 (ant and butterfly), 45 (top), 59 (butterfly).
Ruth (Goodwin) Weaver: 66.
Samuel Hoover: 12, 13 (right), 14 (right), 18 (left), 23, 24, 26, 32 (top), 40 (top left), 41 (black widow), 52 (egg), 57 (left), 59 (Japanese beetle), 68 (bottom right), 83 (right), 92 (right), 98, 102, 110, 112, 113 (all photos), 114, 115 (both photos), 118, 119, 120 (all except cars), 121 (left and top right), 122, 127, 128 (left), 131 (all photos), 133 (all photos), 136, 137, 138 (top left), 139 (top left), 143 (left), 148 (all except straws), 161 (left), 169 (bottom), 170, 172, 178 (left), 196 (bottom all except heart), 211 (left), 221 (both photos), 223, 224, 225 (right), 228, 230, 231 (strings), 232 (left), 252, 257, 258, T–147, T–292.
Samuel Hoover and William Bosco: 245 (right), 248 (brown), 255, 256 (photo), 261 (both top photos).
Thomas Eisner and Daniel Aneshansley, Cornell University: 44 (top right).
William Schueler: 72.
U.S. Department of Agriculture: 56 (alfalfa weevil and cockroach), 59 (boll weevil).
U.S. Geological Survey: 67, 71, 75 (left).

*The royalty-free drawing is modified from its original to better serve this science course.

Contents

Introduction

"Thou art the God that doest wonders" (Psalm 77:14). The study of science reveals the wonders of the Creator's works; hence the title of this science textbook, *God's Wonderful World*. May you, the Christian teacher of science, adore "the God that doest wonders" and seek to impress those wonders on the young minds entrusted to you.

Goals for the Study of Science

1. *Fifth graders should grow in their awareness of the greatness and wisdom of God as seen in the world He created.* Unit 3 discusses God's awesome power as evident in catastrophic natural events. These wonders reveal man's helplessness. Only God can control them.

Other wonders of creation are very small, yet they fill an important role. For example, God made molecules to function in an orderly way to produce many useful compounds. Myriads of microorganisms busily break down dead matter into useful soil.

2. *Fifth graders should grow in their knowledge of the facts and the order of the created world.* In Unit 1, they learn the basics of orderly, scientific study. God's wonderful world is best understood by careful observation and by consistent testing of one's conclusions.

In Unit 2, students study and classify various arthropods. They learn practical information about destructive insects and beneficial insects.

Middle grade students enjoy collecting. Science gives order and purpose to such collections. This year will be a good time to collect insects and make displays of them. Perhaps a student already has a bug collection. As a teacher, you have an excellent opportunity to direct your students into areas that could become special interests for a lifetime.

3. *Fifth graders should become acquainted with practical applications of everyday science.* Unit 4 teaches how simple machines help us in our work. Understanding principles of motion, such as inertia, can aid efficiency. In Unit 5, students learn that even a simple thing like fire is a wonder of chemical change. In Unit 7, students learn principles and uses of sound.

4. *Fifth graders should learn to be good stewards of God's gifts.* Various paragraphs remind students about the importance of conserving God-given resources.

This book also instills safety. It teaches the principle of prevention and cautions students about harmful chemicals, insects, and plants. Unit 6 teaches good care of the human body.

Remember the Textbook Theme

The theme and title of this book is *God's Wonderful World.* The many marvels of science and the numerous quotations from the Book of Psalms emphasize this theme.

Be sure your students understand the relation of their text to the Book of Psalms. You may want to give a series of morning devotional talks from Psalms or from the unit introductions to provide a background for their science study. Tell the students to watch for Scripture verses from Psalms.

The psalmist wrote much about the wonderful handiwork of God in the created world. Let these Bible truths about the natural world set a sober and uplifting tone to your classes. As you study various wonders of God's creation, be sure to give frequent credit to the Creator. We certainly do live in God's wonderful world!

Recommended Teaching Plans

Before the school year begins, work out a schedule that will allow you to complete the book in one year. The textbook contains 48 lessons and 8 tests, which calls for at least 56 class periods. If you have a 36-week school year and schedule two science classes per week, you will have 72 periods for teaching science. This allows 16 sessions for activities, additional review, and unit introductions. You could have one session for the unit introduction and one for activities, but this would not allow 2 sessions for the unit review.

A better plan may be to have two science classes per week for half a year and three science classes per week the remainder of the year, for a total of 90 sessions. This allows 4 extra class periods for each unit.

Many exercises are designed to have answers that are fairly specific and easy to grade. However, exercises of this kind do not promote exploratory thinking. Try to make the class discussions practical and challenging. Ask questions that stimulate insight that goes beyond the facts given in the book. Encourage questions and discussions about the practical applications of the lessons. Be a teacher of science and not just a textbook administrator. Let the textbook be a tool in teaching and not your master.

Provide for Frequent Review

Children learn by repetition and review. Set aside a few minutes for oral review in each class. Rivet main concepts and key vocabulary terms from previous lessons. If you review regularly, you will need to spend less time reviewing just prior to the tests.

Beginning with the second lesson, a short review is included with the lesson exercises. These may also be used as oral review. Even though the coverage of concepts and vocabulary is not thorough, just the recall of several specific facts will be valuable to keep the general subject matter fresh in the students' minds. At the end of the unit, you can use these review questions as a closed book oral review. Notice, however, that some lessons review material from previous units.

How to Use the "Activities"

Do not ignore the suggested activities in the teacher's manual and in the student's book. But do not feel obligated to do everything suggested, for nothing in the study exercises or tests is based on these activities. However, demonstrations and experiments do much to stimulate interest and broaden insight. You may want to do different activities from year to year. You can then report on the results of some activities that were used in the past, to enrich the lesson without actually doing them.

Be sure to do some activities as a class with everyone becoming involved and contributing. Other activities can be assigned to individual students, who can then give a report or demonstration.

You may want to require that each student do one activity of his choice every unit, semester, or year. The research, organization, handicraft, and public speaking involved in such work are valuable beyond the scientific understanding gained through doing the activity.

Glance over the suggested activities, and obtain the needed supplies. Materials for the activities on pages 37, 57, 71, 148, and 152 may need to be purchased from a science supplier. Below are possible sources. Get a current catalog, and place your order. You will be glad to have the supplies when needed. Home Science Tools is excellent for ordering small quantities.

Home Science Tools
546 S 18th St. W, Suite B
Billings, MT 59102
1-800-860-6272

NASCO
901 Janesville Ave.
Fort Atkinson, WI 53538-0901
1-800-558-9595

Frey Scientific
P.O. Box 8101
Mansfield, OH 44901
1-800-225-3739

Pupil's Book Introduction

Your science textbook is called *God's Wonderful World.* When we see something so well made that it astonishes us, it is wonderful to us. God's world is truly wonderful.

"Many, O LORD my God, are thy wonderful works which thou hast done, and thy thoughts which are to us-ward: they cannot be reckoned up in order unto thee: if I would declare and speak of them, they are more than can be numbered" (Psalm 40:5). The world is full of wonderful things that God made.

When we study science, we try to understand the wonderful world God created. In the first unit, you will learn how to study science in an orderly way. Then you will learn wonderful details about small things such as arthropods, molecules, and plantlike organisms. You will study the large and powerful wonders of volcanoes, earthquakes, storms, and galaxies. You will discover useful things about machines and chemicals. You will see how God made your body in a wonderful way. Understanding these things will help you to live wisely in God's wonderful world.

The lessons in this book contain many Bible verses from the Book of Psalms. Because of the many wonderful things in creation, the psalmist was strongly impressed with the greatness of God and he worshiped God in deep humility. Considering God's wondrous works can help you to honor and worship God too.

Many wonders await you as you look closely at the surrounding world. While observing, you should see the reflection of a wonderful Creator. May the Lord bless your year of studying these wonders.

Unit 1

Title Page Photo

God created the earth with much natural variety. The different landforms are only one example of this. Science is the study of these everyday wonders.

Introduction to Unit 1

God created all things. He is the origin of all truth. All His works are consistent with each other. His Word and His works of creation do not conflict. Therefore, His revealed Word helps us in our investigation of the natural world. To ignore God's Word while studying God's world is folly.

People who believe the Bible study science (God's wonders) for three primary reasons: (1) to stimulate themselves to praise God for His marvelous works, (2) to become better stewards of what God has entrusted to their care, and (3) to fortify themselves against errors that would destroy their faith in God. The people of God do not see science primarily as an effort to make the world a more comfortable place. Neither do they see it as an effort to more efficiently lay up treasure on this earth. Of course, science is useful for solving the earthly problems that we face. But science does not have the answers to life's greatest problems, and it must never be lifted to the position of a god wherein we trust. God alone deserves our trust. Science is valuable when it leads us to admire Him more and serve Him better.

Unit 1

Science, the Study of God's Created Wonders

Look around you. What do you see? Birds? Animals? Grass? Trees? Hills? The sky? Where did they come from? We learn from the Bible that God created them. "O LORD, how manifold are thy works! in wisdom hast thou made them all" (Psalm 104:24).

The things we see make us curious. We want to know more about them. As we study God's created wonders, we marvel at the beauty and order that we find. With the psalmist, we say, "When I consider thy heavens, the work of thy fingers, the moon and the stars, which thou hast ordained; what is man, that thou art mindful of him?" (Psalm 8:3, 4). The things we find cause us to praise our Maker. "O LORD our Lord, how excellent is thy name in all the earth!" (Psalm 8:9).

But God not only created the world; He also rules and controls it. He times

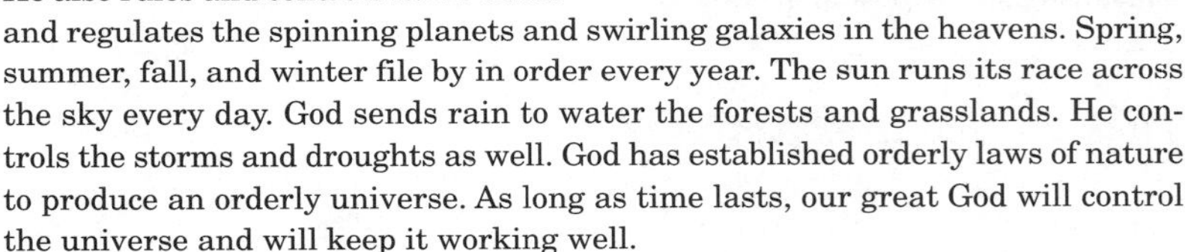

and regulates the spinning planets and swirling galaxies in the heavens. Spring, summer, fall, and winter file by in order every year. The sun runs its race across the sky every day. God sends rain to water the forests and grasslands. He controls the storms and droughts as well. God has established orderly laws of nature to produce an orderly universe. As long as time lasts, our great God will control the universe and will keep it working well.

In science class, we try to understand the wonderful world God created. This study is an exciting adventure. Many wonders await us as we look closely at the world around us; and in looking, we see the reflection of a wonderful Creator.

Story for Unit 1

The Sweet Corn Experiment

"Each of you find a hoe in the garage and come down to the garden. We want to get the sweet corn planted this evening," Mother told Philip and Henry as she headed for the garden. She was carrying several packages of seeds and a hoe for herself.

The boys arrived, breathless from their race to the garden. Father was already at work with the tiller, getting a long strip ready to plant. Mother was positioning a string between two stakes to mark the first row. "Philip, you make the rows, and Henry and I will plant the corn," Mother directed.

"But who is going to cover the rows?" Henry wondered.

"Father has almost finished tilling. Maybe he can cover the rows for us," Mother replied.

"I don't remember how far apart to plant corn seeds," Henry commented as he took the package of seeds Mother offered him.

"Oh, plant three seeds every sixteen inches," Mother answered, measuring the distance with her hands.

Steadily they planted. The loose, black soil was just right for planting corn. It made a beautiful bed for the seeds. Father finished tilling the plot and began covering the rows Mother and Henry had planted.

"You know what James told me on Monday?" Henry asked as he straightened his aching back and surveyed the row he had just planted. "He said

Lesson 1

"By the Word of the LORD"

"By the word of the LORD were the heavens made; and all the host of them by the breath of his mouth" (Psalm 33:6).

Vocabulary

Creation (krē·ā′·shən), God's act of making the entire world, including the stars, the earth, and all living things.

the Flood, the great punishment from God in the days of Noah when the whole earth was covered with water.

miracle (mir′·ə·kəl), an event that cannot be explained by natural laws.

natural law, a law that God made to control the natural world He created.

The Creation

Everything we can see or touch was created by the words of God. The Bible tells us that during the ***Creation,*** God simply spoke and the world came into being.

In six days God created the earth and everything in it as well as the sun, moon, and stars. On the very first day of time, God said, "Let there be light." Those were mighty words. Into the darkness burst forth beautiful light. On the second day, God created the air we breathe and the sky above us. During the next three days, He made the plants, the heavenly bodies, the fish, and the birds. On the sixth day, God created land animals and finally man himself.

Only God, by His word, can create a world from nothing. Imagine yourself in a cave with total darkness. What would happen if you shouted very loudly, "Let there be light"? Nothing

A poster of the Creation

they plant their corn spaced evenly instead of in clumps of three. They think it produces better that way. I think it would be much easier to plant that way, too."

"Yes, Mother, I heard that too," Philip added. "Why don't we plant some that way?"

"What do you think, Father?" Mother asked, turning to see how nearly Father had caught up with them.

Father looked up. "I don't think I ever planted sweet corn that way, but I don't see why we couldn't. Let's try a few rows and see what happens."

Father walked over to Henry. "Instead of planting three kernels every sixteen inches, plant one seed every five inches. That way we will have nearly the same number of plants in a row."

"That will be much easier, I think, than the three-kernel method," Henry said. "It's hard to flip three kernels out of my hand every time. Often I get four on a pile, and sometimes I get only two. This way I need to flip only one at a time out of my hand."

"Father, why can't we plant them closer together? Then we should get more ears of corn from each row. Do you think that would work?" Philip asked.

"We don't want to crowd the plants," Father answered. "Besides, how could we tell whether it helps to space the corn evenly if we had more plants in the one row than in the other?"

"Well, I guess we couldn't tell very well, could we? I didn't think of that," Philip admitted.

"Suppose we plant one row with three-inch spacing and see what happens," Mother suggested. "Then we'll have some rows with 16-inch spacing, some with 5-inch spacing, and one row with 3-inch spacing."

"Sounds okay to me," Father consented. "Let's just be sure to remember which planting method we used for each row."

"We'll be able to tell when the plants are up, won't we?"

"That's probably right. But we're going to have to hurry now. It will be dark in an hour."

Finally all the corn was planted, and the tools were put away. A crimson sun was slipping below the horizon as the four headed for the house to wash up.

"How will our sweet corn experiment turn out?" Henry wondered.

"We'll have to wait until harvest to see," Father replied. "Then our questions will be answered."

To the Teacher

This story illustrates how the scientific method can give answers to everyday questions. Your pupils may wonder which planting method proved to be best. Maybe they will feel the story is incomplete. But this is one of those questions about the natural world for which there is no single, correct answer. The best planting method is the one that produces the highest yield. And the only way to tell for sure is to plant a test plot and see. Not every answer can be found in the classroom encyclopedia.

Of course, a knowledge of some principles does enable us to form an educated guess, or hypothesis. When planted too close, sweet corn competes for moisture and nutrients. Narrow spacing produces many small ears, especially in infertile soils or dry weather. Wider spacing produces the large ears of sweet corn that are so desirable. Therefore, the 3-inch spacing would almost surely yield inferior ears.

But the only way to tell for sure is to try it. And the results of your test will most likely vary with soil fertility and structure, the available moisture, and the variety of corn. Test results may vary from year to year and farm to farm. Compiling the data from many trials and averaging the results should produce guidelines that will work most of the time.

While we do find the scientific method useful in the natural world, we must always reverence the Creator. He is sovereign. The success or failure of our tests is subject to His will. If we increase our yields, He deserves the credit. We may plant, we may water, but only God gives the increase.

would happen; not even the faintest glow of light would appear. Men cannot create by saying words, as God can. We must have things like matches or flashlights to make light. But that is very different from creating light out of nothing by simply speaking the word, as God did. "Let them praise the name of the LORD: for he commanded, and they were created" (Psalm 148:5).

The Flood

After the six days of Creation, time moved on for about fifteen hundred years. Many people lived in the world, and they became more and more wicked. God decided to send ***the Flood*** as a punishment on mankind. Torrents of rain fell for forty days and forty nights. The water kept rising all over the earth until it covered even the highest hills. Noah and his family were the only people who escaped. The animals with them in the ark were the only land animals that were saved. Never before had there been a flood—or even rainfall—and never again would there be a flood like this one.

Moving water has great power to cause changes. Perhaps you have seen deep gullies washed out by streams of water after a hard rain. Sometimes when rivers overflow, trees and whole buildings are washed away.

The Flood of Noah's day had much greater power than any flood you may

A hurricane flood washed this huge tree out. The trunk is nearly 8 feet across.

have seen. Much soil was washed away. As the rocks and mud settled, they formed great layers pressed tightly under tons of water. Have you ever seen layers of rock where a hill was cut away to make a road? Those layers may have been formed during the Flood.

Exactly how much the Flood changed the earth's surface, God has not told us. But many things we observe seem to point to the Flood. Deep canyons may have been cut by the rushing waters of the Flood. Rock layers and deep soil deposits were possibly formed by Flood waters. The Flood may even have caused the land to sink in some places and to rise in others.

God has promised that He will never again send a flood like the one in Noah's day. The floods we have today are quite tiny when compared

Lesson 1

Lesson Concepts

1. The natural world came into existence by the spoken word of God at Creation. (*Note:* The word *creation* is capitalized when it refers to this miraculous event. It is not capitalized when it refers to one created item or to created things collectively.)
2. God also brought about changes in the natural world by the Flood.
3. God created an orderly world that follows natural laws which He established.
4. A miracle is an event that does not follow natural laws.
5. God sometimes sets aside His natural laws to perform miracles.
6. Both the Creation and the Flood are miracles beyond man's ability to understand or explain.

Teaching the Lesson

This first lesson establishes the basis for science: "In the beginning God created the heaven and the earth" (Genesis 1:1). The Creation story should be familiar to the students. Help them to see the connection between the Creation and the study of science. Everything around us (grass, insects, water, hills—yes, everything) is ultimately a result of the Creation. Even laws of nature, such as gravity and inertia, were established at the Creation. Evil, of course, is not a creation of God, but is the result of rebellion against God.

This lesson focuses on two miracles: the Creation and the Flood. Help the students to understand what miracles are. We can define them quite simply as the doings of God that supersede natural laws. Do not try to explain them naturally, for they have no natural explanations. Point the students toward God.

This canyon cuts through many rock layers.

A road cut showing layers of rock

with that one. We cannot explain exactly what happened in the Flood or how God could make a worldwide flood. But we do know there was such a flood because the Bible says so.

Ungodly men have tried to explain how the earth became as it is today. They say that it all just happened in natural ways and that God had little or nothing to do with it. This false teaching is called evolution. But since the Bible tells us about the Creation and the Flood, we know it was God who made the earth as it is today. In the beginning, He created everything in six days. Later, He sent a yearlong flood that drowned all the wicked people and made great changes on the earth.

Natural Laws

God is a God of order. The earth that He created is also very orderly. "And God saw every thing that he had made, and, behold, it was very good" (Genesis 1:31). God made ***natural laws*** to produce order in the earth. With these natural laws, God controls the whole universe. These laws exist wherever you go.

One natural law says that the earth's gravity pulls things toward the earth. When you hold out your pencil and let it go, it falls downward, not upward. Another natural law says that living things always reproduce after their kind. Cats always produce kittens, never puppies. Corn plants always produce corn seeds, never bean seeds.

Discussion Questions

1. How did God create the world?
   Stimulate the pupils' thinking. Lead them to the fact that the question does not have a natural explanation. The creation of the world was a miracle. God spoke, and there it was.
2. Have you ever seen proof of Noah's Flood?
   Help them to understand that the hills they see today are possible proof of the Flood. Particularly point out the strata in mountains that were probably formed as the Flood waters receded.
3. Do miracles happen today?
   While this question is somewhat outside the scope of the lesson, help the pupils to see the miraculous hand of God in such things as healing, answers to prayer, and even life. (What is life? Where does it go when something dies?) Help them to see the hand of God in daily life.

"Thy faithfulness is unto all generations: thou hast established the earth, and it abideth. They continue this day according to thine ordinances: for all are thy servants" (Psalm 119:90, 91). The word *ordinance* is another word for *law*. The earth remains for us because God made natural laws to keep things in order. Man cannot change the laws that God has created to control the universe.

But sometimes God performs deeds that do not follow natural laws. Jesus walked on the water. This was an exception to the natural law of gravity. Elijah prayed, and God sent fire that burned the wet sacrifice, the stones of the altar, and even the water in the trench. In both cases, things happened that cannot be explained by natural laws. Such an event is called a ***miracle.*** Both the Creation and the Flood are miracles that man cannot completely understand or explain.

Living things always reproduce after their kind.

Lesson 1 Answers

Exercises

1. "by the word of the Lord"
2. nothing
3. to punish men for their wickedness
4. a, c, d

Exercises

1. How does Psalm 33:6 say the heavens came to be?
2. In the Creation, God made everything out of ———.
3. Why did God send the worldwide Flood upon the earth He had created?
4. What terrible effects did the Flood bring upon the earth? Choose all the correct answers.
   a. All people except Noah's family drowned in the water.
   b. Everyone got wet when the water rose.
   c. The rushing water caused great destruction.
   d. Many plants and animals perished in the Flood.

5. Which of the following are proofs of the Flood? Choose all the correct answers.
   a. natural laws c. soil and rock layers
   b. canyons d. plant life
6. The floods we have today are
   a. sometimes nearly as large as the Flood.
   b. seldom large enough to do much damage.
   c. small in comparison with the Flood.
   d. sometimes more destructive than the Flood.
7. When scientists disagree with God's Word about how the earth came to be, whom should we believe?
8. God made natural laws
   a. to teach us how to live.
   b. to produce order in the earth.
   c. so that He could do miracles.
   d. so that we cannot understand how things work.
9. When God does something that is impossible according to natural laws, we say that He performs a ———.

5. b, c

6. c

7. God (the Bible)

8. b

9. miracle

Activities

1. Mix one-half cup fine white sand, one-half cup coarse brown sand, one-half cup potting soil, and one-half cup loam (ordinary garden soil). Fill a quart jar half full with this mixture. Add water until the jar is nearly full. Screw a tight-fitting lid onto the jar, and shake it well. Shake it until the sand, water, and soil are well mixed. Then set the jar aside, and let the mixture settle for a few hours. Can you see the layers that are formed?

   That is a little like what happened during the Flood. The water stirred up great masses of dirt and rock. After the water calmed, the dirt settled out and formed layers.
2. During the Flood, millions of animals and shellfish as well as much plant matter were quickly buried deep in mud. Under the great pressure of tons and tons of water, the mud eventually hardened into layered rocks, and the trapped animals and plants became fossils. Perhaps you live in an area where many of these fossils are found. If so, you and your classmates can make an interesting display of rocks formed from real "Flood mud," with fossils of creatures that lived while Noah walked the earth.

Activities

Help your students with some of the activities described in the pupil's book. Hands-on experience helps them to understand the concepts in focus.

Lesson 2

"Come and See the Works of God"

"Come and see the works of God: he is terrible in his doing toward the children of men" (Psalm 66:5).

Vocabulary

astronomy (ə·stron′·ə·mē), the branch of science that deals with stars and other heavenly bodies.

biology (bī·ol′·ə·jē), the branch of science that deals with living things.

chemistry (kem′·i·strē), the branch of science that deals with the makeup of materials and their reactions to each other.

observe, to watch or study carefully.

physics (fiz′·iks), the branch of science that deals with force and energy and with the effect of energy on matter.

science, the observation and study of God's wonderful world.

What Is Science?

Science is the study of God's wonderful world. God wants us to learn about the world He made. He has given us five senses so that we can ***observe*** our world. Our five senses are seeing, hearing, touching, smelling, and tasting. These senses make life more enjoyable and enable us to observe the things around us. We can *hear* the birds singing. We can *see* and *smell* an apple pie and then *taste* its delicious flavor as we eat it. We can *touch* an ice cube and feel how cold it is. Without our five senses, we could know nothing about the world around us.

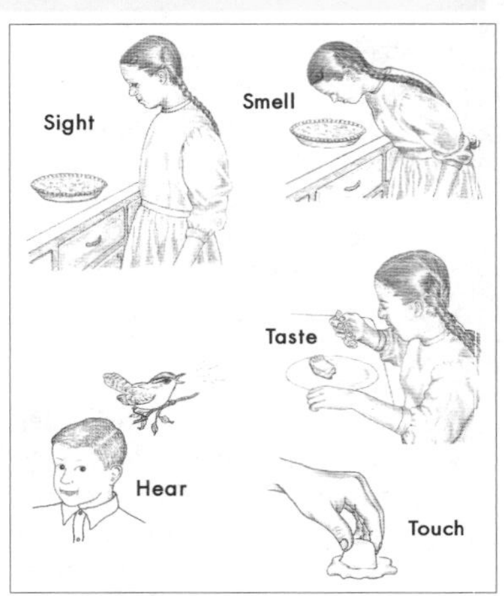

The five senses help us to observe the world around us.

Lesson 2

Lesson Concepts

1. God gave man five senses, along with the intelligence needed, to learn things about the natural world.
2. Science is man's effort to learn about the natural world.
3. The proper study of science will begin with God and inspire us to worship Him. (This is the spiritual benefit of studying science.)
4. The proper study of science provides us with practical knowledge about the natural world and thus is useful in carrying out our God-given responsibilities. (This is the practical benefit of studying science.)
5. Science has been conveniently divided into various branches.
6. The natural world is full of displays of God's wonder-working power.

Teaching the Lesson

Do your students really know what science is? Science is man's attempt to learn about the world we live in. Since God both made this world and placed us in it, His praises belong in every science class. Science is the study of His wonderful world.

God has given us the faculties we need to make observations. (Help the children to understand that their five senses are a gift from God.) He has also given us the ability to think logically so that we can learn from our observations and better serve and glorify Him in this life.

We live in a real world of practical science. The more you bring the real world into science class, the more practical science will become for the students. Do you live in a farming community? Adapt your science classes to that environment. Are you surrounded by industry and the trades? Bring the real world of the

God has also given us the ability to think. With the intelligence that He has given, we can think about the observations we make and recognize the orderly laws that God has created to control the universe.

Take an electric fence as an example. If we observe an electric fence with our sense of touch, we receive a shock. We can repeat this observation again and again. But after several hard shocks, we come to the conclusion that electric fences are unpleasant and even dangerous.

By observing and thinking, we have learned something. We have decided that if we do not want to be shocked, we must not touch the fence. This is something valuable to know about the world in which we live. Much of what men know about God's world has been learned in the same way: by observing and thinking.

Why Study Science?

The study of science revolves around God Himself. God created the whole universe, along with all the laws that control the universe. As we explore the great wonders of creation, our studies inspire us with praise and awe for the great God we serve.

Many of the psalms in the Bible praise God for what we study in science. "Praise ye him, sun and moon: praise him, all ye stars of light. Praise him, ye heavens of heavens, and ye waters that be above the heavens.... Let them praise the name of the LORD: for his name alone is excellent; his glory is above the earth and heaven" (Psalm 148:3, 4, 13). We study science to praise God for the wonders of His creation.

Science also provides us with knowledge that is useful in everyday life. The Lord told man to "have dominion" over the world He created. To do this properly, man must learn how the world operates. We do not study science to become rich or to lift ourselves up in pride. Instead, we study science to be able to do what God has commanded. We study science to honor God and to help us know how to take care of the world He created.

What Does Science Include?

Science is a broad subject that deals with all the natural wonders God created. In fact, the study of science is so vast that it has been divided into many smaller subjects, called branches of science. Like the branches of a tree, the many branches of science grow out of one main trunk—science.

Each branch of science includes a certain area of knowledge. The branch called ***biology*** is the study of living things. The study of what things are made of and how different materials react to each other is the branch called ***chemistry.***

Astronomy, yet another branch, is

students into science class. It will do wonders for their interest and retention. After all, science is practical. Be sure to keep it practical for the students.

Discussion Questions

1. What are your five senses?
   seeing, hearing, touching, smelling, tasting
2. What areas does science include?
   This question can be answered only in a general way. For the sake of interest, steer the discussion toward practical, everyday areas about which the children are knowledgeable.

The dome roof of an observatory gives astronomers a broad view of the night sky.

the study of the stars, planets, and other heavenly bodies. ***Physics*** deals with the laws of motion and energy. It includes the study of levers, pulleys, and gears. Science has many more branches than we are able to consider in this lesson.

Many of these branches break down into even smaller branches. Biology, for example, can be further divided into botany (the study of plants), zoology (the study of animals), anatomy (the study of the human body), and a number of others.

God's Wonderful World

We can see God's handiwork all around us. One example is snow, which is a much greater marvel than you may realize. Snow will melt into water if it is warmed, for a snowflake is made of frozen water. It forms in the clouds when water vapor freezes into tiny water crystals. As these crystals fall through the cold, moist air, they pick up more moisture and the lacy shapes grow larger.

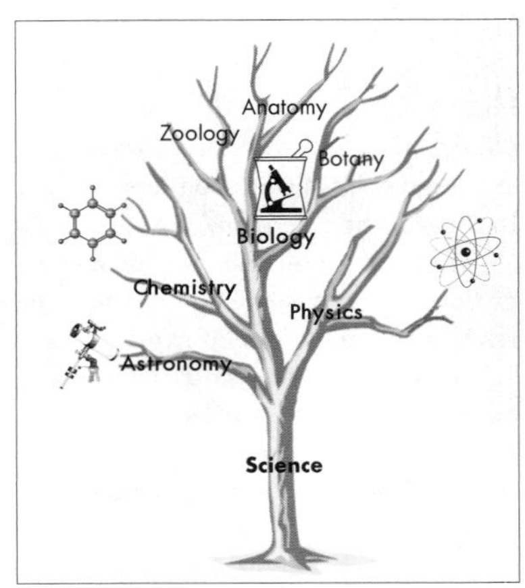

The study of science has many different branches of knowledge.

Snowflakes generally have six sides or six points. Yet even though they all have the same basic shape, no two snowflakes look exactly alike. A man named W. A. Bentley studied snowflakes under a microscope. He published a book that has about three thousand photographs of snowflakes. Each snowflake is distinctly different from all the rest. Sometime when you are out in the fresh, falling snow, catch a few flakes on your coat sleeve and examine them closely. You should be able to see their six sides and their delicate design.

The natural world is filled with such marvels just waiting to be explored. This year you will continue studying the world that God created and learning how to take good care of what God has given to us.

The delicate design of snowflakes shows God's wonderful handiwork.

Exercises

1. Write a good definition of science.
2. God has given us five senses to ——— the natural world, and the ability to ——— so that we can learn from our experiences.
3. Why should we study science? Choose all the correct answers.
   a. to be able to solve some of our everyday problems
   b. to learn more about God's creation
   c. to help us to get rich
   d. to honor and glorify God
   e. to help us care for God's world as good stewards
   f. to impress others with what we can do
4. Name the correct branch of science for each description below. Choose from these words: *astronomy, biology, chemistry, physics.*
   a. the study of what things are made of
   b. the study of how a bicycle can change speeds
   c. the study of the heavenly bodies
   d. the study of living things
   e. the study of how to make rubber
   f. the study of lizards
   g. the study of force and energy
   h. the study of why the moon appears to change its shape
5. True or false? We need to look closely before we can notice any proofs of the Creation.

Lesson 2 Answers

Exercises

1. Science is the observation and study of God's wonderful world.
2. observe, think
3. a, b, d, e
4. a. chemistry
   b. physics
   c. astronomy
   d. biology
   e. chemistry
   f. biology
   g. physics
   h. astronomy
5. false

Review

6. the Creation and the Flood
7. six, nothing, word
8. (Any two.)
   The Flood caused soil and rock layers to form.
   Rushing water cut deep canyons.
   Some lands rose and others sank.
9. A miracle is a happening that cannot be explained naturally. God can do miracles because He can set aside His natural laws if He so chooses.
10. gravity;
    reproduction according to kind

Activities

1. (Sample answers.) botany, zoology, physiology, biochemistry, anatomy, pathology, taxonomy, ecology, psychology, biomathematics, genetics, bacteriology, morphology

Review

6. What two miraculous events did God use to make the earth as it is today?
7. The Creation is truly a marvelous work of God. In ——— days, God created an entire universe out of ——— by simply speaking the ———.
8. Give two ways in which the Flood changed the surface of the earth.
9. What is a miracle?
10. Give one example of a natural law.

Activities

1. The study of God's world is so detailed that the branches of science given in this lesson may each be further divided. For instance, biology is the study of living things. Using an encyclopedia, find and write at least eight subdivisions of the science of biology.
2. Find an object near you to observe. Write a short description of that object, showing that you observed well. Use as many of your five senses as possible.
3. Your skin is a sense organ. With it, you can feel pain, heat or cold, and roughness or smoothness. Are all parts of the skin equally sensitive? Or has God designed some parts to be more sensitive than others?

   To test your skin, you will need the following items.
   (1) A helper. One of your classmates will do.
   (2) A short, thin piece of bent wire. A hairpin or a compass will do.
   (3) Two charts like the one on the next page to fill in as you do the test.

   Use the following procedure.
   a. Tell your helper to close his eyes for the testing.
   b. Set the points of your wire or compass 1 inch apart. (Measure them with a ruler.)
   c. Touch the points lightly to your helper's hand and arm at the locations listed on the chart. Try to touch both points to the skin at the same time.
   d. Fill in the chart with a *1* or a *2* to tell whether your helper felt one point or two.
   e. Continue until the chart is completely filled in.
   f. Next, trade places, and use the other chart to have your helper test your skin.

What do you think? Is the skin equally sensitive at all places? Of the three locations you tested, which was most sensitive? Which was least sensitive?

Distance of Points	Back of Forearm	Hollow of Palm	Tip of Finger
1 inch			
½ inch			
¼ inch			
⅛ inch			
1/16 inch			

4. Though your five senses work very well, God's wonderful world includes millions of creatures that are too small to see without assistance. To observe these tiny creatures, scientists have developed the microscope, a tool that magnifies things hundreds of times or more.

   If you have a microscope, try raising some tiny creatures called protozoa. Pour 3 cups of pond water into a quart jar. Try to include a bit of pond scum or pond weeds. Add a dozen grains of rice, and let the mixture sit for a week in a dark, warm location. Then take a medicine dropper and remove a bit of the scum from the surface of the water. Put this scum on a microscope slide, and view it under a microscope at 100 power. You will observe some of God's creatures that you probably have never seen before.

4. Use these microscopic creatures to demonstrate God's marvelous works to your pupils. **Be sure to supervise the use of the microscope. Otherwise, the children might not see anything, and the microscope could be damaged.**

Lesson 3

The Scientific Experiment

Vocabulary

data (dā′·tə *or* dat′·ə), information collected and organized so that it can be used to make a decision.

experiment (ik·sper′·ə·mənt), a test made to discover new truths about God's creation or to show how something works.

observation, the use of the five senses to study God's creation.

superstition (sü′·pər·stish′·ən), a false idea about why something happens, which is based on fear or ignorance rather than fact.

Superstition

As we observe the world we live in, we find joy in remembering that the One who created this wonderful place is our Father. He cares for us and talks to us in His Word, the Bible. When people who do not trust God or believe His Word observe God's world, their ignorance and fear often lead them to form ideas that are not true. These false ideas are ***superstitions.*** Superstitions do not agree with God's Word or the natural laws God made.

Superstitious people have wrong ideas about why things work. For example, some people think that toads cause warts. They see all the little knobs on a toad's skin and decide that touching a toad will give them warts. And they are afraid of toads because they do not want to get warts. But the idea that toads cause warts is just a superstition. Careful observation shows that people do not get warts from toads.

The Scientific Experiment

Often in our ***observation*** of God's world, we see something that puzzles us. We wonder, "How does this work?" or "What would happen if I should do that?" Sometimes just a little thinking will give us our answer. But at other times, we may need to set up a special

Lesson 3

Lesson Concepts

1. Superstitions result from the poor observations and the fearfulness of those who do not trust God.
2. To draw correct conclusions, we must observe carefully and record accurate data.
3. Scientific experiments should be used to find truth about natural laws, but not about moral and spiritual laws.
4. If the Bible and science seem to conflict, we always believe the Bible.

Teaching the Lesson

What is the practical value of science? Help your students to understand that the scientific method can be used to solve everyday problems. A farmer experiments with different types of feed to try to improve the productivity of his cattle. A housewife experiments to discover how to make better cakes. Even children experiment. Perhaps a boy would experiment to see whether he can haul gravel faster with a bucket or a wheelbarrow. Every time you test something to see how it will turn out, you are doing a sort of experiment. An orderly approach to an experiment generally yields more accurate results.

The key word in today's lesson is *observation.* Accurate observation is an art the child must be taught. By default, we tend to miss many things that are quite close to us. Teach your students to observe their surroundings carefully. Careful observation is a key to arriving at sound conclusions.

Many wrong ideas about God's world are accepted by careless observers. A scientific attitude is not as important to our children as faith is, but the habits of observing carefully, thinking logically, and refusing to form hasty conclusions are as valuable to Christians as to scientists.

How would a careful observer think about the toad/wart question mentioned in the lesson text? He may not take time to do thorough tests on whether toads cause warts, but he would certainly withhold judgment until someone has. And he would realize the difficulty of establishing the truth of the matter. For an enlightening class exercise, you may wish to discuss what sort of experiment would be required to answer the toad/wart question. How would a person go about it?

We could choose one person to touch a toad and see if he gets warts. That is the right idea, but it is not test so that we can observe more closely. This test is called an ***experiment.***

We follow four steps when we experiment.

Step 1: Write down the question you are trying to answer.

It is a good idea to write down the question to keep it clearly before you as you try to answer it.

Step 2: Plan a special test (an experiment) that will help to answer the question.

Part of planning is deciding what information to record as you experiment, and preparing a sheet to write your information on. The information that you gather is called ***data,*** and the sheet you write it on is called a data sheet.

Step 3: Work the experiment, recording your observations on a data sheet.

Step 4: Use the information on the data sheet to answer your question.

An Experiment

To see how these four steps work, let's try an experiment. What shall we investigate? Perhaps in driving through town one day, you observed a tall, silo-like water tower. It is easy to understand why the water is needed. But why is the tank standing up on end? Why is it not buried in the ground or lying on the ground, where it would be easier to fill? Does having the water level high above the ground make the water flow out with greater speed and pressure? That would be important in case of a fire, and it would be convenient at all times. Let's look for some answers. To find them, we will follow our four steps.

A town water tower

Step 1: Write down the question you are trying to answer. Here is the question for our present experiment: *If a water tank has an outlet near ground level, does the water flow from the outlet with more pressure if the water level in the tank is high? or if it is low? Or does the water level make no difference?*

Step 2: Plan an experiment. Of course, we cannot experiment with a real water tower in the classroom. But we can bring the problem down to a workable size by using a small water tower—a clear plastic bottle. For an outlet, we will drill a 1/16-inch hole about ½ inch from the bottom of the bottle and put a piece of tape over it so the bottle holds water.

To keep an accurate record of the water level, we will measure up from the outlet hole and draw marks one inch apart up the side of the bottle.

really accurate enough. The person may be resistant to warts, or perhaps our toad's wart-powers may be unusually low! For a better test, we could have 100 different persons touch 100 different toads. On our data sheet, we could record each person's name, the date he touched the toad, and which hand he used. And we would record when and where the warts develop—if he gets them, of course.

To be really sure of our results, we would need to choose 100 other persons who never touch a toad at all, and observe how many warts they get. This would be the control group.

After doing such a test, we should have a fairly good idea of whether toads cause warts. Of course, we would need to observe for several months or maybe even a year, and it may be good if our toad touchers would touch their toads every day for the first week or so of the test.

Unfortunately, such an experiment requires too many persons, too many toads, and too much time for a fifth grade science class. However, it is something useful to think about. Superstitions begin because some person sees a toad and notices its warty skin. He touches it or picks it up, maybe shivering a little at its ugly appearance. A week later, he gets a wart. "Ah," he thinks, "toads must cause warts." Actually, he has no basis for such a conclusion until he has performed an experiment like the one described above.

Fifth graders may be a bit young to catch the full impact of the limitations of the scientific method, but it is good to expose them to these limitations: (1) science does not have all the answers, and (2) we must experiment only with natural things and never with spiritual things. God's revealed Word is our final authority. We do not experiment with Bible-taught facts, but we accept them by faith.

Next, we will put a three-inch block of wood on a flat, level surface to set the bottle on. Then we will place the end of a ruler against it to measure how far the water squirts out the side of the bottle.

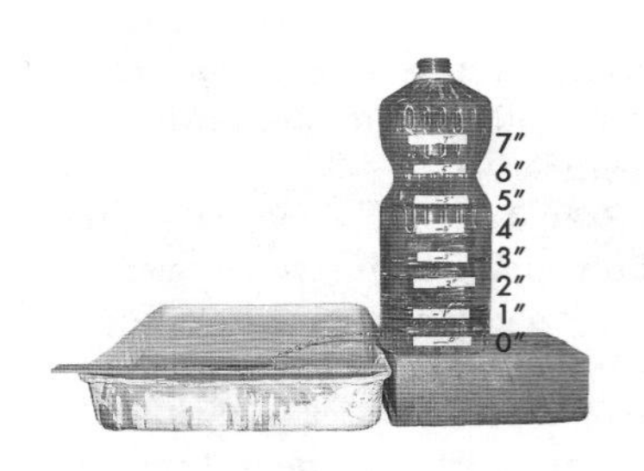

Before we begin, we will need a data sheet to record the data we gather from the experiment. So we will make a chart like the one below.

	Water Level in Bottle				
	7 in.	6 in.	5 in.	4 in.	3 in.
Test 1					
Test 2					
Test 3					

Step 3: Work the experiment. We are now ready to work the experiment. We fill the bottle up to the 7-inch mark and leave the cap off. We pull the tape off the hole to let the water shoot out on to the ruler. Then we must observe how far out the water squirts, and record the information on our data sheet. We must record the distance each time the water is exactly level with one of the inch marks on the side of the bottle. The chart below shows data collected from such an experiment. Repeating the experiment three times helps us make sure the data are accurate.

	Water Level in Bottle				
	7 in.	6 in.	5 in.	4 in.	3 in.
Test 1	$7\frac{1}{4}$ in.	$6\frac{1}{4}$ in.	$5\frac{1}{4}$ in.	$4\frac{3}{8}$ in.	$3\frac{3}{8}$ in.
Test 2	$7\frac{1}{8}$ in.	$6\frac{1}{4}$ in.	$5\frac{1}{4}$ in.	$4\frac{3}{8}$ in.	$3\frac{1}{8}$ in.
Test 3	7 in.	$6\frac{1}{8}$ in.	$5\frac{1}{4}$ in.	$4\frac{1}{4}$ in.	$2\frac{7}{8}$ in.

Discussion Questions

1. This lesson mentions the superstition that toads cause warts. What other superstitions do you know about?

   (Handle this discussion carefully. The purpose here is not to introduce superstitions, but to eliminate those that your pupils may believe. Sample answers are given.)

   If a black cat crosses your path, something bad will happen.

   Bad things happen on a Friday that is the thirteenth day of the month.

   A rabbit's foot will keep you from harm.

   Wishes made on falling stars come true.

2. What is the difference between seeing and observing?

   When you see something, you do not necessarily think about it. When you observe something, you study it closely and try to understand what you are seeing.

3. Discuss the value of recording the data. Is it important to record data, or could we observe just as well without writing anything down?

   If memories were infallible, perhaps we would not need written records. These records are also needed to show other people the results of an experiment.

Step 4: Use the data to answer your question. Now that we have the data from the experiment, we are ready to answer our question. Study the data chart above. Do you know the answer to the original question? *If a water tank has an outlet near ground level, does the water flow from the outlet with more pressure if the water level in the tank is high? or if it is low? Or does the water level make no difference?*

Limitations of Scientific Experiments

We experiment frequently. Father experiments to discover what kind of gasoline makes his car run best. Mother experiments to discover how to make better bread. Jerry experiments to see which kind of sled will slide best in deep snow. We have many questions that experiments help us to answer.

However, we must never question what the Bible says. The Bible is always true. If the Bible says something is true, we need no experiment to prove that it is true. Neither will any experiment ever show the Bible to be wrong. But what about cases where men say the Bible is untrue? For instance, the Bible says that God created the world, but many scientists say the world was formed in some other way. We know these scientists are not telling the truth. Whenever the Bible and scientists disagree, we know that the Bible is true and that the scientists are in error. God's Word is always the highest authority.

Exercises

Use vocabulary words to answer numbers 1–4.

1. A ——— is a false idea about the world, which people form through ignorance and fear.
2. When we want to test something to find out how it works, we often do an ———.
3. The information we gather when we do an experiment is called ———.
4. During an experiment, we do careful ——— to gather the data we need.
5. Write the four steps for doing an experiment. Be ready to write them from memory.
6. Think about the experiment with the water bottle.
   a. As the water level dropped, what change took place in the distance that the water squirted?
   b. Does this show that the pressure decreased? or that it stayed the same?
7. a. What scientists say is (always, sometimes, never) true.
   b. What God says is (always, sometimes, never) true.

Lesson 3 Answers

Exercises

1. superstition
2. experiment
3. data
4. observation
5. (Quiz the students to see if they can write these steps from memory.)
   (1) Write down the question you are trying to answer.
   (2) Plan an experiment that will help to answer the question.
   (3) Work the experiment, recording your observations on a data sheet.
   (4) Use the information on the data sheet to answer your question.
6. a. The distance became smaller.
   b. It shows that the water pressure decreased.
7. a. sometimes
   b. always

8. a. yes
   b. no
   c. no
   d. yes
   e. yes
   f. no

Review

9. Science is the study of God's wonderful world.
10. nothing, six
11. a. physics
    b. chemistry
    c. astronomy
    d. biology
12. Flood
13. (Any two.)
    Science leads us to praise God for His great creation.
    Science helps us to solve everyday problems in life.
    Science helps us to be good stewards of God's world.

Activities

Try one of the experiments in today's lesson yourself before you get to class. This will acquaint you with how the experiment works and will alert you to any potential problems. Being prepared for such problems can save you frustration during class time. It will also reduce the amount of time needed to do the experiment.

8. We must never do experiments to test God's words. Write *yes* for the questions that would be all right to test by doing experiments. Write *no* for the ones that should not be tested.
   a. What kind of gasoline provides the best fuel mileage?
   b. How well does a garden produce if it is planted on Sunday?
   c. How fast will this car go?
   d. Which pair of shoes will last the longest?
   e. How well does fiberglass insulation burn?
   f. How long will a man's hair grow if it is not cut?

Review

9. What is science?
10. God created the world out of ——— in ——— days.
11. Name the branches of science that deal with the following things.
    a. the study of force and energy
    b. the study of materials
    c. the study of the heavenly bodies
    d. the study of living things
12. Many of the rock layers, canyons, and mountains we see today were probably formed during the ———.
13. Give two reasons why we study science.

Activities

1. In the experiment you did in this lesson, you learned that water pressure is greater when the water level is higher. Here is a related question for you: Does increasing the size of the bottle also increase the water pressure? Suppose you were to try the experiment in the lesson again, keeping the water levels the same as before, but this time using a much larger bottle or even a bucket. Would the larger container cause more pressure? Would it cause less pressure? Or would it make no difference?

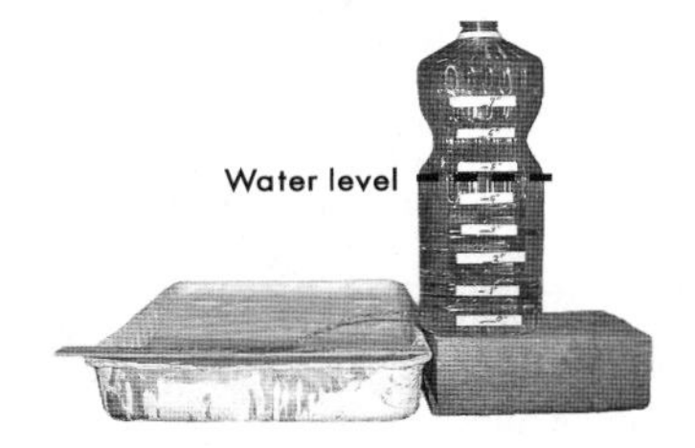

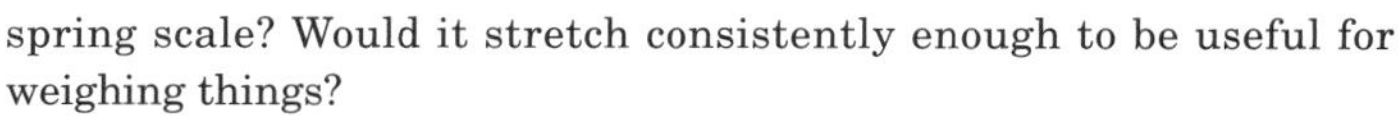

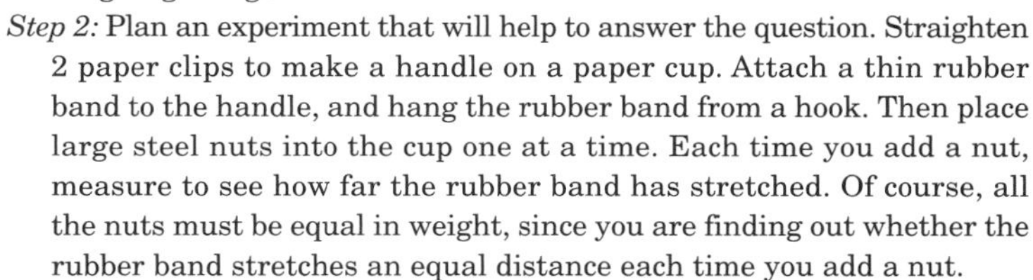

2. As you probably know, many scales for measuring weight use a spring to tell how heavy something is. The spring scale works because if 1 pound stretches a spring ½ inch, 2 pounds will stretch the spring 1 inch. A weight three times as heavy will stretch the spring three times as far.

   Step 1: Observing a spring scale raises a question: Could a rubber band be used to make a spring scale? Would it stretch consistently enough to be useful for weighing things?

   Step 2: Plan an experiment that will help to answer the question. Straighten 2 paper clips to make a handle on a paper cup. Attach a thin rubber band to the handle, and hang the rubber band from a hook. Then place large steel nuts into the cup one at a time. Each time you add a nut, measure to see how far the rubber band has stretched. Of course, all the nuts must be equal in weight, since you are finding out whether the rubber band stretches an equal distance each time you add a nut.

   Part of planning an experiment is making a data sheet. On your data sheet, you will need spaces to record how many nuts you put into the cup, and spaces to record how far the rubber band stretched each time.

   Step 3: Work the experiment, recording your observations on the data sheet. If your rubber band does not stretch when you add the nuts, you should use either a thinner rubber band or heavier nuts.

   Step 4: Use the information on your data sheet to answer your question. What do you think? Does a rubber band stretch consistently enough to use it for a spring scale? Why or why not? What possible problems do you see?

3. Plan experiments that will help to answer the following questions.
   a. Is honey a satisfactory substitute for the sugar in a chocolate cake?
   b. Which can stop more quickly—a bicycle with hand brakes? or a bicycle with a coaster brake?

3. a. Bake two cakes, one with honey and one with sugar. Be sure to keep all other ingredients the same, use the same oven, and have the same person mix both. And try it more than once.
   b. Be sure to have two bicycles of the same size and with tires of the same type and condition. If the tire or wheel size is different, different amounts of rubber will contact the road.

Have the same person ride both bikes, or else have two riders of the same weight. Weight affects stopping time.

Do all testing on the same stretch of road, under the same conditions. The speed of the bicycles must also be the same.

Do the experiment several times. Switching the two riders for different tests may eliminate some variables.

If you are deciding only between the brakes of two bicycles, the above procedure will be sufficient. If you are trying to establish a general rule, you will need several bicycles of each type.

Lesson 4

Unit 1 Review

Review of Vocabulary

astronomy	data	natural law	physics
biology	experiment	observation	science
chemistry	the Flood	observe	superstition
Creation	miracle		

The first story we read in the Bible is the account of the _1_. This account tells how God made the world and everything in it. A _2_ is a rule that God set up to provide order in creation. Man cannot change such a law; only God can do that. When God performs a deed that does not follow natural laws, we say He performs a _3_. One example of this is _4_ that God used to punish sinful people.

The study of God's creation is called _5_. There are various branches of study to help us understand the different parts of nature. The study of the stars and other heavenly bodies is called _6_. Pulleys, levers, and other machines are studied in _7_, and living things are studied in _8_. We learn how different materials react with each other in the branch called _9_. If we carefully _10_ the things around us, we shall learn many lessons about God's wonderful created world.

Living on God's wonderful earth, we find many things to observe and study. Sometimes we do an _11_ to help us understand how something works. We must use careful _12_ with our five senses so that we can gather accurate _13_ to answer our questions. We must also be sure that all our ideas agree with God's Word. Through fear and ignorance, some people believe things that disagree with God's Word or with natural laws. Such an idea is a _14_.

Multiple Choice

1. The universe and everything in it came into existence
   a. in six weeks' time.
   b. in one instant.
   c. in six days.
   d. in thousands of years.

Lesson 4 Answers

Review of Vocabulary

1. Creation
2. natural law
3. miracle
4. the Flood
5. science
6. astronomy
7. physics
8. biology
9. chemistry
10. observe
11. experiment
12. observation
13. data
14. superstition

Multiple Choice

1. c

Lesson 4

The review exercises are designed to clinch the concepts of the unit and prepare students for the unit test. You can assign the two review sections as one assignment, or you can separate them and use the extra time for drilling the vocabulary words and reviewing the concepts.

Do not assume that your students have fully mastered the concepts just because they have completed all the lessons. Children tend to forget even what was carefully drilled. Use these review exercises as one more opportunity to rivet the concepts being taught.

2. God sent the Flood
   a. because the world needed to be cleaned.
   b. because there were too many people on the earth.
   c. because He needed to punish men for their sinfulness.
   d. because He no longer cared about mankind.

2. c

3. The great Flood of Noah's day
   a. made slight changes in most of the earth's surface.
   b. made great changes in a small part of the earth's surface.
   c. destroyed all living things but did not change the earth's surface.
   d. most likely produced canyons, rock layers, and possibly even mountains.

3. d

4. Nature is orderly because
   a. God created it that way.
   b. it slowly became better over the years.
   c. scientists have learned much about it and have improved it in many ways.
   d. the Flood made it that way.

4. a

5. The laws that God created to control nature
   a. can be changed by man.
   b. change frequently.
   c. can be changed only by a miracle.
   d. do not affect people.

5. c

6. Which of the following is an example of a natural law?
   a. The speed limit is 55 miles per hour.
   b. Stop, look, and listen before crossing a busy highway.
   c. "Be sure your sin will find you out."
   d. Gravity pulls objects toward the earth.

6. d

7. The most important reason for studying science is
   a. to find enjoyment in life.
   b. to discover ways to save time and money.
   c. to learn how to help others.
   d. to learn to praise the Creator and to care for His creation.

7. d

8. Science is divided into many different branches because
   a. everyone has different ideas about how things work.
   b. some people believe God created the world and others believe in evolution.
   c. there are so many different things to study in the created world.
   d. people are interested in different things.

8. c

9. b
10. c
11. d
12. d
13. c

9. When we study ordinary things around us,
   a. we find that very few of them are really interesting.
   b. we learn to appreciate the wonderful world God created for us.
   c. we seldom find anything that we did not know before.
   d. we discover how intelligent men really are.
10. Superstitions
   a. agree with the Bible but not with natural laws.
   b. agree with natural laws but not with the Bible.
   c. do not agree with the Bible or natural laws.
   d. agree with the Bible and natural laws.
11. Incorrect ideas are formed when
   a. we gather data accurately.
   b. we perform the experiment several times to compare results.
   c. we read and believe what the Bible says about science.
   d. we form ideas without close observation and careful thinking.
12. Which of the following is the second basic step in doing an experiment?
   a. Decide what you are trying to find out.
   b. Use the data to answer the original question.
   c. Make observations and record the data.
   d. Plan an experiment.
13. Which one of the following is **not** an observation?
   a. You tasted the cookie dough and decided that it was good.
   b. You smelled the smoke and knew someone was burning leaves.
   c. Your uncle told you that his goat has only one horn.
   d. You heard the baby crying and knew he was awake.

Unit 2

Title Page Photo

This crayfish is an arthropod that belongs to the crustacean class. Notice that it lives in the water. Why do crayfish live there? What advantages do they gain from the water?

Introduction to Unit 2

We tend to overlook some of the most common and plentiful things around us. Stars, stones, and insects are all likely on the "neglected wonders" list. In this unit we want to take a direct look at arthropods, including insects, crustaceans, and arachnids. The wonders we observe in the world of arthropods are awesome. These living gems take on a new luster if we will only stop and observe them.

In this unit you will find it easy to bring the student's world into focus in the classroom. Anywhere you live, you will be able to find a host of insects and a substantial array of arachnids. Most climates also have specimens in the crustacean category. Bring these creatures to class, and have students do the same. Firsthand observation is always better than even the best drawings or photographs.

Be sure to emphasize the beneficial aspects of the arthropods, not just their harmful effects. Recognize them as a blessing from God, and seek to impress that concept on your students as well.

Unit 2

Wonders Among the Arthropods

Crayfish

Did you know that spiders and grasshoppers have no bones? Instead of having skeletons inside their bodies as we do, these creatures have stiff shells on the outside. You have probably noticed that the grasshopper has a hard outer crust. The grasshopper and many other creatures belong to the arthropod family. God has created the arthropods without bones.

Arthropod means "joint-footed." As the name suggests, these creatures have many joints in their legs. In this unit you will study some details about these many-jointed, boneless creatures that God made. "O LORD, how manifold are thy works! in wisdom hast thou made them all: the earth is full of thy riches. So is this great and wide sea, wherein are things creeping innumerable, both small and great beasts" (Psalm 104:24, 25).

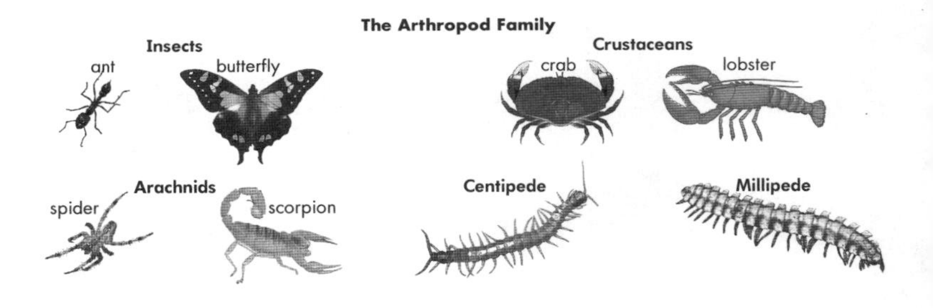

Story for Unit 2

The Giant Spider Hunt

I had wanted to catch and mount a tarantula (tə·ran′·chə·lə) for a good while. I first saw one of those giant spiders when I was a small boy visiting the Dominican Republic with my parents. Ever after that, I dreamed of getting a giant spider for myself. Now here I was, visiting the Dominican Republic again eleven years later. Today I would go tarantula hunting. I had seen a small one the day before. Its leg span was only about 2½ inches across. Would I be able to find a big, hairy one today? I hoped so!

I dug through a trash barrel and found a gallon-size tin can. I carried it out to an old rock heap that I had noticed at the back of the mission property. Surely some tarantulas would be hiding there.

I set down my can and started turning rocks. The sun moved high into the sky and poured its heat down mercilessly. My fingers were getting sore from the rough, heavy rocks, but I kept on working. Where were the tarantulas? Did they hear me coming and keep moving farther back into the pile? I hoped not.

Finally I flipped a large rock, and there sat one of the giant spiders. But he was just a small giant, with a leg span of 3½ inches. With a stick, I scooped him into my can and carried him to the house. I put him into a small cardboard box and set him in the freezer to kill him. I might want this one if I didn't find any others.

Back out I went. I wanted a *big* giant. Maybe if I dug through the rocks a while longer, I would find him. Rock after rock I flipped and piled aside. Hosts of tiny insects scampered away, but not a sign of another tarantula appeared. Twenty minutes later, I found another 3½-incher.

Once it was safe in the freezer, I went back to the

Lesson 5

The Armored Crustaceans

Vocabulary

abdomen (ab′·də·mən), the back part of the body of an arthropod.

antennae (an·ten′·ē), *singular* **antenna** (an·ten′·ə), a pair of long, thin feelers on the head of an arthropod.

arthropod (är′·thrə·pod), an animal with two or three main body parts, with jointed legs, and without bones.

crustacean (krə·stā′·shən), an arthropod with a crusty shell surrounding its body.

exoskeleton (ek′·sō·skel′·i·tən), the stiff outer shell of an arthropod, which protects it and gives shape to its body.

gills (gilz), the feathery organs that water animals use to get oxygen from the water.

thorax (thôr′·aks′), the middle part of an animal having three main body parts, to which the legs are attached.

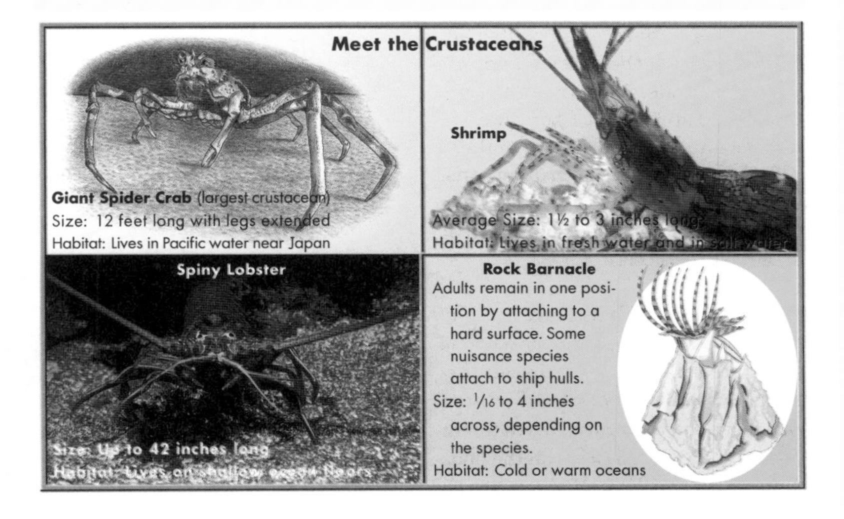

hunt. I was disappointed. Already I had moved half the rock pile. It was backbreaking work, and I still had not found my trophy.

I had nearly finished digging when I remembered that I would need to pile all the rocks back in place after I quit hunting. But I *had* to find a nice, big tarantula to show my brothers back home.

I had dug another half an hour when suddenly I saw a big, hairy leg sticking out from under a rock. Quickly I dug away all the other rocks from around that one. I wanted a big area of open space around there so that I could catch the giant spider before it scampered under another rock.

Finally I had a space cleared. I bent down and flipped the rock. A beautiful, hairy tarantula crouched there before me, ready to jump. I put the can down near the spider and tried to herd him into it with a stick. When my stick got too close, the giant tarantula reared up on its hind legs and jumped for my stick. It attacked the stick viciously and then scampered under another rock close by.

Once more, I dug around the rock to clear a space. I flipped the rock and quickly laid my can directly behind the tarantula. With my stick, I carefully coaxed it to back into the can. Then I dropped the stick and righted the can.

The giant tarantula covered most of the bottom of the can. It reared to jump out of the can. What should I do? I did not have a lid! Quickly I shook the can sideways and flipped the spider off balance. Wiggle, jiggle, I carried him safely all the way to the house. I found a cardboard cereal box and herded him into it. Carefully I taped the flaps shut and slipped him into the freezer.

I somehow managed, before lunchtime, to pile all the rocks back as they had been. When the other boys came in for the noon meal, I pulled out my spider to show them. We measured the giant. His leg span was 5 inches wide and 5½ inches long. His chunky body measured 1 inch by 2½ inches.

I was satisfied.

Note: To avoid problems such as importing new diseases, customs officials normally do not allow living creatures to be brought into the United States. We need to recognize and respect this restriction when we travel to foreign countries.

Mark and his brother were wading in the creek one summer day. "Here is a crayfish!" Mark shouted excitedly. "Let's see if we can catch it." But the crayfish was quicker than the boys. Soon the mud was stirred up, and the water was too cloudy for them to see the little creature.

The boys were trying to catch an animal called a ***crustacean.*** Crustaceans are ***arthropods*** with stiff outer shells. In fact, *crustacean* means "creature with a shell." This group of arthropods includes crayfish as well as creatures like lobsters, crabs, shrimp, and barnacles.

Animals like dogs, cats, and rabbits have bones inside to support their bodies, but crustaceans are different. Instead of an inside skeleton, a crustacean has a hard shell called an ***exoskeleton*** on the outside and soft body parts inside. Since an exoskeleton is hard and stiff, it must be jointed so that the animal can move around.

To learn more about crustaceans, we will look closely at the familiar crayfish. This creature is not a fish at all, but a crustacean.

The Common Crayfish

It is easy to see that the crayfish is a crustacean. It has a hard shell covering much of its body, and jointed body parts that allow it to move freely. Notice the two large claws on the ends of the two front legs. If the crayfish were not such a small creature, it would look dangerous.

Like other crustaceans, the crayfish has three main body parts: head, ***thorax,*** and ***abdomen.*** Each of these three main parts has a number of smaller body parts attached to it. Crayfish have many interesting parts in their strangely shaped bodies.

Head. At the front of the crayfish, you will find its pointed head. Two pairs of ***antennae,*** one short and one long, curve out from the very point of the head. The crayfish uses its long antennae for feeling, smelling, and tasting. They are the crayfish's fingers, nose, and tongue.

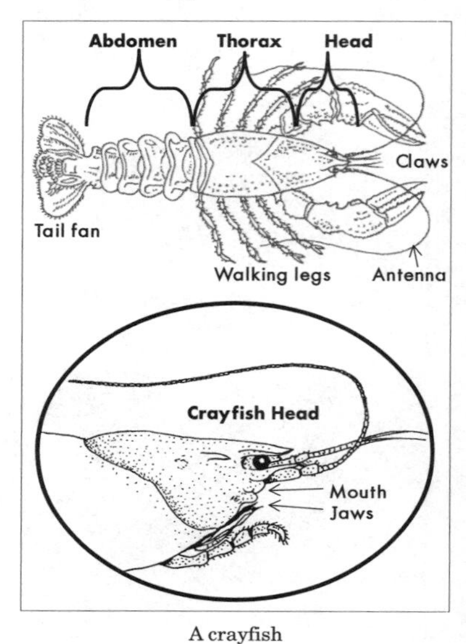

A crayfish

Teaching the Lesson

Have your students ever seen a crustacean? Perhaps you live close to the ocean and are familiar with several marine crustaceans. Be sure to focus on them in the lesson. Or maybe you are fortunate enough to have a stream near your school. Are there any crayfish in it? If so, your students would benefit from a little field trip to see a crustacean in its native habitat.

Help the pupils to understand that these creatures are real, not just some kind of "textbook animals." It would be profitable to dissect a crayfish or similar crustacean as described in the activities below.

For information about science supply companies, see the introduction of this book.

Lesson 5

Lesson Concepts

1. Lobsters, crabs, shrimp, barnacles, and crayfish are all classed as crustaceans.
2. Animals with exoskeletons have jointed bodies.
3. Typical crustaceans have three main body parts.
4. Crustaceans take in oxygen through their gills.

Just at the base of the antennae, you can see two knobby eyes. They are at the ends of two movable eyestalks so that the crayfish can move them around.

A tiny mouth, two strong jaws, and two pairs of mouthparts are also on the head. The mouth is a small slot on the bottom side of the head. Below the mouth slot are two small jaws. The crayfish uses its jaws to grind food before it pushes it on into its mouth. Next to the jaws are two pairs of mouthparts. The first pair moves food to the mouth, and the second pair pumps water through the gills, which allow the crayfish to breathe underwater.

Thorax. The second main body part is the thorax. The head and the thorax are fastened together under one stiff piece of exoskeleton. Only a small groove in the crayfish's shell shows where the head stops and the thorax begins.

On the underside of its thorax, the crayfish has three kinds of limbs. At the front end of the thorax are three pairs of small feet used for holding food. Next are two large claw feet. They end with large pincers that the crayfish uses to capture food. The next four pairs of legs are used for walking. Each of the walking legs has seven joints. Since each joint bends in a different way, the crayfish can move its legs in almost any direction.

Inside its thorax are the ***gills*** of the crayfish. It uses these little feathery organs to get oxygen out of the water. The second pair of mouthparts helps to keep water flowing through the gills.

Abdomen. The last main body part of the crayfish is its abdomen. The abdomen is not stiff like the head and thorax. It has six segments, which allow it to bend easily. Each segment is covered by its own plate of exoskeleton.

A large muscle fills most of the abdomen. The crayfish uses this muscle to snap its abdomen rapidly forward under its thorax. As it snaps its abdomen forward, a tail fan on the very end of its abdomen spreads out and catches water like a paddle. The powerful stroke of its abdomen sends the crayfish streaking backward through the water.

On the bottom side of the abdomen are five pairs of tiny limbs called swimmerets. The female crayfish uses these swimmerets to hold the eggs she lays until they hatch. After the eggs hatch, she uses the swimmerets to help hold several hundred baby crayfish until they are old enough to live on their own.

God has created the crayfish with many different parts. One of the most amazing things about all these parts is that if a leg or an antenna breaks off, the crayfish is able to grow a new one! Truly, the crayfish is a wonderful creature that God has made.

Discussion Questions

1. What are some characteristics of crustaceans?
   (No specific list is given in the lesson; have pupils find them for themselves.)
   three main body parts: head, thorax, and abdomen;
   a crusty shell-like exoskeleton covering soft insides;
   jointed legs;
   five pairs of legs (including claws);
   most live underwater
2. In what two ways does a crayfish move about?
   walking—a slow, awkward movement;
   swimming backward—a quick, agile movement accomplished by snapping the abdomen rapidly forward under the thorax

Lesson 5 Answers

Exercises

1. (Answers from lesson.) lobsters, crabs, shrimp, barnacles
2. exoskeleton
3. Joints in the exoskeleton allow a crustacean to bend its body.
4. head, thorax, abdomen
5. c. head
   d. grinding food
   e. head
   f. moving food to the mouth
   g. head
   h. pumping water through the gills
   i. thorax
   k. thorax
   l. catching food
   m. thorax
   n. walking
   o. abdomen
   p. holding eggs and baby crayfish

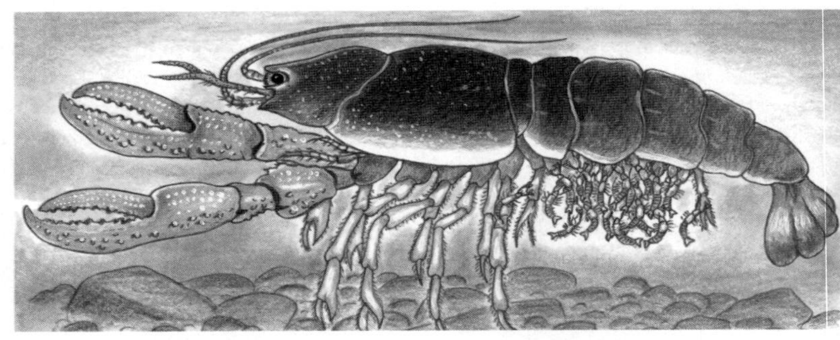

The female crayfish uses its swimmerets to hold baby crayfish.

Exercises

1. List four crustaceans besides the crayfish.
2. Instead of having bones inside, a crustacean has a hard outer shell called an ———.
3. Since a crustacean has a stiff outer shell, how is it able to move?
4. List the three main parts of a crustacean in order from the front to the back.
5. Copy and complete the following chart, which gives the parts of a crayfish from front to back. Under *Location,* write *head, thorax,* or *abdomen* to tell where those parts are located. A few spaces are filled in for you.

Part	Location	Use
Antennae	a. head	b. feeling and tasting
Jaws	c.	d.
First pair of mouthparts	e.	f.
Second pair of mouthparts	g.	h.
Three pairs of small feet	i.	j. holding food
First pair of legs (claws)	k.	l.
Last four pairs of legs	m.	n.
Swimmerets	o.	p.

6. What do the gills do for the crayfish?
7. How does a crayfish move rapidly through the water?
8. What happens when a crayfish loses a leg?

Review

9. The branch of science dealing with living things is called ———.
10. We use our five senses to ——— God's world.
11. List the four steps of an experiment in the correct order.
12. An event that does not follow ——— ——— is called a miracle.
13. The information collected while performing an experiment is called ———.

Activities

1. Use an encyclopedia to answer these questions about crayfish.
   a. How do crayfish benefit many people?
   b. What harm do crayfish cause?
   c. What foods do crayfish eat?
2. Ask your teacher for help in dissecting a crayfish.
3. Catch a live crayfish, and watch it swim around in a pail of water. Notice how rapidly it can move when it snaps its abdomen forward. Observe the different parts of the crayfish and how they work together.
4. If you have an aquarium or a large glass jar, you could keep your crayfish in the classroom for a while. Place some sand and gravel from a stream in the bottom of the aquarium. Add some water plants and a rock large enough for the crayfish to hide under. Fill the aquarium half full of water. Small minnows make the aquarium more pleasant and also provide a fresh food supply for the crayfish. Your crayfish will eat bits of earthworm, bits of raw meat, insects, and almost any other kind of food. Put food into the aquarium every other day or so, and remove what has not been eaten.
5. Consider this question: How can an animal with an exoskeleton grow? Doesn't its hard suit of armor keep it from getting larger? And yet crayfish must grow. Baby crayfish are tiny when they hatch, and in a stream you may find crayfish of all sizes. Can you find the answer in an encyclopedia or a nature reference book? If you keep a crayfish in the classroom long enough, you might see for yourself what happens.

6. Crayfish use gills to get oxygen out of the water.
7. It snaps its muscular abdomen rapidly forward under its thorax. This powerful stroke sends it streaking backward.
8. It grows a new one to replace the one it lost.

Review

9. biology
10. observe
11. (1) Write down the question you are trying to answer.
    (2) Plan an experiment that will help to answer the question.
    (3) Work the experiment and record your observations.
    (4) Use the data to answer your question.
12. natural laws
13. data

Activities

1. a. Many people eat crayfish, especially in the Scandinavian countries and the southern United States.
   b. Some crayfish weaken dams or dikes by their burrowing.
   c. minnows, insect larvae, worms, snails, tadpoles, plants
2. Get a large preserved crayfish from a scientific supply company. Obtain a pair of small, sharp scissors and a small scalpel. You can get both of them from a scientific supply company or buy them locally. A farm supply store would be a good place to look for a scalpel. Perhaps a local farmer would even have one you could borrow. Dissection does not require elaborate equipment, but you do need a steady hand and a little practice.

   Find a dissection manual at a local bookstore or library, and follow it closely. Show the students the gills attached to the base of the legs and extending up under the carapace. Show them the intricate mouthparts, the knobby eyes, the jointed legs, and the grinding mechanism inside the stomach. Cut and lift away the armor plating of the abdomen, and show them the large muscle that a crayfish uses to snap its abdomen forward for reverse locomotion. Many students are fascinated by this kind of demonstration.
5. Crayfish periodically shed their exoskeletons as they outgrow them. The newly molted crayfish is soft until it secretes a new exoskeleton. Soft-shell crayfish are especially vulnerable to predators and are prized by fish as well as fishermen.

Lesson 6

Spiders and Other Arachnids

Vocabulary

arachnid (ə·rak′·nid), an arthropod that has four pairs of legs and two body parts.

predator (pred′·ə·tər), an animal that eats other animals for food.

spinnerets (spin′·ə·rets′), the organs used by a spider to make silk for its web, found at the back of its abdomen.

web, a silk net woven by a spider to catch prey.

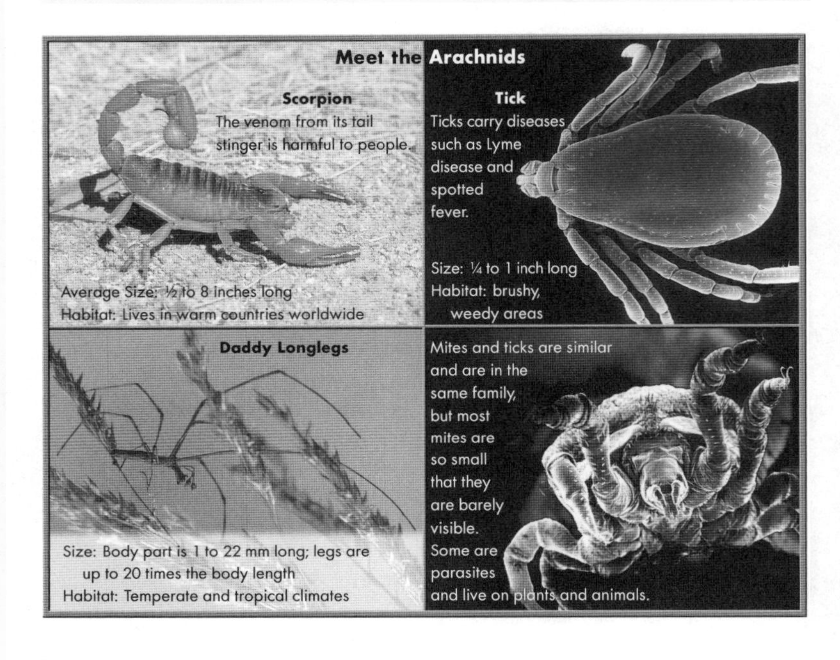

Lesson 6

Lesson Concepts

1. Scorpions, mites, ticks, and daddy longlegs are similar to spiders.
2. Spiders have two main body parts and eight legs.
3. Different spiders make different webs.
4. The spider's weaving of a web is a marvel of creation.
5. Spiders are useful as insect predators.

Teaching the Lesson

Many people consider spiders to be pests. Their webs might often be a nuisance, especially to housekeepers, but that hardly qualifies spiders for the label "Pests." They are very effective predators, and they do us a valuable service by destroying harmful insects. Some large spiders catch mice, lizards, and frogs.

Spiders' webs are interesting. Instead of destroying them as soon as you find them, take time to observe them. Some spiders weave tangles, but others weave very elaborate webs. Point out the ornate designs to your students. Help them to see the beneficial and beautiful side of spiders. God had a definite plan in mind when He made these creatures.

The strength of spider silk is amazing. The only man-made material whose strength comes close is steel, but the strongest silk has a tensile strength five times greater than that of a steel wire of equal weight.

Another group of joint-footed arthropods is the ***arachnids.*** We can identify an arachnid by several characteristics. Arachnids have two main body parts and four pairs of legs. None of them have wings.

The arachnids include scorpions, mites, ticks, and daddy longlegs. Most common of all are the spiders. In hot places and cold places, in clean places and dirty places, we can find spiders almost anywhere in the world, even in kings' palaces (Proverbs 30:28).

Body Parts of Spiders

Spiders, like all arachnids, have two main body parts. The front part is the combined head and thorax. The back part is the abdomen.

Fastened to the underside of the front body part are eight legs for walking. Each leg has seven parts connected by joints that allow the legs to bend.

Unlike crustaceans, spiders do not have jaws to chew their food. Instead, a spider has two sharp mouthparts that are connected to its poison glands. When a spider bites an insect, it pumps poison into the insect to kill it. It also pumps digestive juice into the insect. The digestive juice softens the inside parts of the insect so that the spider can suck them out. After the spider eats, all that is left of the insect is an empty shell. As you can see, a spider has no need for chewing mouthparts.

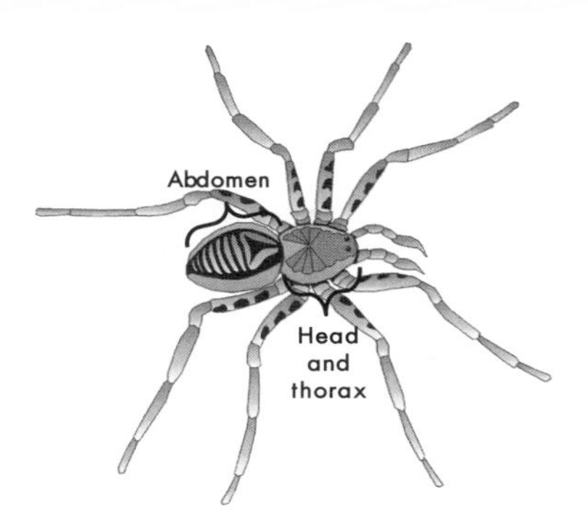

Did you ever observe a spider's eyes? Spiders usually have eight eyes, and some are quite large. Most spiders that hunt their prey can see well. They spot their prey and pounce on it when it is close enough.

Spiders that build ***webs*** and wait for their prey to get trapped cannot see very well. They have an excellent sense of touch and can feel an insect vibrating the web even if they cannot see it very well.

At the rear of a spider's abdomen are silk-spinning organs called ***spinnerets.*** Most spiders have six spinnerets. The spinnerets are connected to tiny glands that produce liquid silk inside the spider. All spiders have glands to make three kinds of silk, and some can make five different kinds. Using its six spinnerets and several kinds of silk, a spider can produce a thick thread or a thin one, dry silk or sticky silk—whatever is needed.

Tensile strength is the longitudinal (lengthwise) stress that a substance can bear without tearing apart. In some parts of the world, the webs of large spiders are used as fishnets and are said to be strong enough to hold a one-pound fish.

Collect spiders. Glass gallon jugs will hold most small spiders quite well. Be sure to keep a good supply of small flying insects on hand for them to eat. Larger spiders can be difficult to keep in captivity. They need more room for their larger webs. Capitalize on the spiders that insist on building webs in or around your school building. These will provide excellent lesson material for you because you can observe them right where they are, webs and all.

A field guide will help you to identify the spiders that you and the pupils collect. One of these should be as near as the library.

Discussion Questions

1. What do spiders use their webs for?
   Spiders use their webs to trap insects that they need for food.
2. What are some ways that spiders are different from crayfish?
   Spiders have four pairs of legs instead of five as crayfish do.
   Spiders have two main body parts, and crayfish have three.
   Spiders do not have chewing mouthparts as crayfish do.
   Spiders are land creatures, and crayfish are water creatures.
   (Spiders also have a softer exoskeleton than crayfish do.)
3. How are spiders helpful to man?
   They kill many harmful insects.

Spider Webs

To catch the insects they eat for food, many spiders build webs. One of the most beautiful webs is the orb web built by the black and yellow garden spider. At the right is a diagram that shows the steps in spinning an orb web. This little garden spider can make a beautiful web in 30 to 60 minutes.

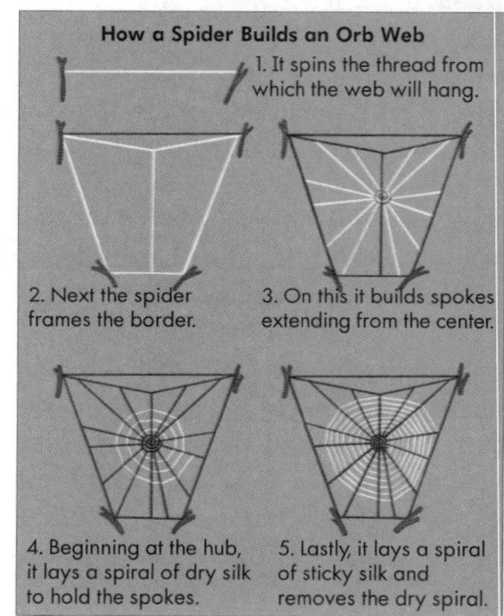

After the web is finished, the garden spider moves to the center and waits. When an insect flies into the web and gets caught, the spider feels the vibration and hurries over to its prey. With a quick bite, the spider kills

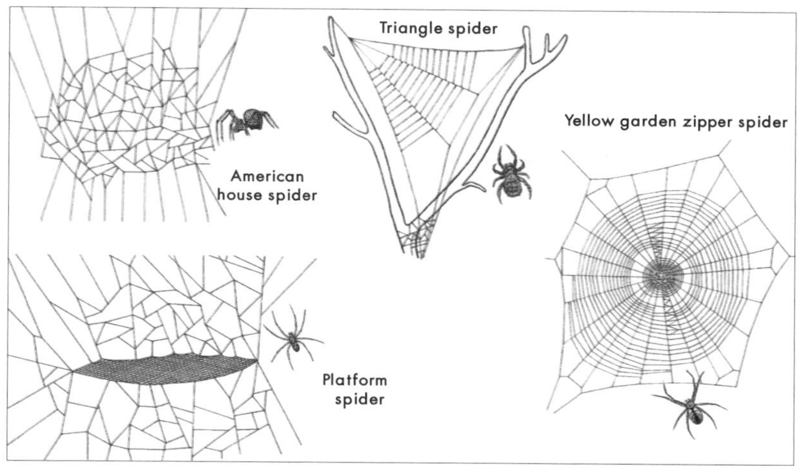

A spider is one of the "four things which are little upon the earth, but... exceeding wise" (Proverbs 30:24).

the insect. It often waits to eat the insect until later when the digestive fluids have dissolved the inside of the insect.

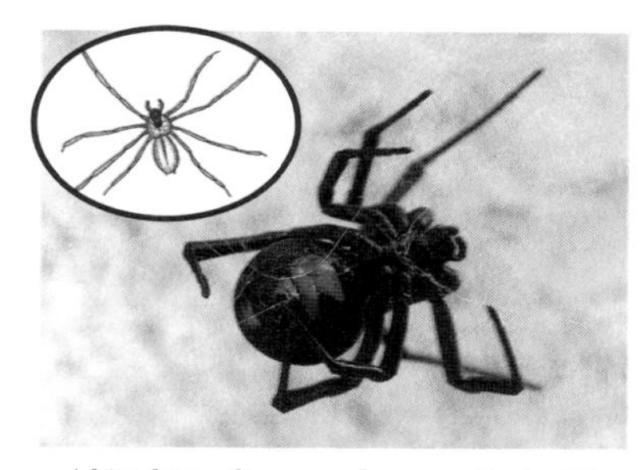

A bite from a brown recluse or a black widow is very harmful to humans. Black widows have a reddish hourglass shape underneath the abdomen. The brown recluse has a violin design on its back.

Spiders are ***predators*** that kill many harmful insects. Even though their webs can be a nuisance, spiders are a blessing that God has created to keep harmful insects under control.

Not all spiders build webs, and not all webs are beautiful. The common house spider simply spins a big tangle of silk. Whenever something gets trapped in the web, the spider climbs through the tangle to bite and kill its prey. Different kinds of spiders all make webs a little differently. Some webs are beautiful, and others look untidy; but each is a witness to the ability that God put into a spider's tiny brain.

A spider uses its silk for other

purposes too. Wherever a spider goes, it trails a fine silk thread. If danger threatens, it can jump out of its web and let itself down by this natural safety rope. When the danger is past, it climbs safely back up to its web.

Spiders also use silk to make nests. Many spiders wrap their eggs in a special silk sac. The sac protects the eggs until they hatch, and it also helps to keep the young spiders safe. When the young spiders are old enough to survive on their own, many of them spin streamers of silk that catch the wind and carry them to new homes.

Spider silk is very strong. It is stronger than cotton and wool, and it is even stronger than steel. A strand of spider silk is up to five times stronger than a steel wire of the same weight! Spiders show the wisdom of God in some very remarkable ways.

Exercises

1. Arachnids have ——— main body parts, ——— pairs of legs, and no ———.
2. Name three arachnids other than spiders.
3. Which spiders cannot see very well—the web builders or the ones that hunt their prey?
4. By which of the five senses does a spider tell when an insect is in its web?
5. A spider pumps two liquids into its prey. Name each kind of liquid, and tell what it does.
6. Spiders use their ——— to spin silk threads.
7. A spider's main reason for building a web is
   a. to have a place to hide from enemies.
   b. to trap insects for food.
   c. to have a place to lay eggs.
   d. for something enjoyable to do.
8. Write the letters of the steps below in the order a spider follows to build a web.
   a. Stretches spokes of dry silk.
   b. Weaves a spiral of sticky silk.
   c. Spins a border.
9. Because they eat other animals, spiders are called ———.
10. All spiders
    a. build beautiful spiral-shaped webs.
    b. build webs to catch insects.
    c. can produce more than one kind of silk.
    d. are harmful pests that should be killed.

Lesson 6 Answers

Exercises

1. two, four, wings
2. (Any three.) scorpions, mites, ticks, daddy longlegs
3. the web builders
4. touch
5. One liquid is a poison that kills the insect.
   The other is digestive juice that softens its inside parts.
6. spinnerets
7. b
8. c, a, b
9. predators
10. c

11. Give one way that spiders use their silk besides building webs.
12. Spider silk is
    a. always sticky to help hold insects.
    b. stronger than cotton and steel.
    c. used only to capture insects.
    d. used by people to make silk cloth.

Review

13. How many main body parts do crustaceans have?
14. The stiff outer shell of an arthropod is called an ———.
15. Crustaceans breathe by means of ——— located in the thorax.
16. True or false? Both spiders and crayfish are arthropods.
17. True or false? Both spiders and crayfish are crustaceans.

Activities

1. Study the different types of webs. Look for webs in your garden, lawn, flower bed, or barn. Can you discover which kind of spider made each web? Make drawings of the webs to show to your classmates.
2. Find a spider web with a spider in it. Catch several flies or other small insects, and throw them into the web. Watch the spider run over and kill its prey. Do this with several different kinds of spiders. Notice the different ways that spiders behave.
3. Mount a spider web. You will need a piece of black construction paper, a spray can of white paint, and scissors.

   First, locate a nicely shaped web, and chase the spider off. Spray the web lightly on both sides with the white paint. Using short bursts of spray and directing the spray at an angle will keep the web from tearing.

   Before the paint dries, ease the black construction paper close to the web. Carefully touch the paper to the web, trying to touch the whole web at once. After the wet web is stuck to the paper, do not move the paper until you snip the strands of silk extending beyond the paper.

   Try to find several webs with interesting shapes. If possible, label the web with the kind of spider that made it.
4. Use an encyclopedia to answer these questions.
   a. How does a bolas spider catch its prey?
   b. Since spiders breathe air, how can the water spider live underwater?
5. In this lesson, you learned that spiders prey upon insects. But things are sometimes reversed, and the hunter becomes the hunted one. Find out how the mud dauber wasp feeds spiders to its young, and write several sentences about it.

11. (Any one of these.)
    as a safety rope to escape from danger;
    to make a nest and egg sac;
    to travel to new homes
12. b

Review

13. three
14. exoskeleton
15. gills
16. True
17. false

Activities

1–3. Help your students observe spiders and their webs as described here.

4. a. The bolas spider spins a strand with a ball of sticky silk at the end. It throws this at its victim and entangles it in silk. (Compare the bolas used on the pampas of South America.)
   b. The water spider spins a web under the water. Then it makes many trips to the surface, capturing air bubbles and taking them underwater to inflate its web.
5. The mud dauber wasp stings a spider to paralyze it. Then she takes it to her mud nest, lays an egg on the spider, and seals the entrance of the cell. When the egg hatches, the larva has plenty of fresh food to eat.

Lesson 7

Grasshoppers and Other Insects

Vocabulary

adult, a creature that is fully mature.

compound eye, a large eye made of many tiny simple eyes working together.

egg, a round or an oval body out of which young animals hatch; the first life stage of insects.

hearing membrane (mem′·brān′), a tiny organ on the abdomen of some insects, which enables them to hear.

life cycle, the stages that an insect goes through in becoming an adult.

nymph (nimf), a young insect in the stage between egg and adult, looking much like an adult but smaller.

Lesson 7

Lesson Concepts

1. Insects have three main body parts and six legs.
2. The grasshopper is a very complex animal.
3. Many animals with exoskeletons have life stages.

Teaching the Lesson

Insects are the most abundant creatures on earth. Both in number of individuals and in number of species, they far exceed the larger animals. It should not be difficult to find specimens for this lesson. Be sure to include the common insects. Though we tend to highlight the rare and unusual species of insects, we must not overlook the ladybugs, ants, and even roaches and other pests that might be around.

This lesson centers on grasshoppers, so by all means try to catch several for observation. Show your students the different body parts of a live grasshopper. Gently lift the wings to see if you can find the hearing membranes. Watch the grasshopper jump. Help the students calculate how far they would be able to jump if they could jump proportionately as far. (See Activity 3.) If your area lacks grasshoppers, be sure to find some other insect common to your area, which will make this lesson come alive for your students.

A drone bumblebee

Another group of arthropods is the insects. This group has more different kinds of creatures than any other group of arthropods. Butterflies, houseflies, beetles, bees, grasshoppers, and many more insects are in this group. Although insects differ widely, they all have four characteristics.

1. They have three main body parts: head, thorax, and abdomen.
2. They have six legs.
3. They have one pair of antennae.
4. They breathe through tiny holes in the thorax or abdomen.

Different kinds of insects look different, yet they have these similarities. Let's look more closely at the common grasshopper to see what an insect is like.

The Common Grasshopper

The grasshopper, like all insects, has three main body parts: head, thorax, and abdomen. The head and thorax are not joined into one stiff piece like those of a crayfish. The abdomen is soft and segmented, and it connects to the thorax just behind the last pair of legs.

Head. Look closely at the grasshopper's head. You will see that it has several moving parts. Two wiggly antennae are fastened to the front of its face. Like a crayfish, the grasshopper uses its antennae to feel and smell the things around it.

At the bottom of its head, the grasshopper has chewing jaws surrounded by other mouthparts. As it chews, its jaws move from side to side instead of up and down as our jaws do. Its mouthparts help to hold the grass and other plants that it eats.

The grasshopper has two large ***compound eyes*** and three tiny simple eyes. The compound eyes are found on the top front corners of the grasshopper's head. These eyes are quite

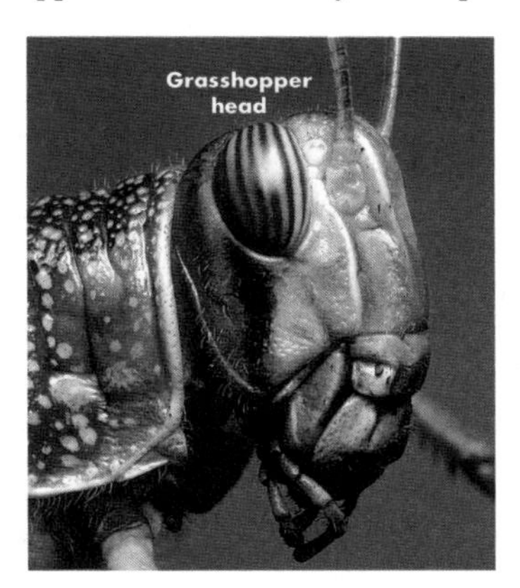

Discussion Questions

1. How are insects different from spiders?
   Insects have chewing mouthparts, and spiders do not.
   Insects have three main body parts, and spiders have only two.
   Insects have six legs, and spiders have eight.
   (Also, insects usually have wings, but spiders never do. Some insects eat plants and some are predators, but all spiders are predators.)
2. How did grasshoppers get their name?
   Grasshoppers got their name from the way they can hop around. With their powerful hind legs, they can jump great distances.
3. How can you tell if a grasshopper is a nymph or an adult?
   If it is small and its wings are not fully developed, it is a nymph. If it is full size and has functional wings, it is an adult.

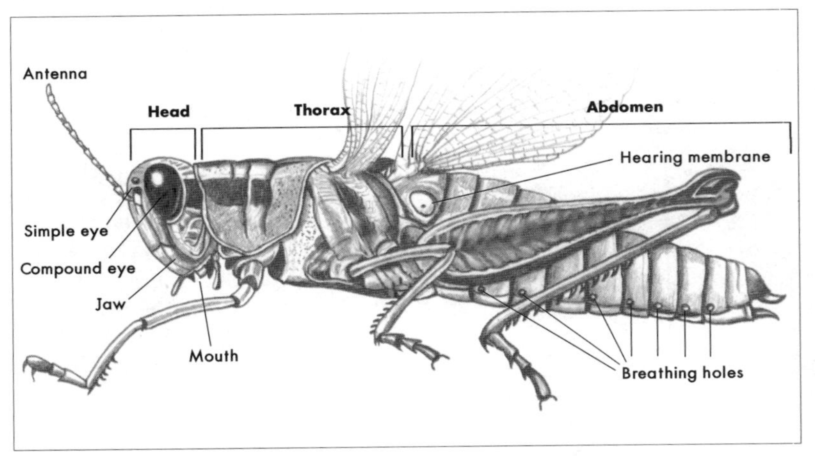

A grasshopper

large and are made of thousands of small, six-sided lenses. Each part faces a slightly different direction, allowing the grasshopper to see in many directions at once. The three simple eyes are small, one-part eyes located on the front of the head.

Thorax. The grasshopper has two pairs of wings fastened to the top of the thorax. The front two wings are stiff and narrow. When the grasshopper is not flying, the two front wings cover the two back wings to protect them. The back wings are thin and fanlike, and the grasshopper uses them for flying.

The six legs of the grasshopper are fastened to the bottom of its thorax. The front two pairs of legs are small and thin, but the back pair is powerful. These two back legs enable the grasshopper to hop distances up to twenty times its length. A grasshopper can jump straight up to a height ten times its body length. If a five-foot boy could jump that well, he could jump straight up as high as a four-story building. *Grasshopper* is a good name for it.

Abdomen. The abdomen of the grasshopper is soft, flexible, and segmented. The first ten segments each have a pair of tiny holes that the grasshopper uses for breathing. Muscles inside the abdomen force air in and out of little air sacs connected to the holes. The sacs absorb the oxygen that the grasshopper needs.

On the first segment of the abdomen, the grasshopper has two ***hearing membranes.*** These tiny patches are the grasshopper's ears. They are up under the wings on either side so that the wings protect them.

On the very end of her abdomen, the female grasshopper has two stiff prongs that she uses to lay ***eggs.*** The grasshopper uses the prongs to drill a hole in the soil, she pushes her abdomen down into the hole, and she lays from 2 to 120 eggs in the hole. After capping the hole with a sticky material, she leaves the eggs to hatch on their own.

Life Cycle of Insects

God has designed insects to grow in a special way. Because of their stiff exoskeletons, it would be hard for them to grow as many other animals do. God created insects with a ***life cycle*** of several stages. All insects begin life as eggs. Some kinds, such as houseflies and butterflies, hatch into wormlike larvae. Others, like the grasshopper, hatch into ***nymphs*** that look much like full-grown grasshoppers but are smaller. After several changes, these young, partly formed insects become ***adults.***

Grasshoppers generally lay their eggs in late summer or fall. The following spring, the eggs hatch into nymphs. Grasshopper nymphs grow for 40 to 60 days before they become adults. As they grow, their exoskeletons become too small for them. Every time they outgrow their exoskeletons, they simply shed the old ones and form new ones that are larger.

Nymphs shed their exoskeletons from five to seven times before they become adults. After they have shed their exoskeletons for the last time, they usually have wings and are adults. In late summer, they lay eggs to produce the next year's grasshoppers.

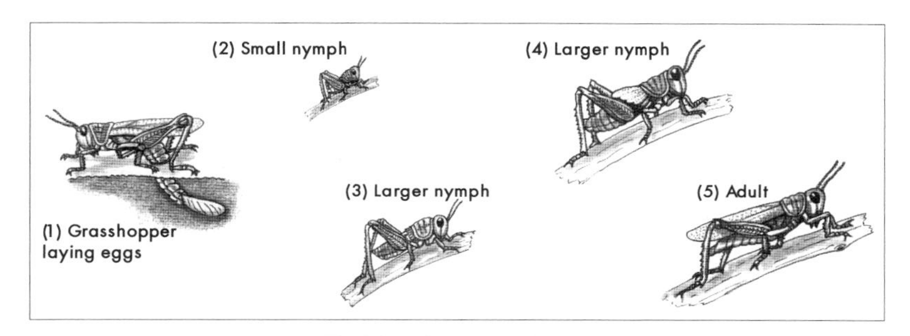

The life cycle of a grasshopper

This enlarged view of a robber fly's head shows thousands of six-sided lenses in the compound eyes. The fly uses its funnel-shaped mouth as a straw to drink liquids, its only food.

Lesson 7 Answers

Exercises

1. They have three main body parts: head, thorax, and abdomen.
   They have six legs.
   They have one pair of antennae.
   They breathe through tiny holes in the thorax or abdomen.
2. a
3. grass and other plants
4. It enables a grasshopper to see in many directions at once.
5. The first pair is a protection to cover the second pair when the grasshopper is not flying.
   The second pair is used for flying.
6. holes, thorax, abdomen
7. a. hearing membranes
   b. on the first segment of the abdomen, under the wings
8. egg, nymph, adult
9. As the nymphs grow, their stiff exoskeletons become too small and must be shed.
10. false

Exercises

1. List the four characteristics of all insects.
2. Grasshoppers use their antennae to
   a. feel and smell.
   b. taste and smell.
   c. taste and hear.
   d. feel and hear.
3. What kind of food do grasshoppers eat?
4. How does a compound eye help a grasshopper?
5. Grasshoppers have two pairs of wings. Briefly describe the purpose of each pair.
6. Insects breathe air through tiny ——— in the ——— or ———.
7. a. What hearing organs do grasshoppers have instead of ears?
   b. Where are these organs located?
8. What are the three stages in the life cycle of a grasshopper?
9. Why must insect nymphs shed their exoskeletons as they grow to adulthood?
10. True or false? After a grasshopper nymph hatches from an egg, it sheds its exoskeleton two times and is then an adult.

Review

11. a. How many main body parts do crustaceans have?
    b. How many main body parts do spiders have?
    c. How many main body parts do insects have?
12. a. How many legs do spiders have?
    b. How many legs do insects have?
13. a. Crayfish use (gills, lungs, tiny holes) to breathe.
    b. Insects use (gills, lungs, tiny holes) to breathe.
14. Crustaceans, spiders, and insects all belong to the group called ——.
15. True or false? Spiders produce threads of silk with their spinnerets.
16. In what way are spiders helpful?

Activities

1. Catch several grasshoppers. Can you tell whether they are nymphs or adults? Check whether they have wings. If they do, they are adults. If they lack wings, they usually are only nymphs.
2. Try feeding grass to a grasshopper. Hold him between your thumb and forefinger of one hand, and feed him with the other. Push the edge of a blade of grass up against his jaws. Does he start chewing? He may spit out some dark green fluid instead. But if you have enough patience, usually you can get him to nibble at the grass. Watch all his tiny mouthparts move as he eats. Isn't it fascinating to see?
3. Measure how far your grasshopper can jump. Have your teacher help you calculate how far you could jump if you were able to jump as far for your size as your grasshopper can.
4. Use an encyclopedia or a Bible dictionary to answer these questions.
   a. What use do people of the Middle East and parts of Africa make of grasshoppers?
   b. What Bible character used grasshoppers in this way?
5. Have you ever met an insect that was older than you? Use an encyclopedia to learn about the insect called a cicada. How old is the cicada when it emerges from the ground?

Review

11. a. three
    b. two
    c. three
12. a. eight
    b. six
13. a. gills
    b. tiny holes
14. arthropods
15. True
16. Spiders kill many harmful insects.

Activities

3. Use a proportion as follows to calculate the jumping distance.

$$\frac{\text{length of grasshopper}}{\text{distance jumped}} = \frac{\text{height of student}}{\text{distance he could jump}}$$

Substitute measurements for words, and you have the following:

$$\frac{2''}{40''} = \frac{60''}{?''}$$

Multiply the two numbers across the corner (*40* and *60*) and divide that product by the third number (*2*). Your answer (*1,200 inches* or *100 feet*) will be the distance the student could jump.

4. a. Those people eat grasshoppers. Grasshoppers are candied in China; dried, powdered, and mixed into bread in Arabia and northern Africa; enjoyed by the Filipinos; and considered a delicacy by some people in Europe and the United States.
   b. The Bible says that John the Baptist ate locusts and wild honey (Matthew 3:4). Locusts are a kind of grasshopper. Grasshoppers and locusts are specifically mentioned as being suitable for food in Leviticus 11:22.
5. The cicada emerges from the ground 17 years after tunneling underground. Children often enjoy finding the empty shells they leave behind on tree trunks and similar places when they fly off as adults.

Lesson 8

Metamorphosis of the Butterfly

Vocabulary

chrysalis (kris′·ə·lis), the hard, smooth case of the pupa stage of a butterfly, in which the caterpillar changes to a butterfly.

larva (lär′·və), *plural* **larvae** (lär′·vē), the wormlike second stage in the life of an insect that passes through four life stages.

metamorphosis (met′·ə·môr′·fə·sis), a great change that certain insects go through as they grow to adulthood.

molt (mōlt), to shed an outer covering, such as an exoskeleton.

pupa (pyü′·pə), the third stage in the life of an insect with four life stages, in which it is inactive and changes into an adult.

All insects have a life cycle with several stages. In the last lesson, you noticed the three stages in a grasshopper's life: egg, nymph, adult. Now you will study the four stages that every butterfly goes through. Butterflies do not have nymph stages as grasshoppers do. They change from caterpillars to butterflies in a process called ***metamorphosis.*** Metamorphosis completely changes wormlike caterpillars into beautiful butterflies.

Four Stages of Butterflies

Egg. When a female butterfly is ready to lay her eggs, she carefully chooses a place to put them. She must find a place that will provide a good food supply for the caterpillars when they hatch. She lays the tiny eggs in clusters on the undersides of leaves. A sticky substance holds them in place. Some butterflies lay only one or a few eggs at one place; others lay a mass of eggs. Eggs that are laid in the fall may not hatch until the following spring. But other eggs may hatch within several days after they are laid.

Larva. Instead of hatching into a nymph as a grasshopper egg does, a butterfly egg hatches into a wormlike ***larva.*** We call it a caterpillar. Some caterpillars look ugly, but others have beautiful colors and interesting patterns.

A caterpillar is a peculiar creature. Although it is an insect, it looks more like a worm. The caterpillar has a head followed by a long segmented body. The head has two short antennae, chewing

Lesson 8

Lesson Concepts

1. A larva is the wormlike stage of the butterfly and many other insects.
2. During the larva stage, the caterpillar eats enough food to carry it through the pupa stage.
3. A marvelous metamorphosis takes place in the change from a larva to an adult butterfly.
4. A butterfly in the adult stage lays eggs for the next generation.

Teaching the Lesson

Few people find it hard to appreciate butterflies. Their brilliant colors and carefree flitting are inspiring to the observer. Perhaps even more fascinating than their appearance is their life story. They are nature's classic example of the ugly duckling. From a scrawny, destructive worm to a beautiful, useful butterfly is the delightful story of butterfly metamorphosis.

Are any butterflies still in season in your area? Catch a few, and mount them to show to your students. Have your students look for caterpillars. They will enjoy watching them grow and change into pupae. When the butterflies emerge, catch and mount them. Keep a collection of butterflies and moths from year to year. Such collections are excellent interest generators.

mouthparts, a spinneret for making silk, and a dozen simple eyes.

The first three segments behind the head make up the caterpillar's thorax. Extending from the underside of each segment in the thorax are two short legs. That gives the caterpillar six legs just as adult insects have.

The rest of the caterpillar's body is its abdomen. The caterpillar's abdomen has as many as five pairs of false legs. False legs are short, non-jointed limbs that look like legs and help the caterpillar to walk. They are not considered true legs because they disappear when the caterpillar changes into a butterfly.

The caterpillar is a big eater. God has given it chewing mouthparts with which it eats a large amount of leaves and other plant matter. It needs to store up food for the pupa stage during which it cannot eat. Some caterpillars eat enough vegetation that they become pests. Cabbage butterfly caterpillars are one example of such pests.

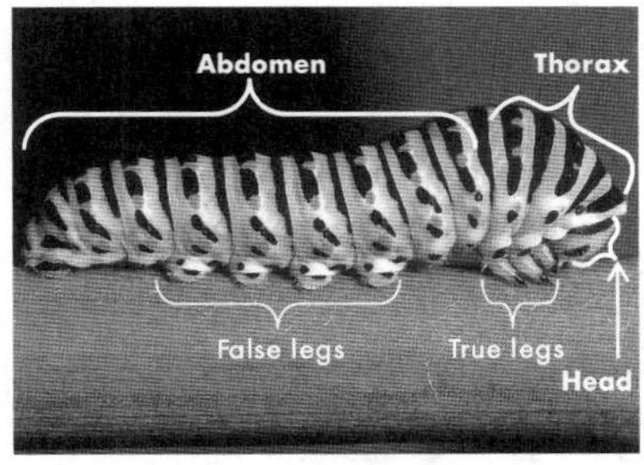

This caterpillar tilts its head down to eat.

They can quickly strip the leaves from young cabbage plants and completely ruin older plants.

The caterpillar grows very fast because of all the food it eats. Like the grasshopper nymph, it soon outgrows its exoskeleton. When its exoskeleton gets too tight, the caterpillar ***molts,*** shedding its old exoskeleton and growing a new one. After molting four or five times, the caterpillar goes into its next life stage—the ***pupa*** stage.

Pupa. As it gets ready for its last molt, the caterpillar deposits a sticky liquid from a spinneret on its head onto a leaf or twig. The sticky liquid hardens into a silk pad that looks much like a spider web. The caterpillar grasps this silk pad with special claws on the end of its abdomen and hangs there upside down. Its exoskeleton splits open and falls off.

Now it no longer looks like a caterpillar. Its three body parts are not easy to see. Its new exoskeleton hardens into a case called a ***chrysalis.*** The chrysalis hangs from the silk pad until the caterpillar completely changes into a butterfly. This takes several weeks or months, depending on what kind it is.

God works a special miracle inside the chrysalis. The wormlike caterpillar slowly changes into a beautiful butterfly. This mysterious change is called metamorphosis. Even though many people have studied metamorphosis,

The emphasis of this lesson is on the miracle of metamorphosis. Impress upon your students the miraculous nature of the dramatic change a caterpillar makes as it matures. Be sure to credit the hand of God as you observe and study His delightfully designed butterflies.

Note: Insects that pass through four life stages undergo complete metamorphosis. Those with fewer stages (like the grasshopper) go through incomplete metamorphosis. In this course, *metamorphosis* refers only to complete metamorphosis.

Discussion Questions

1. Why do butterflies not have nymph stages?
   Butterflies have larva and pupa stages instead of nymph stages. God created them that way.
2. List some differences you can see between caterpillars and butterflies.
   Butterflies have large, colorful wings, and caterpillars have none.
   Butterflies have compound eyes as well as simple eyes, and caterpillars do not.
   Butterflies' antennae are generally much longer and more graceful than caterpillars' antennae.
   (Have a caterpillar and a butterfly for the class to observe.)
3. How are butterflies helpful? (This is not mentioned in the lesson, but it is a point worth making.)
   As butterflies suck nectar from flowers, they pollinate the flowers. Butterflies are not only beautiful but also useful.

A monarch butterfly lays eggs on milkweed leaves. A milkweed patch may contain all four stages pictured here.

there is still very much that we do not understand about it. Truly, this is a miracle from God.

Adult. When the metamorphosis is complete, the thorax of the new butterfly swells and cracks the chrysalis shell. The chrysalis splits open, and the head and thorax of the butterfly come out first. For several minutes, the butterfly struggles to free its legs, wings, and abdomen from the chrysalis.

This butterfly has come out of its chrysalis. It has spread its wings to dry.

When it is completely out of the chrysalis, the butterfly still does not look very pretty. Its wings are soft, wet, and crumpled. For the first half hour or so after crawling out of the chrysalis, the butterfly fans its wings to help dry and stiffen them. At the same time, it pumps blood throughout its body to help harden the wings, legs, and exoskeleton. Only then is it ready to flit around and enjoy the sunshine.

Butterflies are different from caterpillars in many ways. Butterflies have large, colorful wings, and caterpillars have none. Butterflies have compound eyes as well as simple eyes, and caterpillars do not. Butterflies' antennae are generally much longer and more graceful than caterpillars' antennae.

Differences Between Butterflies and Moths

Butterflies	Moths
1. Most butterflies fly in the daytime.	1. Most moths fly at dusk or at night.
2. Most butterflies have knobs at the tips of their antennae.	2. Most moths have feathery antennae without knobs.
3. Most butterflies have slender, hairless bodies.	3. Most moths have plump, furry bodies.
4. Most butterflies rest with their wings held upright over their backs.	4. Most moths rest with their wings spread out flat.

This heliconian butterfly lives in the tropics. Like the monarch, its body is poisonous to predators because the larva feeds on poisonous plants.

This is an underwing moth. When in danger, it frightens its enemies by suddenly uncovering its vivid underwing.

Lesson 8 Answers

Exercises

1. metamorphosis
2. egg, larva, pupa, adult
3. She must be sure to lay her eggs on plants that will provide food for the larvae when the eggs hatch.
4. caterpillars
5. They eat the leaves of cabbage plants, harming or even killing the plants.
6. d
7. chrysalis
8. b
9. (Any three. Other answers may also be acceptable; be sure they include both sides of each comparison.)
   Butterflies have wings; caterpillars do not.
   Butterflies have compound eyes; caterpillars do not.
   Butterflies' antennae are longer and more graceful than caterpillars' antennae.
   Butterflies have sucking mouthparts for drinking nectar; caterpillars have chewing mouthparts for eating vegetation.
10. adult

Review

11. head, thorax, abdomen
12. egg, nymph, adult
13. through tiny holes
14. True
15. false

Instead of chewing mouthparts like those of caterpillars, butterflies have sucking mouthparts. Butterflies suck nectar out of flowers instead of eating plants as caterpillars do. Nectar gives butterflies the energy they need to fly.

Most adult butterflies live about a month or six weeks. Some kinds may live almost a year. The adult butterflies lay the eggs, and the caterpillars that hatch become the next generation of butterflies.

Exercises

1. The amazing process through which a caterpillar changes into a butterfly is called ———.
2. List the four life stages of a butterfly in order from first to last.
3. Why must the female butterfly be careful where she lays her eggs?
4. Instead of hatching as butterfly nymphs, butterfly eggs hatch as larvae called ———.
5. In what way do cabbage butterfly caterpillars cause damage?
6. In what way are caterpillars and spiders alike?
   a. Both have eight legs.
   b. Both can see quite well.
   c. Both have spinnerets on their abdomens.
   d. Both can produce silk.
7. A butterfly spends one stage of its life in a stiff, smooth case called a ———.
8. The change inside a chrysalis is
   a. from the larva stage to the pupa stage.
   b. from the pupa stage to the adult stage.
   c. from the egg stage to the pupa stage.
   d. from the egg stage to the larva stage.
9. List three differences between caterpillars and butterflies.
10. The ——— stage is the one in which eggs for the next generation of butterflies are laid.

Review

11. What are the three parts of an insect's body, from front to back?
12. List the life stages of the grasshopper.
13. How does a butterfly breathe—with gills or through tiny holes?
14. True or false? Spiders, crayfish, grasshoppers, and butterflies are all arthropods.
15. True or false? Arthropods use their antennae for seeing.

Activities

1. Find and capture several caterpillars to study. Put your caterpillars in a glass gallon jar. Since caterpillars spend most of their time eating, be sure to give them plenty to eat. Gather leaves from the same plant you found the caterpillar on. That way you can be sure you are providing the food it needs as it grows and molts. Also keep the leaves fresh and crisp; perhaps you can put their stems in water. How many times will your caterpillar molt before it changes into a pupa?

   A good place to look for monarch caterpillars is on milkweed plants. The monarch caterpillar is interesting to watch because it is very active and grows rapidly. In the pupa stage, it forms a pale green chrysalis with small specks of gold. It stays in the pupa stage two or three weeks before crawling out as a ragged-looking butterfly. After an hour of fanning, it is ready to fly away as a beautiful black-and-orange butterfly.
2. See if you can find answers to these questions.
   a. Why do caterpillars eat more than adult butterflies?
   b. How do monarch butterflies survive the cold northern winters?

Activities

1. Keep several caterpillars in your classroom for the students to observe. Catch the caterpillars as young as you can; then follow their maturing process carefully as they grow. It would be interesting to record on a calendar every several days how much the caterpillar has grown. You could also indicate when it entered the pupa stage and when the butterfly finally emerged.
2. a. The caterpillar needs to store enough food to last through the pupa stage and to provide energy for the great change from larva to adult. The pupa appears to be resting, but great changes are happening inside the chrysalis, and these changes require energy.
   b. In the fall, North American monarch butterflies gather in flocks and migrate to the southern United States or to Mexico.

Lesson 9

A World Abounding With Insects

Over one million kinds of insects have been discovered. They are crawling, jumping, and flying around us every day, especially during summer. Insects can be found in hot deserts, on cold mountains, under ice, on land, in water, in the air, and underground. Insects are interesting and beautiful to study and collect. Here are some pictures of the many kinds of insects.

You can use an insect guidebook to identify insects. With words and pictures, it helps you to tell the differences between common insects. For example, butterflies have thin antennae with knobs on the ends, and moths have feathery antennae. With this information, you can quickly tell a butterfly from a moth. Also, the caterpillar of a moth spins a cocoon of silk fibers when entering the pupa stage. The caterpillar of a butterfly forms a hard, smooth chrysalis.

An encyclopedia can tell you many fascinating details about insects. The larva and adult stage of a firefly have an interesting similarity—both produce light. The Atlas moth from Asia and Australia is one of the largest silk-producing moths. Its wingspan exceeds 10 inches.

Lesson 9

Lesson Concepts

1. The earth abounds with insects, many of which are beautiful and useful.
2. God created insects for our welfare and enjoyment.
3. Attractive, educational displays can be made by collecting and mounting insects.

Teaching the Lesson

The lesson title is an understatement. There are billions upon billions of insects on the earth. In the water, in the air, in the ground, among the leaves and stems of plants—this planet literally teems with insects.

Most of the insects around us are harmless, and many of them are helpful. Only a relatively small number of species are harmful. The boll weevil, the Colorado potato beetle, the mosquito, the housefly—these are some of the bad characters in the insect world. They cause damage in the millions of dollars every year, as well as much suffering.

But we tend to focus on the bad insects and overlook the good ones. A British scientist is said to have speculated that British soldiers might have to go without beef if it were not for the spinsters of England. The spinsters kept cats, the cats ate mice, and the mice made holes. Empty mouseholes made ideal homes for

Collecting Insects

Insect collecting is a very common and educational hobby. It provides a way to study this part of God's creation, and it yields a colorful collection to be enjoyed and added to for years to come.

Catching insects can be quite simple. Some insects can be caught with your fingers. Others move rapidly and must be caught in other ways. If you are quick enough, you can use a net to scoop flying insects out of the air. To catch many different insects in a short time, you can sweep the net through small bushes or tall grass.

Put the insects you want to keep into jars. If the insects are very tiny, a good tweezers can be a useful tool.

To keep insects alive, put them into clear jars. Be sure to close the top with a lid that has tiny holes in it. A piece of nylon stocking also makes a good cover. Use a rubber band to fasten the nylon around the neck of the jar. Put some of the plant where you found the insect into the jar for its food. With a set of jars containing different insects, you can have an insect zoo.

To keep the insects for a collection, you will need to kill them. The best and safest way to kill insects is to freeze them in a freezer. Just place the insect, jar and all, into a freezer for at least half an hour. Stinging insects, such as wasps and bumblebees, should be frozen overnight to make sure they are dead.

Mounting Insects

After removing an insect from the freezer, you will want to mount it. There are two common ways to mount insects. One way is to place them on a layer of cotton covered with glass. Spread a layer of cotton evenly over the bottom of a box, such as one for a

A simple insect cage

Mounted insects can make very interesting displays.

bumblebees, which pollinated the clover fields that produced hay for cattle. Though the story is not entirely logical (bumblebees do not nest only in mouseholes), it may not be as far off the mark as it appears. When red clover was introduced to Australia and New Zealand, the plants did not produce seed until bumblebees were imported to pollinate the flowers.

The goal of this lesson is to help the children appreciate the many beautiful and beneficial insects that are all around. One way to enjoy butterflies and other colorful insects is to catch and mount them. Carefully arranged insect displays are educational as well as attractive. Develop in your students an appreciation for such displays, and encourage them to start a personal collection.

You might want to make a butterfly net to use for catching butterflies. A properly made net is almost indispensable for butterfly collecting and will provide many years of service.

One of the best ways to preserve a collection of butterflies is to mount them under glass. You will want to store them in a cool, dark place to keep them from fading. Butterflies and other insects can also be mounted in a shoebox with a piece of ceiling tile (or similar material) on the bottom. Run a pin through the body, and stick it into the tile. Beside the insect, pin a label with the pertinent information.

For mounting butterflies, special insect pins are more satisfactory than standard straight pins. (Check with a science supplier.) These super thin, rustproof pins have very sharp points that make it easier to pierce insects with a minimum of downward pressure.

thin book. Then arrange the insects on the cotton and put a piece of glass on top of them. Write the name of each insect (and perhaps where it was found) on a label, and place it beside the insect.

Another way to mount insects for a collection is to push a pin through the thorax of the dead insect. Slide the insect about three-fourths of the way up the pin. Then stick the pin into a piece of corklike material, such as a piece of ceiling tile. Also use a pin to hold a label with the insect's name on it. Be sure to arrange the insects to make an attractive display.

Collections of butterflies and moths are especially attractive. You must handle these insects carefully to keep the wings from being damaged. As much as possible, handle butterflies with tweezers so your fingers do not rub the coloring from the wings.

Immediately after thawing a frozen butterfly, gently lay it upside down on a block of soft wood. (Working with it upside down helps to protect the coloring on the upper side of the wings.) Push a pin through the thorax to anchor the butterfly to the wood. Then use tweezers to spread the wings one at a time, and hold them down with ½-inch strips of paper pinned to the board. Be sure to spread the wings evenly and attractively.

Place the butterfly in a warm, dark place for several days to dry. The butterfly will be dry enough after the abdomen is stiff when you nudge it with a pin. Now you can remove it from the

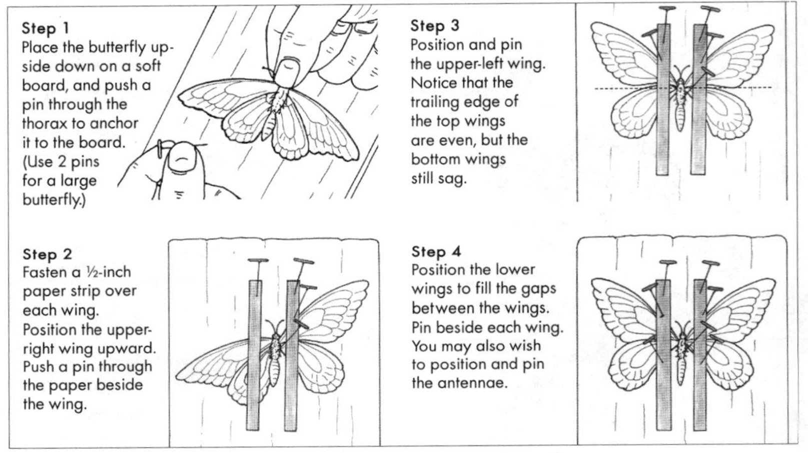

Spreading a butterfly's wings

Discussion Questions

1. What is the value of mounting insects?
   It is an attractive way to preserve insects for future enjoyment and an excellent way to become acquainted with God's wonderful insects.
2. Where could we find insects near our homes?
   (Many answers could be given; try to keep them as practical as possible.)
3. Why is it better to freeze an insect than to kill it with poison?
   (Sample answers.)
   Poisons can be dangerous and may be difficult to obtain.
   Nearly everyone in North America has access to a freezer; this is hardly true of poisons.
   Note: One drawback of freezing is that a freezer is not portable. You cannot take it with you on your hike.

board, and the wings will stay spread open. The butterfly is ready to be mounted with a pin or between cotton and glass. Keep the collection in a dark place so that the colors do not fade.

Insects: Friend or Foe?

One of the plagues that God sent upon Egypt was a great swarm of locusts, a kind of grasshopper. These destructive insects ate all the green plants in Egypt. They were pests indeed. You may think that all insects are pests. Beetles eat leaves off garden plants. Flies and mosquitoes carry diseases. Bees and wasps have painful stings. But actually, only a few of the many insects are really harmful to us.

Bees and some other insects pollinate garden plants and fruit trees, making it possible for them to bear fruit. We get some products from insects, such as honey, wax, silk, shellac, and some dyes. The ladybug helps us by eating other insects that are pests. Insects provide food for birds and other animals. Some insects help to get rid of dead animals. We are thankful that God created insects and made them so beautiful, helpful, and interesting.

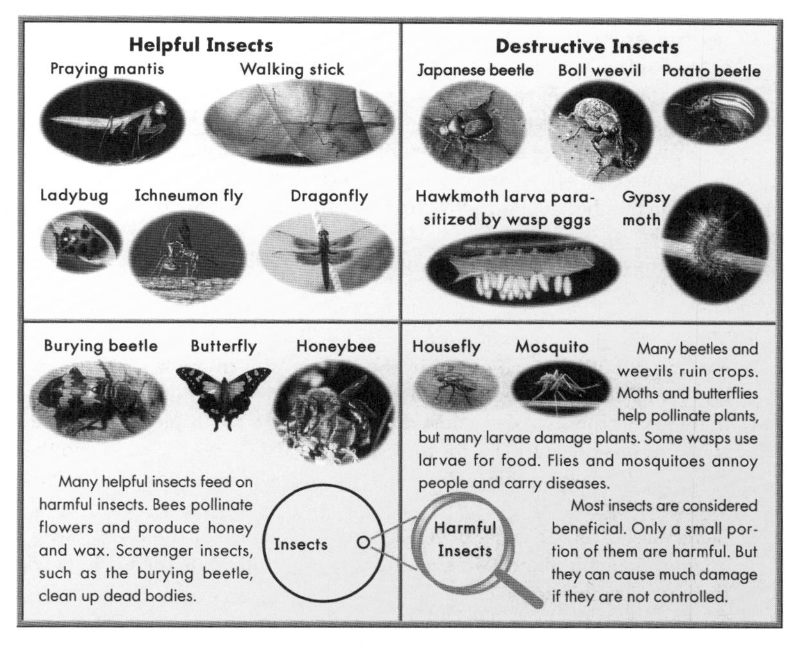

Lesson 9 Answers

Exercises

1. more than one million kinds
2. in hot deserts and on cold mountains;
   on land and in water;
   in the air and underground
3. Butterflies have thin antennae with knobs on the ends, but moths have feathery antennae.
   The caterpillar of a butterfly forms a hard, smooth chrysalis, but the caterpiller of a moth spins a cocoon of silk fibers.
4. false
5. net
6. freeze it
7. cotton, glass, pin
8. to avoid rubbing the coloring from their wings
9. (Any two.)
   Beetles eat leaves off garden plants.
   Flies and mosquitoes carry diseases.
   Bees and wasps have painful stings.
10. (Any two.)
    Bees and some other insects pollinate garden plants and fruit trees.
    We get products such as honey, wax, silk, shellac, and dyes from insects.
    The ladybug eats other insects that are pests.
    Insects provide food for birds and other animals.
    Some insects help to get rid of dead animals.

Exercises

1. How many kinds of insects have been discovered?
2. Name some contrasting places where insects are found.
3. What are two differences between butterflies and moths?
4. True or false? Insects are often difficult to find.
5. To catch insects that fly or move rapidly, it is handy to have a ———.
6. What is the best way to kill an insect without damaging it?
7. Insects can be mounted by putting them on a layer of ——— under a piece of ———, or by pushing a ——— through each insect and sticking it into some corklike material.
8. Why must you be especially careful when handling moths and butterflies?
9. What are two ways that insects are harmful?
10. What are two ways that insects are helpful?

Review

11. What organs do grasshoppers and crayfish use for touching and smelling?
12. Because exoskeletons do not grow, animals with exoskeletons must ———, or shed this outer covering, so that they can grow.
13. Name the stages in the life cycle of a butterfly in correct order.
14. Which statement is correct?
    a. Exoskeletons are soft and flexible, allowing arthropods to move freely.
    b. Exoskeletons have many joints, allowing arthropods to move freely.
15. Give the number of main body parts and the number of legs that each creature has.
    a. arachnids b. insects

Activities

1. Make a simple terrarium out of a clear gallon jar. Lay the jar on its side, and put about one inch of soil in the bottom. Put in a few rocks and several handfuls of dried leaves. Catch a dozen crickets to put in it, and feed them several pieces of crushed dog food. Soak a small piece of sponge in water, and set it in the jar for the crickets to drink from. If you keep them in a cool, dark place, they should sing for you. Add a handful of fireflies for nighttime sparkle.
2. Some insects have a sense of smell that is much keener than ours. Read in an encyclopedia about two such insects, and tell what they can do with their remarkable sense of smell.
3. Find out why the color rubs off a butterfly's wing.

Review

11. antennae
12. molt
13. egg, larva, pupa, adult
14. b
15. a. two main body parts and four pairs of legs
    b. three main body parts and three pairs of legs

Activities

2. (Sample answers.)
   Some male moths can find other moths from miles away by following scent trails.
   A certain parasitic fly can smell a grub through two inches of solid wood when it is seeking a good place to lay its eggs.
3. The butterfly's wing is covered with tiny colored scales that rub off easily.

Extra Activity

You may want to supervise an insect-mounting activity or a net-making project. To make an insect net, see the instructions on page T–291.

Lesson 10

Unit 2 Review

Review of Vocabulary

abdomen	compound eyes	larva	predator
adult	crustacean	life cycle	pupa
antennae	egg	metamorphosis	spinnerets
arachnid	exoskeleton	molt	thorax
arthropod	gills	nymph	web
chrysalis	hearing membrane		

Animals in the __1__ group have two or three main body parts and jointed legs. Within this group is the __2__, whose name means "creature with a shell." Its hard shell is called an __3__. One such animal is the crayfish, which has three main body parts. The head is the front body part. On it are the eyestalks, the mouth, and two pairs of feelers, or __4__. The head and the second body part, called the __5__, are fastened together in one stiff piece of armor. Under the exoskeleton of the second body part are __6__ that the crayfish uses to breathe. The last main body part of the crayfish is the __7__. It uses this taillike part to swim.

The __8__ is a kind of arthropod with only two main body parts. The most common one is the spider. Spiders have __9__ on the tips of their abdomens. With these, the spider can spin a beautiful silk __10__ to trap its food. Since the spider kills and eats insects and other tiny animals, it is a __11__.

Grasshoppers belong to the group of arthropods called insects. The grasshopper has three stages in its __12__. It begins as an __13__ laid by a female. In the second stage, called the __14__ stage, the young grasshopper looks like an older one but is not able to fly. After shedding its exoskeleton for the last time, the grasshopper enters the __15__ stage. The __16__ of the grasshopper help it to see in many directions at once. Instead of ears, it has small patches on its abdomen that enable it to hear. Such a patch is called a __17__.

The butterfly has four stages in its life cycle. It begins as an egg that hatches into a wormlike __18__ called a caterpillar. The caterpillar eats much vegetation and grows very rapidly. It must __19__, or shed its exoskeleton, several times. Later, the caterpillar forms a case called a __20__ and enters the __21__ stage.

Lesson 10 Answers

Review of Vocabulary

1. arthropod
2. crustacean
3. exoskeleton
4. antennae
5. thorax
6. gills
7. abdomen
8. arachnid
9. spinnerets
10. web
11. predator
12. life cycle
13. egg
14. nymph
15. adult
16. compound eyes
17. hearing membrane
18. larva
19. molt
20. chrysalis
21. pupa

Lesson 10

Remember that children learn by repetition and review. So take a brief journey back through the chapter with your students. Comment on the pictures, and drill major concepts and facts as you go. Notice in particular the differences between crustaceans, arachnids, and insects. Those facts will be useful for them to know later on in life.

22. metamorphosis

Multiple Choice

1. c
2. c
3. a
4. b
5. b
6. c

Inside its case, the caterpillar changes into a butterfly through an amazing transformation called __22__. Finally the butterfly comes out, fans its wings, and flies away.

Multiple Choice

1. Animals with exoskeletons
   a. can perform only a few simple motions.
   b. have soft coverings that allow them to move.
   c. have many joints that allow them to bend and move about.
   d. are hard and stiff and must stay at one spot all their lives.
2. The main body parts of a crustacean are
   a. head and thorax.
   b. head and abdomen.
   c. head, thorax, and abdomen.
   d. head, thorax, and pincers.
3. A spider has
   a. two main body parts and eight legs.
   b. three main body parts and eight legs.
   c. two main body parts and six legs.
   d. three main body parts and six legs.
4. Which of the following is true?
   a. All spiders spin webs.
   b. All spiders can produce silk.
   c. All spiders can see well.
   d. All spiders are pests.
5. A spider's web
   a. is barely strong enough to hold insects.
   b. is made of very strong fibers.
   c. contains only sticky fibers.
   d. always has a beautiful design.
6. Insects have
   a. three body parts and eight legs.
   b. two body parts and eight legs.
   c. three body parts and six legs.
   d. two body parts and six legs.

7. The stages in a grasshopper's life are
   a. egg, larva, adult.
   b. egg, larva, nymph, adult.
   c. egg, nymph, adult.
   d. egg, larva, pupa, adult.

7. c

8. The stages in a butterfly's life are
   a. egg, larva, adult.
   b. egg, larva, nymph, adult.
   c. egg, nymph, pupa, adult.
   d. egg, larva, pupa, adult.

8. d

9. The change from a larva to a butterfly takes place
   a. during the last time the larva molts.
   b. gradually as the larva grows.
   c. during the pupa stage.
   d. after the pupa stage.

9. c

10. Most kinds of insects
    a. destroy crops.
    b. are not harmful and may be helpful.
    c. spread disease.
    d. are quite intelligent.

10. b

11. Catching and mounting insects
    a. requires great skill.
    b. requires expensive equipment.
    c. requires little or no skill.
    d. can be done with simple, homemade equipment.

11. d

12. The best way to kill an insect for a collection is to
    a. pour a chemical on it.
    b. squash it with a stick.
    c. push a pin through it.
    d. freeze it.

12. d

Unit 3

Title Page Photo

The grandeur of a waterfall cannot be captured completely by a photograph. The sense of depth and the sound of falling water is lost. God's earth is full of such marvelous wonders.

Introduction to Unit 3

The primary objective of this chapter is to inspire wonder and praise to God for His awesome works. God has created the earth with enough wonders large and small, common and extraordinary, to keep us constantly amazed and busy learning for several lifetimes. But we sometimes forget what a wonderful place God has made for us. The common becomes commonplace, and we lose our sense of wonder. This hardly pleases God, who rejoices in His works.

Stop and ponder the special design that went into the earth. Do not be satisfied with a glance; take a deeper look, and you will find fresh inspiration.

Make this chapter practical for your students. Every community on earth has its own unique set of wonders. Not everyone experiences earthquakes, nor does everyone have beautiful waterfalls or mountains next door. But everyone is blessed with created wonders. Help your students to discover and appreciate the natural wonders that surround them. If the big wonders are missing, look for the small wonders. Learn to praise God for what you do have, and examine it more closely than ever before. You will be surprised at your discoveries.

Unit 3

Wonders of the Earth and Sky

"Praise ye the LORD from the heavens.... praise him, all ye stars of light.... Praise the LORD from the earth, ye dragons, and all deeps: fire, and hail; snow, and vapours; stormy wind fulfilling his word: mountains, and all hills" (Psalm 148:1, 3, 7–9). When we look around us at the great works of creation, we marvel at God's power. Mountains, storms, waterfalls, earthquakes, volcanoes—all these are proofs of a divine Designer and Creator. The vast beauty and number of the stars inspire wonder as well. Who has ever counted all the stars? Who has ever named them all? Only our great and wonderful God is wise enough to do that.

Many people today believe that all these wonders just happened to come into existence. They say that nobody made them; they just slowly came into being. But the Bible tells us the truth about their origin. It tells us that God created heaven and earth in just six days. We believe the Bible because it is the Word of God. We believe it even if many men today say it is not true.

As we explore the wonders of creation, let us look for things that show God's hand at work. We can find these things in all parts of nature. We must be sure to praise and thank our Maker for all He has given to us.

Story for Unit 3

Trail to the Sky

I was finally going to climb a volcano! For years I had dreamed of doing that. I was visiting my uncle and his family in Guatemala and was helping with a building project in the area. But one pleasant Saturday in July, we took a vacation from our work and headed for Volcano Agua (ä′·gwä).

That eventful Saturday, we awoke soon after 3:30 in the morning. By 5:00 we had packed a lunch and were riding in two vehicles, heading toward the volcano. We approached the mountain and began the steep ascent on a severely bumpy road. The two four-wheel-drive vehicles toiled along the trail for an hour or so.

The early morning was dark and foggy. As our vehicles ground along, we noticed a neighboring volcano, Pacaya (pä·kä·yä′), glowing through the haze ten miles to the west. Occasionally we saw hazy streaks through the gloom as it spewed hot rocks into the air.

The fog began to melt in the light of morning. The air grew thin as we ascended, and the gasoline engine in one vehicle began to balk in the thinning air. By now, the road was a rocky trail that wound its steep and rutted bed in tight switchbacks up the mountain. Finally we had to park our vehicles because water from a recent rain had torn a giant gash across the trail. We had driven nearly a third of the way to the top. Now we detoured around the gully and began climbing on foot. The trail narrowed until it became only a small footpath.

It was a grueling climb. We gasped for breath in the thin air of the high elevation. My legs felt fine, but my lungs could not supply me with enough oxygen. I

Lesson 11

Majestic Mountains

Vocabulary

earth's crust, the outer layer of the earth, about five miles thick under oceans and twenty-five miles thick under continents.

glacier (glā′·shər), a large sheet of slowly moving ice, made from many layers of snow piling up in cold mountain valleys.

lava (lä′·və), hot, molten rock that comes to the earth's surface through a volcano or other crack in the earth's crust; also, the same rock after it cools and hardens.

magma (mag′·mə), hot, molten rock located below the earth's surface.

volcano (vol·kā′·nō), an opening in the earth's crust from which hot lava, ash, and gases erupt; also, a cone-shaped mountain formed by the erupting material.

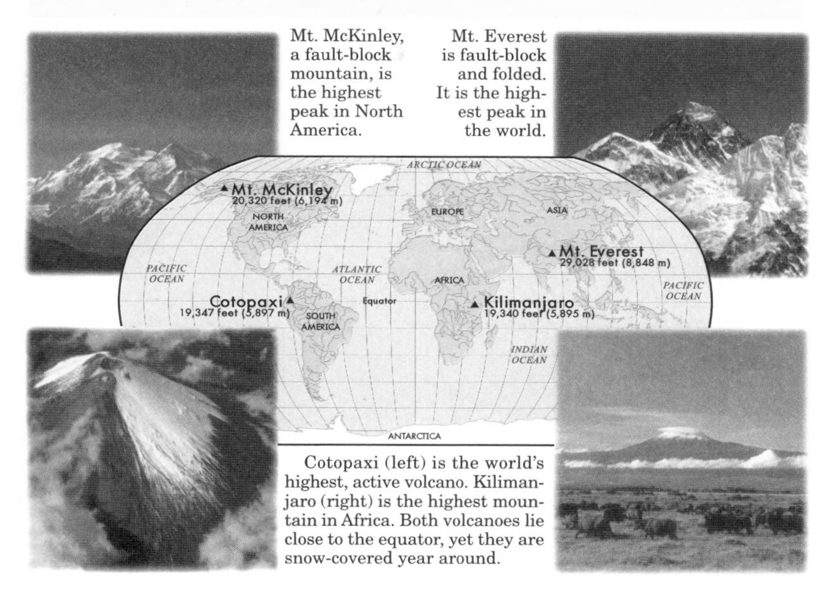

Mt. McKinley, a fault-block mountain, is the highest peak in North America.

Mt. Everest is fault-block and folded. It is the highest peak in the world.

Cotopaxi (left) is the world's highest, active volcano. Kilimanjaro (right) is the highest mountain in Africa. Both volcanoes lie close to the equator, yet they are snow-covered year around.

had to slow to a creeping trudge to avoid getting a severe headache. Back and forth, back and forth, wound the trail. Because of the winding, the trail was not too steep to climb, but it was much longer than if it had gone straight up the side of the mountain.

We paused to catch our breath. A cold breeze and the burning sun fought to control the temperature. Our abandoned vehicles were no more than tiny specks below us. We had an eagle's view of the valley below—a breathtaking scene. But the top was still quite a distance away. We couldn't stop and rest too long!

After what seemed like miles of toilsome climbing, we approached the notch of the crater on Volcano Agua. This crater had once been whole and full of water. But one day in 1541, a large piece broke out of the north lip of the crater, and the lake poured violently into the valley below. The water left great destruction in its path and completely wiped out Ciudad Vieja (sē′·ù·däd′ vyā′·hä), the capital of Guatemala at the time.

Now we were standing in the bottom of the notch that had broken out centuries before. A half-hour climb took us from there to the opposite lip of the crater, the highest point on the volcano. The sight nearly took my breath. The houses in Antigua (än·tē′·gwä), a town five miles north of the mountain, were tiny specks clustered together, glittering occasionally as the sun reflected from the steel roofs.

To the west, the twin peaks of Volcano Acatenango (ä·kä′·te·näng′·gō) and the lone peak of Volcano Fuego (fwā′·gō) rose slightly higher than our level. They were ten short miles away. Smoke rolled from Fuego, an active volcano.

I turned my gaze toward the south. On the horizon, I could see a band of blue that marked the Pacific Ocean, forty miles away. We had to make our time count because light, fluffy clouds were rolling in over the land. The morning breeze had brought in moisture-laden air from the ocean and had pushed it up over the land, where clouds were forming as the air rose and cooled. But we were high up on the volcano, much higher than the clouds.

Nearer and nearer drew the clouds below us. Soon the small puffs that had first formed near the coast became a fluffy mat of white cotton, rushing ever closer. The wind pushed them steadily nearer until the volcano looked like an island in a fluffy white sea. Then the clouds began to ascend the volcano slope toward us. In a matter of minutes, the white mass had churned up the side of the mountain and was swirling past us, piece by wispy piece. We seemed to be way up in the sky where rain and other weather come from. And, actually, we were!

We ate the hearty lunch that we had packed and drank the black coffee that some Guatemalans had given us. And since we could not see through the clouds anyway, we knew it was time to leave our sanctuary of the sky and trek down to earth again.

Slipping and sliding in the loose, sandy volcanic soil and pebbles, we reached the vehicles in a fraction of the time it had taken us to climb to the top. We crawled wearily into our vehicles and made our slow, bumpy way back home. We were sunburned and windburned. Our bodies were aching and terribly tired. But we didn't care. Standing up there in God's wonderful sky had been a priceless experience.

"I will lift up mine eyes unto the hills, from whence cometh my help" (Psalm 121:1). The psalmist thought about God as he looked at the mountains surrounding him. Majestic mountains provide part of the beauty in the world that God has created.

God made hills and mountains of many sizes, from small hills a hundred feet high to mountain peaks towering more than five miles above sea level. The highest mountain is Mount Everest in the Himalayas of southern Asia. It rises about 5½ miles (8.9 km) above sea level.

God has also created great variety in the shapes and features of mountains. Some are grassy. Others are covered with trees. The peaks of many high mountains are rocky and barren because of the cold. These mountain peaks are covered with snow all year long. Hundreds of wild animals live on some mountains, and only a few live on others. Once again, we are inspired by the psalmist who wrote, "O LORD, how manifold are thy works! in wisdom hast thou made them all: the earth is full of thy riches" (Psalm 104:24).

"The Mountains Were Brought Forth"

God created all the majestic mountains on earth. Exactly how or when He did this great work, we are not sure. Perhaps He created some mountains when He formed the earth. Others we know He has brought forth later, even in recent years.

It seems that God formed the mountains we see today in two basic ways. Some mountains developed when material from deep within the earth erupted and piled up on the earth's surface. Others apparently formed when the earth's surface crumpled in certain places, raising the land into peaks.

Volcanic mountains. The mountains formed of material from deep within the earth are called volcanic mountains or simply ***volcanoes.*** The center of the earth is hot enough to melt the rocks and metals that are in it. This hot material is covered by 5 to 25 miles (8 to 40 km) of cooler rock and soil called the ***earth's crust.***

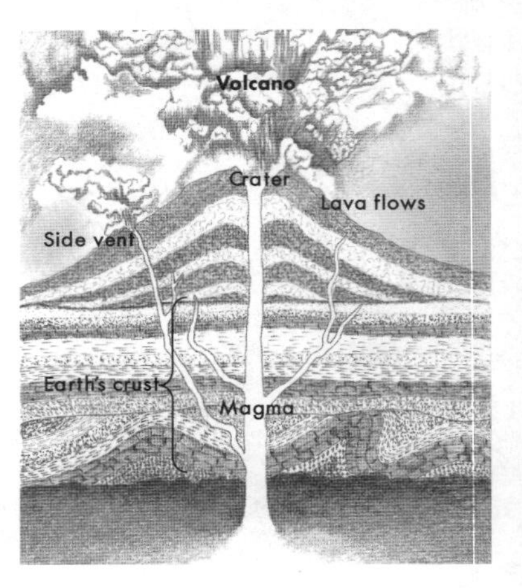

Lesson 11

Lesson Concepts

1. The mountain ranges are the handiwork of God.
2. There are two basic kinds of mountains: *(a)* those formed when the crust of the earth was raised and *(b)* those formed when material from beneath the earth's crust erupted.
3. The center of the earth is very hot.
4. The presence of mountains affects the weather in various ways.
5. The higher the altitude, the colder the temperature.
6. The snow accumulated on high mountains over many years may form a slow-moving river of ice, known as a glacier.
7. The meltwater from mountain snowfall and glaciers provides valuable water for many areas.

Teaching the Lesson

Mountains are giant displays of God's power. Their very existence calls for a skillful Designer and a powerful Creator. Their triangular shapes suggest solidity and power. Their snowcapped peaks remind us of the dignity of a hoary head. We appreciate mountains.

You will not be able to collect mountains as you collect seashells or butterflies, but you can collect beautiful photographs of mountains. Prepare for this lesson by finding several photographs in magazines and flyers. Your students may enjoy supplying pictures as well. Your collection will be more beneficial if it is displayed on a bulletin board, along with the names of the mountains and a map showing their locations. It will be more informative if the mountains are classified as folded, volcanic, or fault-block mountains.

The idea that mountains were formed by movements of the earth's crust should be presented as theory

Some parts of the earth's crust have weak spots, where cracks form in the crust. Great pressure forces hot, melted rock called ***magma*** to flow up through these cracks. When the magma reaches the earth's surface, it is called ***lava.*** This lava piles up and builds a volcano. It usually takes many years to build a mountain out of lava flows.

Some volcanoes grow more rapidly. In the violent ones, magma is not able to flow freely out through the cracks of the volcano. Tremendous pressure builds up underground until finally it can no longer be held back. With great explosions, such volcanoes hurl millions of tons of rock and ash into the sky. The ash, rock, and lava come down around the mouth of the volcano. As this material piles up, the cone-shaped volcano grows rapidly.

Rivers of hot lava flowed down the sides of Paricutín and buried nearby villages, including the lower part of this church.

During the 1940s, scientists were able to watch the birth and development of a violent, rapidly growing volcano. In February 1943, a farmer in Mexico saw smoke coming from a crack in his cornfield. By the next morning, the newborn volcano had started erupting. It grew rapidly, with rumblings and explosions that could be heard at a distance of 50 miles (80 km). One week after it started to form, the new volcano was over 400 feet tall (120 m). Within the next year, it was over 1,000 feet high (300 m) and had buried two small villages. One buried village, called Paricutín, stood about 1 mile away (1½ km). The new volcano was named Paricutín after this village.

Many volcanoes are no longer active and have stopped growing. Paricutín erupted for about nine years and then became inactive.

Fiery, explosive volcanoes often make great changes on the surface of the earth. But we do not need to fear them, for God is watching over us. "God is our refuge and strength, a very present help in trouble. Therefore will not we fear, though the earth be removed, and though the mountains be carried into the midst

rather than established truth. If your students research mountains in standard reference books, they will meet the same terms as those in the lesson (*volcanic, folded, fault-block*); but the process will be set in a time frame of millions of years. Be sure your students understand that God formed the mountains and that the Bible nowhere teaches that He took millions of years to do it.

If you live in a mountainous region, be sure to point out how the mountains affect your local climate and economy. Any tidbit that brings the lesson into your students' world will make it more meaningful to them.

For information about science supply companies, see the introduction of this book.

Discussion Questions

Bring your mountain pictures to class, and discuss the questions in the Activities section.

of the sea" (Psalm 46:1, 2).

Folded and fault-block mountains. Volcanoes are formed from hot materials forcing their way up from deep within the earth. Other mountains were seemingly formed by a crumpling of the earth's crust. How this happened we do not know, but perhaps God used the Flood to form these mountains. It may have happened something like this: As the waters went down, much shifting and moving of the earth's crust probably took place. Some parts of the crust were pushed down, and some were pushed up.

The upward or downward movement in some places would have caused huge ripples in the earth's crust. This is how the mountains known as folded mountains appear to have developed. In other places, it seems that the movement caused great cracks in the crust, and one section was thrust above another. This is how the mountains known as fault-block mountains may have formed. Of course,

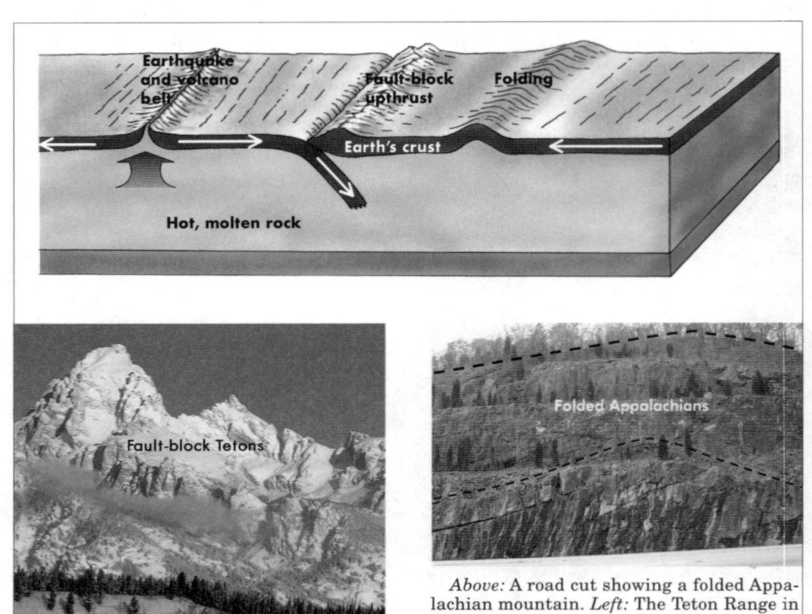

Above: A road cut showing a folded Appalachian mountain. *Left:* The Teton Range in Wyoming is an upthrusted fault block.

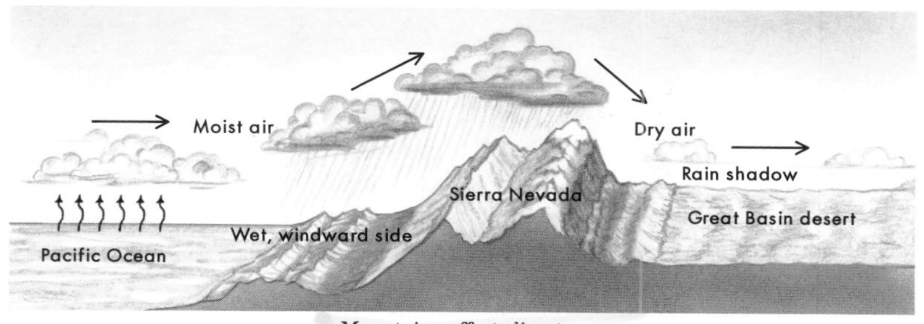

Mountains affect climate.

wind, water, and ice in the years since the Flood have changed and rounded the mountains until they have the shapes that we see today.

Mountains and Climate

Temperature. Have you ever traveled from the bottom to the top of a mountain on a warm day? If so, you probably noticed that it became cooler as you climbed higher. For every 1,000 feet (300 m) of increase in altitude, the temperature generally drops 3 or 4 degrees F (2 degrees C). That is why many tall mountains have snow on their peaks even in the middle of summer.

Rainfall. When warm, moist air from the ocean blows over a mountain range, an interesting thing happens. As this air pushes up over the mountain, it cools and forms clouds in the same way that your warm, moist breath forms a cloud on a cold morning. Much rain and snow falls from these clouds on the windward side of a mountain range (the side from which the winds usually blow). This makes the windward side of the mountain a lush, moist place.

The other side of the mountain is quite different. By the time the ocean air pushes up over the top of the mountain, it is cold and has dropped most of its moisture. As it moves down the opposite side of the mountain, it warms again and becomes very dry. A dry area called a rain shadow forms on this side of the mountain. A good example of this is the Sierra Nevada in the southwestern United States. They are very wet on the west side and very dry on the east side.

Mountains with snowy peaks help to provide moisture in another way. Some of the snow melts during summer, and this provides a good source of water for many areas, especially in rain shadow regions. Each winter a new supply of snow is

A glacier in Alaska

stored in the mountains to supply the surrounding areas with water.

On some mountains, the snow becomes so thick that over a period of years it packs into huge beds of ice called ***glaciers.*** Mountain glaciers become so heavy that they begin to slide slowly down the mountain valleys like huge rivers of ice. In one day a glacier may move only several inches or as much as 50 feet (15 m). The glaciers melt as they reach the warmer regions of lower altitudes, and the meltwater flows into streams and rivers that run through the valleys and flatlands below.

Exercises

1. Mountains that are formed of material coming from beneath the earth's crust are called ———.
2. Below the earth's crust,
   a. it is so hot that rocks burn.
   b. it is cold, but the rocks are soft and liquid.
   c. it is hot enough to melt rocks and metals.
   d. there is a layer of lava.
3. Molten rock that pushes up through the earth's crust is called ——— until it reaches the earth's surface. Then it is called ———.
4. Volcanic peaks are made of what three materials?
5. What great Bible event did God use to change the surface of the earth and form many mountains?
6. What two kinds of mountains were probably formed by crumpling of the earth's crust?
7. What happens to the air temperature as you climb a tall mountain?
8. Explain in your own words why some mountains receive much rain on one side and very little rain on the other side.
9. How do snowcapped mountains help to provide water for surrounding areas?

Lesson 11 Answers

Exercises

1. volcanoes
2. c
3. magma, lava
4. ash, rock, lava
5. Noah's Flood (Genesis Flood)
6. folded and fault-block mountains
7. It becomes cooler. (The temperature drops 3 or 4 degrees F per 1,000 feet of elevation.)
8. As warm, moist air moves up one side of a mountain, the air cools and its moisture falls as rain. By the time the air reaches the other side of the mountain, it is very dry.
9. Some of the snow melts during summer, and the water flows down the mountains.

Review

10. What two great events did God use to make the earth as it is today?
11. What is science?
12. What are the four steps in performing an experiment?
13. Fill in the blanks.
    a. Crustaceans have ——— main body parts.
    b. Arachnids have ——— main body parts and ——— pairs of legs.
    c. Insects have ——— main body parts and ——— pairs of legs.
14. Give three ways that insects help us.

Activities

1. Find pictures of mountains. Can you tell by their shapes whether they are volcanic, fault-block, or folded mountains? Find pictures of the Rocky Mountains in western North America. Which kind are they? What kind of mountains are nearest to where you live?
2. Use Play-Doh to show how God may have formed mountains by the folding of the earth's crust. Use a rolling pin to form several ¼-inch layers of different colors. Push the ends together to produce a folding effect. Notice how the layers are buckled as you see them in road cuts.
3. The largest glacier in the world is not a mountain glacier but the icecap covering the continent of Antarctica. Find out how thick this icecap is.
4. In the United States, a major volcanic eruption occurred on May 18, 1980, at Mount St. Helens in the state of Washington. Find answers to the following questions about this eruption.
    a. In the eruption, the top of Mount St. Helens was blown off. How much of its height did the mountain lose?
    b. How many square miles of forest were leveled in the explosion?
    c. How many people were killed?

This lava column on Mount St. Helens was one mile wide and extended twelve miles up into the atmosphere.

Review

10. the Creation and the Flood
11. Science is the study and observation of God's wonderful world.
12. (1) Write down the question you are trying to answer.
    (2) Plan an experiment.
    (3) Do the experiment, and record the data.
    (4) Use the data to answer your question.
13. a. three
    b. two, four
    c. three, three
14. (Any three.)
    by pollinating garden plants and fruit trees;
    by helping to get rid of dead animals;
    by adding variety to our lives;
    by providing useful things like honey, dye, and shellac;
    by eating other insects that are pests

Activities

3. The icecap of Antarctica is probably one or two miles thick in many places. However, Antarctica has not been completely explored, and even thicker parts of the icecap may yet be found. This kind of glacier is called a continental glacier. The weight of the ice causes such tremendous pressure that the ice slides slowly outward from the interior until it reaches the sea and breaks off as icebergs.
4. a. The mountain lost more than 1,300 feet (396 m) of its height.
   b. About 200 square miles (518 km²) of forest were leveled.
   c. Sixty people were killed.

   The *National Geographic* of January 1981 has some good photographs of the Mount St. Helens eruption. You can also obtain information about this eruption from the Institute for Creation Research.

   Institute for Creation Research
   P.O. Box 59029
   Dallas, TX 75229
   800-628-7640

Extra Activity

Build a model volcano. Using plaster of Paris or clay, form a cone-shaped mountain 6 to 8 inches high. Leave a crater on the top, about 1½ inches across and 1½ inches deep. To make your volcano erupt, fill the crater with fine ammonium dichromate crystals, which can be purchased from a science supply house. When the chemical is ignited, the ash will flow down the mountain in a manner very similar to the way lava flows from a real volcano.

This is a very impressive demonstration. However, it must be done under adult supervision and with good ventilation. The ammonium dichromate crystals are somewhat expensive. You could buy a whole pound (or 500 grams) and share the cost with other schools. Four ounces (or 100 grams) should be enough material for several eruptions.

Lesson 12

Awesome Earthquakes

Vocabulary

earthquake, a shaking of the earth's crust believed to result from the shifting of rock layers.

fault, a crack in the earth's crust where the rock layers shift from time to time.

seismograph (sīz′·mə·graf′), an instrument that detects and records tremors in the earth's crust.

tremor (trem′·ər), a shaking or vibrating, as of the earth's crust.

tsunami (tsü·nä′·mē), a large ocean wave caused by an earthquake or a volcano.

"Then the Earth Shook"

"Then the earth shook and trembled; the foundations also of the hills moved and were shaken" (Psalm 18:7).

About three o'clock on the morning of February 4, 1976, the peaceful Guatemalan countryside was suddenly wakened from sleep. For half a minute, the ground quaked and trembled. In those brief thirty seconds, an ***earthquake*** caused great destruction. Buildings crumbled, and hillsides slid away. Twenty-three thousand people died. More than seventy thousand others were injured. Over one million persons lost their homes.

The Guatemala quake caused parts of this road to slide away.

Many Guatemalans died in collapsed houses.

Lesson 12

Lesson Concepts

1. Earthquakes apparently result from shifts in the rock layers of the earth's crust.
   a. When frequent small tremors occur, they appear to relieve the stress at fault lines little by little, thus preventing major earthquakes.
   b. When stress at a fault line builds up until a major shift occurs, the result is a severe earthquake.
2. The force of an earthquake can wreck buildings, change the course of a river, and cause huge ocean waves called tsunamis.
3. The strength of earthquakes is measured on the Richter scale.
4. Scientists monitor earthquakes and tremors with seismographs.

Teaching the Lesson

Has your area ever been rippled by an earthquake or a tremor? If so, you can easily relate today's lesson to your students' experience. If not, you will need to put forth effort to help them understand just what does happen. Yes, the ground shakes and things fall over. But can your students actually visualize that? Imagine a severe earthquake in your classroom. What would happen? Desks would jostle around, and books would fall off shelves. Students would have difficulty walking, especially amid the shifting and falling debris. Even modern school buildings could be reduced to shambles.

Perhaps you could stage an earthquake drill, which is a regular part of school life in earthquake-prone Japan. In such a drill, all the children quickly huddle under their desks or evacuate the building. These things are real for people living in quake zones.

Portray quakes as awesome manifestations of God's

Earthquakes are nothing new to the people of Guatemala. Many quakes strike Guatemala every year, but most of them are only small ones called ***tremors.*** Some of these tremors can hardly be felt, while others are strong enough to knock jars off shelves. Only rarely does the land of Guatemala suffer an earthquake as strong as the one in 1976.

Cause of Earthquakes

Many scientists believe that the earth's crust contains a number of huge plates of rock. These plates float upon the magma within the earth, and they are constantly moving. Little by little, they slide past each other, perhaps at about the rate your fingernails grow. At the edges where the plates meet, there are cracks called ***faults*** in the earth's crust. It is along these faults that earthquakes occur. Stress builds up as the plates try to slide past each other, and from time to time they slip. With every slip, the ground trembles.

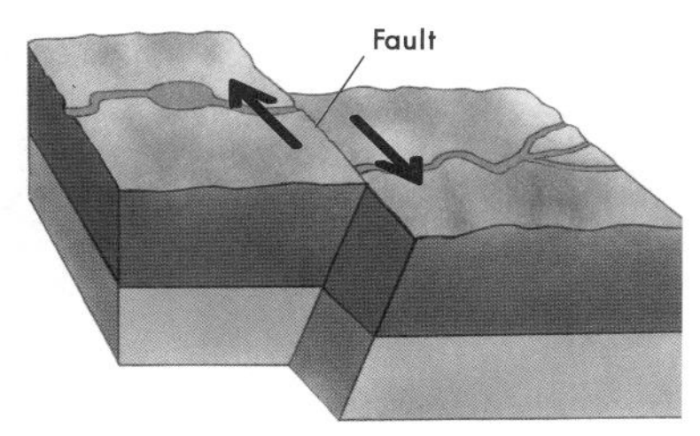

Some plates slide past each other. The diagram on page 68 shows that plates also slide apart or even slide toward each other.

As long as many small tremors occur, there usually is no large earthquake. But sometimes two plates jam, and the small shifts and tremors stop for a while. More and more stress builds up as the plates try to slide past each other. Finally the stress becomes so great that the plates suddenly shift—perhaps 20 feet (6 m)—in one tremendous jolt. It is these great shifts in the rock layers that cause serious earthquakes like the one that shook Guatemala in 1976.

Effects of Earthquakes

An earthquake may actually cause the ground to ripple in waves. These waves cause great damage to buildings as the earthquake twists and shakes their foundations. In many earthquakes, falling buildings kill more people than any other danger caused by the shaking.

Earthquakes often cause fires in cities. With gas lines broken and electric wires down, fire can spread quite rapidly through the ruins. The fire may rage uncontrollably because streets are often so full of rubble that firefighters cannot get through. Besides, with underground water lines broken, little water is available to fight fires. In cities, fire may be the greatest danger brought by earthquakes.

When a major earthquake occurs under the ocean, it causes great shock

power, not merely as thrilling escapades of nature. The toll of human life that they exact makes them serious. However, do not make your students overly fearful. Although earthquakes are sobering events, the hand of God is plainly evident. We put our trust in Him.

Photographs of earthquake destruction are indispensable in such a lesson. If possible, get snapshots taken by someone you know who has been in a quake zone during or after an earthquake. *National Geographic* photos are also excellent. The issues of July 1964, June 1976, and May 1990 are good sources.

Measurement of earthquake energy is not as precise as it may appear in the pupil's text. Different references will report different magnitudes for the same quake. Some of the variation stems from the use of differing scales, and some is just a matter of disagreement among scientists. The figures given in the lesson follow the majority of sources. Do not be surprised if you find differing figures in some reference books.

Discussion Questions

1. Discuss what a tremor must feel like.
   Your students have probably felt the ground shake as a large tractor-trailer rumbled past. Or perhaps they have felt a herd of cattle running close by. A small tremor is somewhat similar to these. Larger quakes, of course, are much stronger!
2. Why do scientists not prevent earthquakes instead of only trying to predict them?
   Earthquakes are much too powerful for man to control. Only the power of God can control an earthquake.

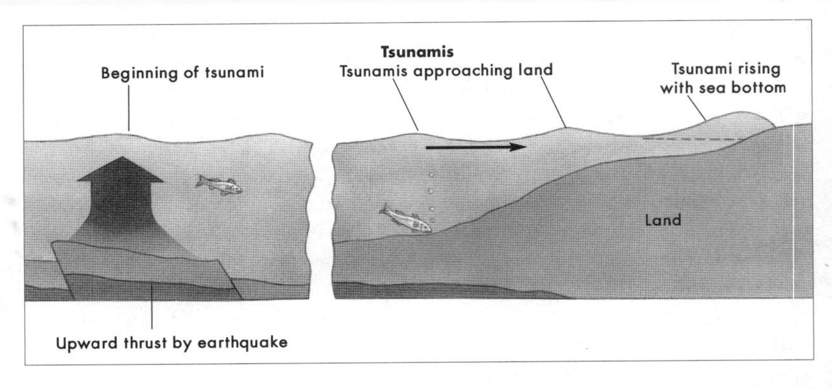

waves that travel swiftly through the water. These shocks are no danger in the open sea, for they cause surface waves only a few feet high. But when the shocks reach shallow water, they cause ***tsunamis*** that tower as high as 30 to 100 feet (10 to 30 m) or even higher. These gigantic waves rush in over land, trees, and buildings, and then they carry the rubble out to sea as the water runs back. Some tsunamis have completely destroyed towns and villages on the seashore.

Even towns far from the earthquake may be endangered by tsunamis. Tsunamis may travel 500 miles per hour (800 km/h) through the open sea and strike coasts hundreds or even thousands of miles away from the quakes that caused them.

Earthquakes change many things. Bridges collapse, hillsides slide away, roads break, and buildings crumble. Sometimes the course of a river even changes to a new location because of shifts in the earth's crust. Surely, earthquakes display the great power of God.

Detecting and Recording Earthquakes

Since it is impossible to prevent earthquakes, people try to predict them instead. Scientists use ***seismographs*** to detect and record even the faintest tremors. They study these tremors to

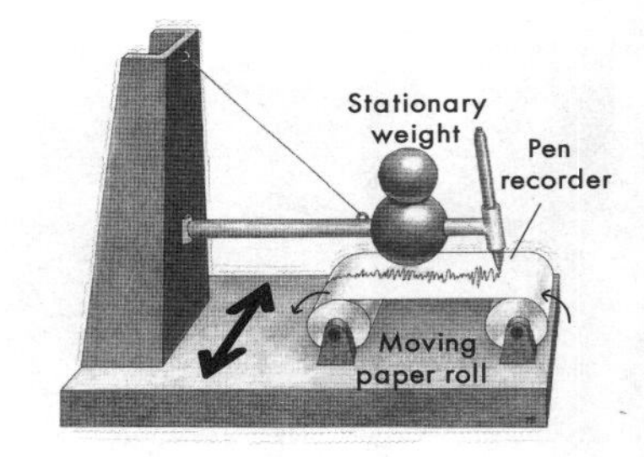

A seismograph

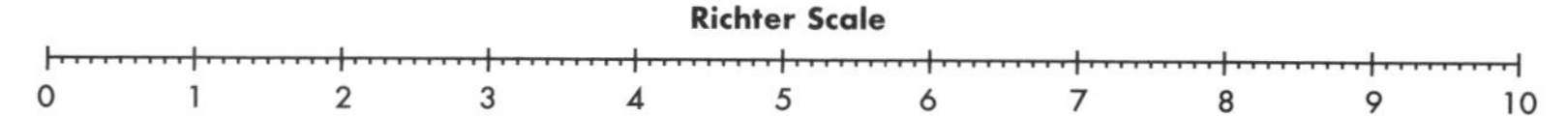

For each whole number higher on the Richter scale, the quakes are ten times as great as the number before.

see if they can find patterns that will give a clue about where and when the next big earthquake will strike. So far, they have not been very successful.

Scientists use the Richter (rik′·tər) scale to describe the strength of earthquakes recorded with their seismographs. A slight tremor may register 2.0 on the Richter scale, and an extremely powerful earthquake may measure between 7.0 and 9.0. For each whole number higher on the Richter scale, the shock waves of earthquakes are ten times greater. A quake that registers 6.0 on the Richter scale has waves ten times greater than one that measures 5.0. The earthquake that struck Guatemala in 1976 registered 7.5 on the Richter scale.

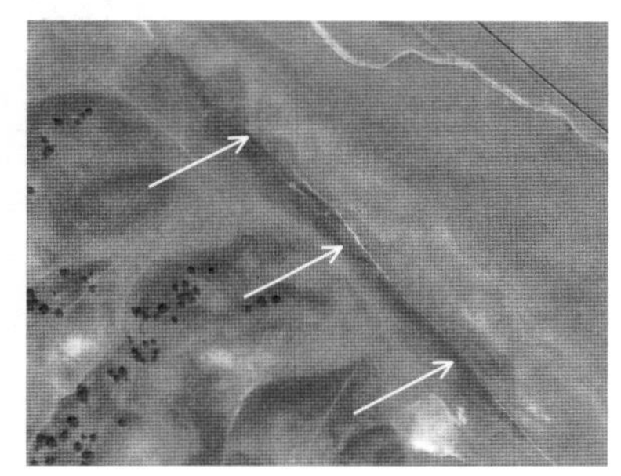

An aerial view near Cholame of a ravine caused by the San Andreas Fault

Notable Earthquakes

On the afternoon of October 17, 1989, a major shift occurred along the famous San Andreas (san·an·drā′·əs) Fault in California. San Francisco trembled for fifteen seconds as it was struck by an earthquake registering 7.1 on the Richter scale. Most of the tall buildings had been specially designed so that they could survive an earthquake. But many buildings still crumbled. Large sections of a two-level highway bridge fell, crushing the cars below. Fires from broken gas lines raged in many places throughout the city, but firefighters were able to put out the flames before any of them were out of control. In spite of modern buildings and equipment, 63 people were killed and 3,800 were wounded. The

The San Andreas Fault extends about 600 miles. California has many smaller faults too.

losses of property amounted to six billion dollars ($6,000,000,000).

Twenty-five years earlier, in March 1964, a terrible earthquake rocked Anchorage, Alaska. This earthquake registered 8.6 on the Richter scale and killed 131 people. Some parts of Anchorage dropped 30 feet in a few seconds, resulting in the destruction of many buildings.

Much damage in the Alaskan earthquake came from the tsunamis it caused. In the coastal village of Chenega, one-fourth of the people were killed as the huge waves crashed down upon them. All that the people could do was try to get out of the way. Only God could create the mighty forces in these earthquakes.

The Alaskan earthquake collapsed bridges and sank streets. Half of the building above sits on stable ground, but the other half sits on an area of town that sank at least one story.

Exercises

1. True or false? If many small shifts and tremors occur at a fault, a major earthquake is likely at any time.
2. Where two sections of the earth's crust slide past each other, the crack between them is called a ———.
3. Explain what causes an earthquake.
4. At a fault, what may be a sign that stress is building up for a major earthquake?

Lesson 12 Answers

Exercises

1. false
2. fault
3. In an earthquake, two sections of the earth's crust shift with a sudden jolt.
4. absence of small quakes in the area

5. Which two results of an earthquake are especially dangerous for people in cities?
6. When an earthquake strikes, towns along a seashore often suffer the greatest damage from
   a. cracks in the ground.
   b. fires.
   c. tsunamis.
   d. shock waves in the ground.
7. What do scientists use seismographs for?
8. A slight tremor might measure (2.1, 5.6, 8.8) on the Richter scale, and an extremely violent one might measure (2.1, 5.6, 8.8).
9. True or false? Modern cities are built so well that very little damage occurs during an earthquake.
10. According to their rating on the Richter scale, which earthquake described in this lesson was the strongest?

Review

11. In what two main ways has God formed mountains since the Creation?
12. How do mountain glaciers help to provide water for surrounding valleys?
13. How can you tell that the center of the earth is very hot?
14. Give two ways in which mountains affect climate.

Activities

1. Perhaps you know someone who has experienced an earthquake. He may be willing to tell the class what an earthquake looks like, feels like, and sounds like.
2. Look at the earthquake pictures that your teacher brings. Which things suffer more damage in an earthquake—man-made buildings? or things that God created, like trees and mountains?
3. a. Some of the worst earthquakes in history have occurred in China. Use an encyclopedia to learn how many Chinese people were killed in the earthquake of 1556.
   b. Why were so many people killed?
4. Volcanoes as well as earthquakes produce tsunamis. In an encyclopedia, read about the destruction of the island of Krakatoa (or Krakatau), and answer the questions below.
   a. In what year was Krakatoa destroyed by a volcano?
   b. How high were the tsunamis that resulted?
   c. How far did the tsunamis travel?

5. falling buildings and fire
6. c
7. to detect and record earthquakes
8. 2.1, 8.8
9. false
10. the earthquake in Anchorage, Alaska, in 1964

Review

11. God formed volcanoes of materials that erupted from within the earth.
    He formed other mountains by the crumpling of the earth's crust, perhaps at the Flood.
12. The glaciers slide slowly down into the valleys and melt as they reach lower altitudes.
13. Molten rock called lava forces its way out of the earth in volcanoes.
14. High mountain altitudes cause cooler temperatures.
    Mountains cause rainfall (and rain shadows) when they cause moist air to rise.

Activities

2. The June 1976 *National Geographic* has an article on the Guatemalan earthquake of February 1976. The July 1964 and May 1990 issues of *National Geographic* also have good earthquake pictures.
3. a. The number of people killed was 830,000.
   b. China has a large population. The quake occurred at night, causing many houses to collapse on the people inside. Also, thousands of people lived in caves dug into unstable cliffs of loess—a fine, windblown silt—and these crumbled when the powerful earthquake struck. Its strength is estimated today at 8.3 to 9.8 on the Richter scale.
4. a. Krakatoa was destroyed in 1883.
   b. The tsunamis were 100 to 120 feet high (30 to 36 m). About 36,000 persons drowned.
   c. Sources vary. Some say 8,000 miles (13,000 km), though no damage is mentioned at that distance. *Nature on the Rampage,* published by National Geographic, says that the waves traveled around the globe two or three times.

Lesson 13

Magnificent Waterfalls

Vocabulary

energy, the power to do work, occurring in forms like heat, light, and sound.

flow, the amount of water running through a pipe or over a waterfall.

head, the distance that water falls at a waterfall or dam.

hydroelectricity (hī′·drō·i·lek·tris′·i·tē), electricity generated by waterpower.

kinetic energy (kə·net′·ik), the energy possessed by a moving thing; the energy of motion.

potential energy (pə·ten′·shəl), energy that is ready to be used; stored energy, such as the energy of water behind a dam.

The Water Cycle

Have you ever wondered where rivers get all their water? A river can flow for years and never run dry. When God created the earth, He designed it with a wonderful system called the water cycle. This system waters plants all over the world and purifies the water at the same time.

The water cycle begins out over the oceans, where winds absorb great amounts of evaporating water. The winds then carry this moist air over an area of land. As the land forces the air upward, the air becomes cooler, its moisture condenses into clouds, and the clouds produce rain that falls to the ground. Some rainwater is used immediately by plants. Some sinks into the ground to provide water for wells and springs. Some does not enter the soil, but runs off into streams and creeks. All runoff water eventually flows into rivers, which carry the water back out to the ocean. Then it is ready to evaporate again and be carried back over the land to fall as fresh rainwater.

Waterfalls occur in the river part of the water cycle. When a flowing river

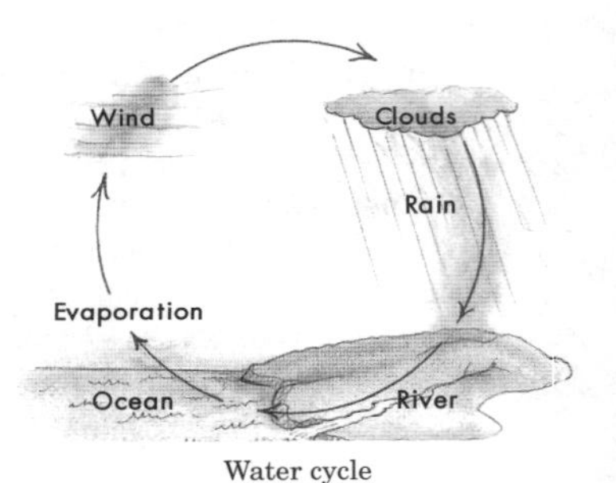

Water cycle

Lesson 13

Lesson Concepts

1. Rivers and waterfalls are part of the water cycle.
2. Energy is the ability to move something.
3. The energy of waterfalls comes from water that was lifted by evaporation and carried by winds, both of which are powered by the sun.
4. The greater the head and the flow, the greater the energy in a waterfall.
5. Waterfalls are good sites for hydroelectric plants.
6. Dams are often built to provide hydroelectricity and water for irrigation.
7. Great waterfalls display God's ability to produce power and beauty.

Teaching the Lesson

Waterpower is an age-old source of energy. For years, men have harnessed moving water to operate machinery or to generate electricity. Moving water is a continuously renewing resource, making it an excellent source of energy. A river can be harnessed year after year without diminishing its flow.

The idea that waterpower is a form of solar energy might be a bit deep for your students. Help them to see the connection between the water cycle and solar energy. If they can grasp that concept, then the fact that waterfalls are a part of the water cycle should help them get the connection between the sun and waterpower.

Dams enter the picture here because of the power production they make possible. Essentially, dams create man-made waterfalls. The water passing over or through a dam has the force of a waterfall and can generate electricity as well.

reaches a cliff and tumbles over the edge, it produces a waterfall.

Energy in a Waterfall

Falling water has much force. That makes waterfalls a good source of ***energy.*** Of course, all the energy in a waterfall comes from water that was lifted by evaporation and carried by winds. And since heat from the sun causes both evaporation and winds, a waterfall actually receives its energy from the sun.

The energy of a waterfall is ***kinetic energy*** because the water is in motion. As water falls, much energy is released. We can use this energy to help us do work.

Two things determine how much energy a waterfall has. One thing is the ***head,*** or the distance that the water falls. The higher a waterfall is, the greater its head will be. The other thing that determines energy is the ***flow,*** or the amount of water running over the waterfall. The larger the river is, the larger will be its flow. Therefore, the farther the water falls and the larger the river is, the more energy a waterfall has.

In the past, people harnessed the energy of falling water by using water wheels. Today they use the falling water to spin turbines (tėr′·binz), which make better use of the water's energy than water wheels do. These turbines usually drive generators to make electricity. Electricity that is generated by waterpower is called ***hydroelectricity.***

If a river has no waterfall and drops too gradually to spin a turbine, people

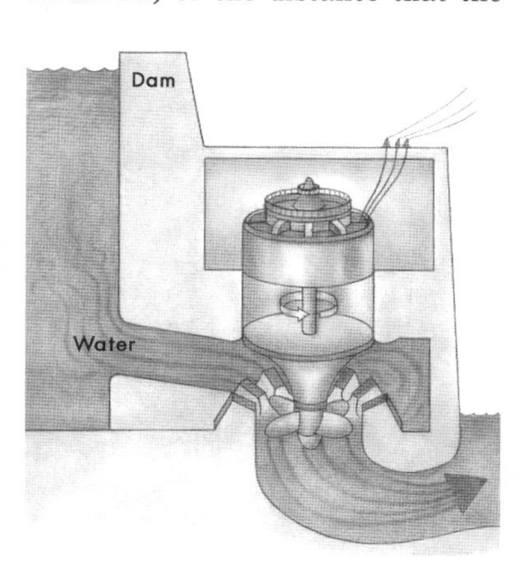

Hoover Dam

Pictures of Niagara Falls in the East, Yosemite Falls in the West, or any other magnificent falls will add sparkle to the lesson. Better yet, is there a waterfall near your school? Take your students to see it. A real waterfall, even a small one, is more impressive than a picture of a large one. Perhaps there is a hydroelectric plant in your area. By all means, tell your students about it, and arrange a tour if feasible.

Discussion Questions

1. In what way does a waterfall receive its energy from the sun?
   The sun provides the heat that causes both evaporation and winds, by which water is lifted and carried over the land to fall as rain.
2. Why is a waterfall a good place to build a hydroelectric plant?
   The falling water provides energy for generating electricity.

sometimes build a dam. The dam raises the water level and forms a man-made lake. This lake has great ***potential energy,*** or stored energy. The water is waiting to run downward through pipes and spin the turbines to generate electricity. Two good examples of dams are the Grand Coulee Dam in Washington and the Hoover Dam on the border of Nevada and Arizona. Millions of kilowatts of hydroelectricity come from these dams, as well as irrigation water for millions of acres of farmland.

God's Magnificent Waterfalls

One of the most famous waterfalls in North America is Niagara Falls. These beautiful falls are on the Niagara River between Lake Erie and Lake Ontario. Goat Island divides the river just upstream from the falls. About one-tenth of the water flows around the east edge of the island and spills over the American Falls. The rest of the water flows around the west edge and thunders over the Horseshoe Falls. With a width of 2,200 feet (670 m), the Horseshoe Falls is more than 2½ times as wide as the American Falls. About 48 million gallons (180 million liters) of water flow over the falls every minute, plunging 170 feet (52 m) to the rocks and whirlpools below.

Not all the water of the Niagara

Horseshoe Falls, part of Niagara Falls

River flows over the falls today. Instead, some of it flows through canals and huge pipes to power stations in the river gorge. Together, the power stations at Niagara produce more than four million kilowatts of hydroelectricity. That is enough to supply hundreds of thousands of homes with electricity.

The highest waterfall in the world is Angel Falls in Venezuela. This waterfall is not nearly as wide as Niagara Falls, but it is much higher. Heavy rains fall on the mountain plateau, and the water flows into huge crevices and underground streams. Near the top of a high cliff, these streams spout out below the canyon rim. The water falls over half a mile (807 m) to the rocks below. From there, it descends another 564 feet (172 m) to the lower river. The water drops 3,212 feet (979 m) in all.

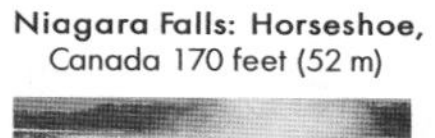

These pictures compare the heights of several well-known waterfalls. Niagara Falls has a small head and a large flow. Angel Falls, the highest waterfall in the world, is nearly nineteen times higher, but its flow is fairly small.

Exercises

1. The water in rivers comes from rainfall, but where do the rain clouds get their moisture?
2. Since falling water is in motion, we say it has ——— energy.
3. What two things determine the amount of energy in a waterfall?

Lesson 13 Answers

Exercises

1. from the ocean
2. kinetic
3. head and flow

4. c
5. the seventy-foot falls (since its head is greater)
6. more energy than normal; because of greater flow
7. turbines
8. to generate hydroelectricity
9. dams
10. the Horseshoe Falls; because much more water flows over it

Review

11. before—magma; after—lava
12. sudden, violent shifting of plates in the earth's crust
13. tsunamis
14. falling buildings, fires, tsunamis
15. fault

Activities

1. The three highest dams in the world are as follows:
   Rogun, 1,099 feet (335 m), Tajikistan
   Nurek, 984 feet (300 m), Tajikistan
   Grand Dixence, 935 feet (285 m), Switzerland
2. Dams are always much thicker at the bottom than at the top because of the tremendous pressure at the base.
3. Temporary dams are used to divert the water of the river away from the dam site. The entire stream may be diverted through a tunnel or canal, or half the riverbed may be used for the water while the dam is being constructed across the other half.

4. Head is best described as
   a. the amount of water that runs over the falls.
   b. the width of the falls.
   c. the height of the falls.
   d. the depth of the river above the falls.
5. Suppose a river flows over a fifty-foot waterfall and then a seventy-foot waterfall. Which falls would you expect to have more energy?
6. When the river flowing over a waterfall is unusually high after a rainstorm, would the falls have more energy than normal or less energy than normal? Why?
7. Instead of water wheels, today people generally use ——— to harness the energy in waterfalls.
8. For what purpose is waterpower most commonly used today?
9. When no waterfalls are available, people build ——— to harness waterpower.
10. At Niagara Falls, which has more kinetic energy—the American Falls? or the Horseshoe Falls? Why?

Review

11. What is molten rock called before it reaches the earth's surface? What is it called afterward?
12. What causes earthquakes?
13. When earthquakes occur under the ocean, they cause powerful waves called ———.
14. What three things resulting from earthquakes are harmful to people?
15. A ——— is a crack in the earth's crust where the rock layers shift from time to time.

Activities

1. Use a recent encyclopedia or almanac to discover which three dams are the highest ones in the world. Give their heights and the countries where they are located.
2. Find out which is always the thickest part of a dam—the top or the bottom. Why is this?
3. You may wonder how men can construct a dam across a large river. What do they do with the water while they are building? See if you can find some answers in a reference book.

Extra Activity

If it is feasible, take your students to see a waterfall. Showing them a falls and explaining *head* and *flow* in terms of what they see will make the concepts more concrete for them. Even a small waterfall in a local creek would be helpful.

Lesson 14

Mighty Oceans

Vocabulary

food chain, a series of creatures in which smaller members are eaten by larger members, and these larger members are eaten by still larger members.

ocean current, a channel of moving water in the ocean, influenced by winds and by water temperature.

plankton (plangk′·tən), tiny plants and animals drifting in the ocean and used as food by fish.

wave, a moving ridge of water such as is common on the ocean surface.

"The floods have lifted up, O LORD, the floods have lifted up their voice; the floods lift up their waves. The LORD on high is mightier than the noise of many waters, yea, than the mighty waves of the sea" (Psalm 93:3, 4).

The oceans are vast stretches of water extending for thousands of miles, much farther than you can see. Small lakes or ponds may sometimes be glassy smooth, but the oceans are continually in motion.

The Movement of the Oceans

Two kinds of motion are found in the oceans: waves and ocean currents. ***Waves*** are the familiar ridges that constantly move across the surface of the

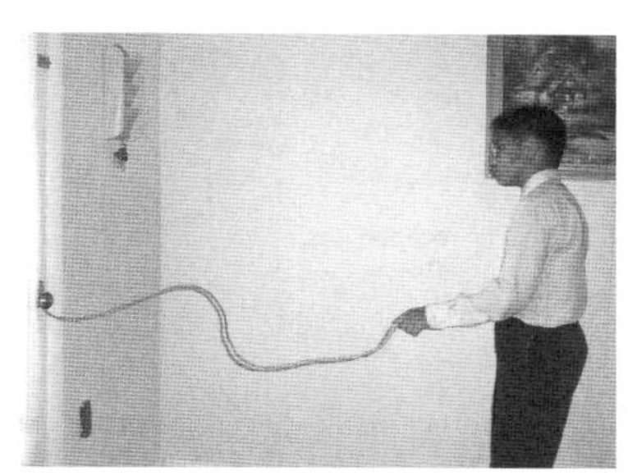

Waves in the ocean do not move water, just as waves in a rope do not pull the rope out of your hand.

Lesson 14

Lesson Concepts

1. The oceans cover about 70 percent of the earth's surface.
2. Oceans provide huge evaporation basins for the water cycle.
3. Oceans provide storage for heat.
4. Ocean currents distribute heat or cold and modify the climate.
5. Oceans are a major source of food.
6. Oceans are connecting links for transportation between continents.

Teaching the Lesson

Oceans cover more than two-thirds of the earth's surface, but these great stretches of water are unfamiliar territory to most people. Familiar or not, oceans have at least some influence on almost everyone. Ocean temperature and moisture influence our weather. Ocean algae supply more oxygen than land plants. Oceans provide food, fertilizer, and chemicals. This lesson should help your students to see the hand of God in the blessings that the mighty oceans bring to us.

Some concepts in today's lesson are difficult to illustrate. Heat distribution and marine influence on climate are not easily captured in photographs. But do your best to make the lesson practical. Pictures of sea creatures and ocean vessels will spark some interest. Perhaps you have some students who are fond of seafood. Oysters, shrimp, tuna, sardines, anchovies—

water. They vary in size from small ripples to towering tsunamis many yards high. Although waves themselves move across the surface of the water, they cause only an up-and-down motion. Waves do not carry water along.

Ocean currents are like rivers moving slowly through the ocean. But ocean currents are much greater than rivers. They may carry as much water as a thousand rivers, and they are thousands of miles long. Yet ocean currents are not visible as waves are, because they hardly disturb the ocean's surface. Currents flow steadily and slowly day and night, constantly stirring and mixing the water of the whole ocean.

Oceans cover about 70 percent (seven-tenths) of the earth's surface. The continents are like great islands surrounded by all this water. These masses of land separate the water into four different oceans: the Atlantic Ocean, the Pacific Ocean, the Indian Ocean, and the Arctic Ocean. With all these waters combined, the earth has about 130 million square miles (335 million km^2) of ocean.

Oceans and Climate

The oceans greatly affect the climate of the continents. Large amounts of water evaporate from the oceans and form clouds. Winds carry the clouds over land, where the moisture falls as rain or snow.

Oceans also influence the temperature of the continents. A body of water does not warm or cool as quickly as land does. So in summer, the land is warmer than the ocean, and winds blowing across ocean water help to cool the land. In winter, the land is colder than the ocean, and winds from the ocean blow warmer air over the land. In this way, the oceans help to make both summer and winter temperatures

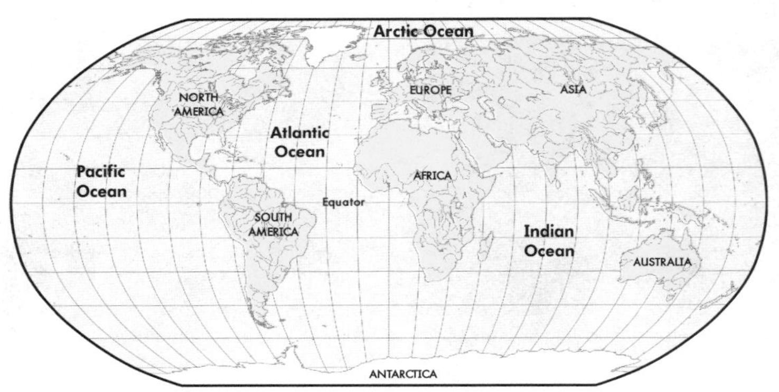

there are many sea dishes that people enjoy even if they live far from the ocean. If you can connect the ocean with a favorite food, you have established a useful link.

Discussion Questions

1. Why is more than two-thirds of the earth covered by oceans?
   People might think it would be better if there were more land and less sea. But God made no mistake when He made the oceans as large as they are. The oceans are huge evaporation basins that provide moisture to water the land. They are also huge "heat reservoirs" that help to moderate the climate. With more land and less sea, the land would be drier, the summers would be hotter, and the winters would be colder.
2. In what ways do oceans provide food for man?
   Oceans are filled with tiny forms of algae and animals that become food for larger species. Man harvests many of these larger species for food. Herring, cod, tuna, sardines, anchovies, lobsters, shrimp, oysters, mackerel, and many other marine species are used directly for human consumption.

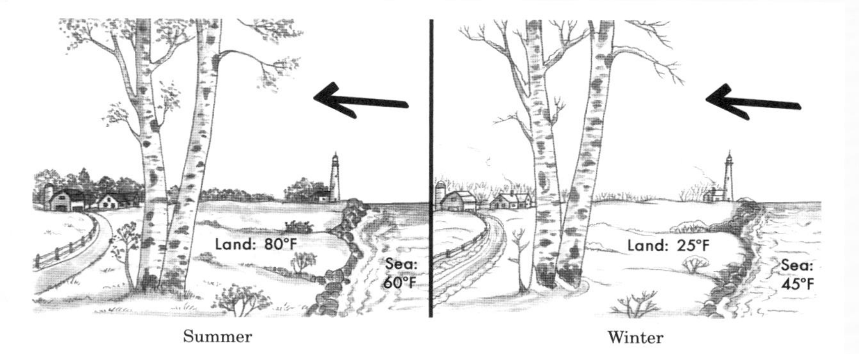

Summer

Winter

milder, especially on the land bordering the oceans.

The currents in the ocean affect temperatures on land. The Gulf Stream, a warm ocean current, carries warm water from the Gulf of Mexico along the east coast of the United States. This causes the climate there to be warmer than it would be otherwise. Then the Gulf Stream turns east and divides into several currents that flow across the North Atlantic Ocean to Europe. The

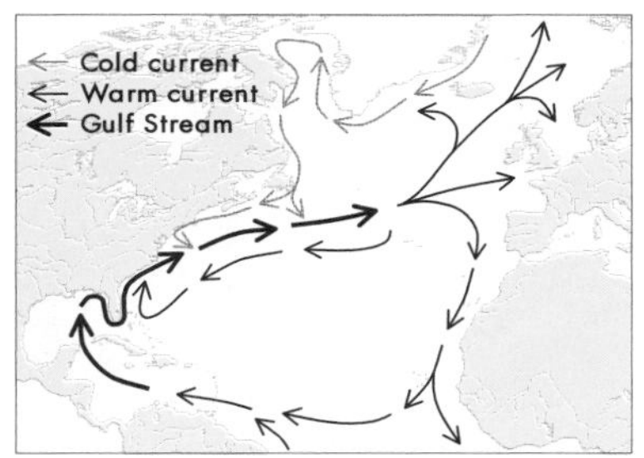

North Atlantic Ocean currents

coasts of Europe are warmed by these currents, with the result that much of western Europe has a mild climate as well.

The Labrador Current affects the climate of northeastern Canada. This current brings cold water from northern Greenland and other Arctic regions to the Labrador coast of Canada. Because of this cold current, the harbors of Labrador are packed with ice for about half the year. The climate of this area is very cold.

Oceans and Food for Man

Within the ocean, a system called the ***food chain*** supplies food for animals and people. The first link in the food chain is algae, which are tiny living things floating in the open sea. Algae make food by using water, carbon dioxide, and sunshine, just as plants do. The next link in the chain is tiny creatures that feed upon the algae. These tiny creatures and the algae together are called ***plankton.*** The plankton is eaten by small fish, the next link in the food chain. Larger fish and other ocean animals that eat the smaller animals are the next several links in the chain. Man becomes a part of this food chain when he harvests fish and other seafoods from the mighty ocean.

In addition to providing food, the oceans have made a way for people to travel all over the world. Men have built boats and ships of all sizes to sail from one place to another. The ocean is surely an important part of God's creation.

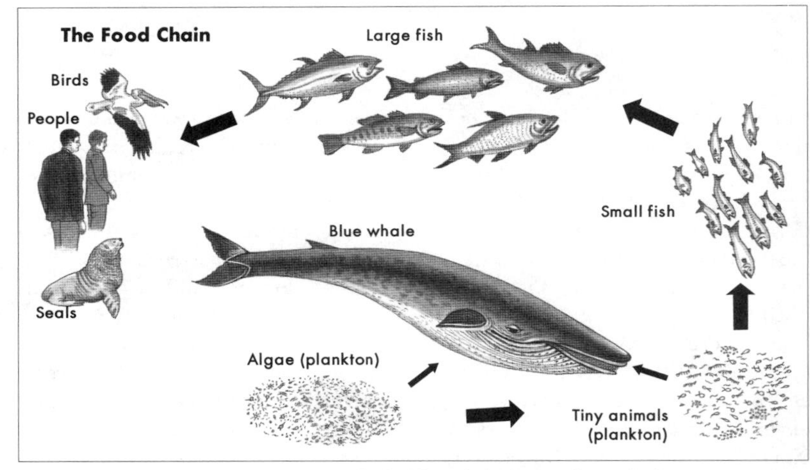

The largest creature on earth, the blue whale, feeds only on plankton.

Exercises

1. Write *waves* or *currents* for each description.
   a. They move water from one place to another.
   b. They simply move water up and down.
   c. They are very large and move slowly.
   d. They can be seen quite easily.
   e. They cannot be seen.
2. Oceans cover ——— percent of the earth's surface.

Lesson 14 Answers

Exercises

1. a. currents
   b. waves
   c. currents
   d. waves
   e. currents
2. 70

3. According to the answer in number 2,
   a. there is as much ocean as land.
   b. there is more ocean than land.
   c. there is more land than ocean.
4. Which part of the water cycle takes place over the ocean?
5. True or false? Ocean temperatures change more rapidly than land temperatures.
6. How do currents warm the climate of the east coast of the United States?
7. What makes the Labrador Current so cold?
8. In which group are the members of the food chain arranged in the correct order?
   a. fish, man, plankton, algae
   b. fish, algae, plankton, man
   c. algae, plankton, fish, man
   d. algae, fish, plankton, man

Review

9. In what two main ways has God formed mountains since the Creation?
10. Name and describe the two things that determine the amount of energy in a waterfall.
11. What is waterpower most commonly used for in modern times?
12. How do mountain glaciers help to provide water for surrounding valleys?

Activities

1. What is the deepest known point in the oceans? Find it on a world map. How deep is it?
2. Experiment with wave motion by tying a light rope to a doorknob and shaking the other end up and down. If you pull the rope tightly enough, you can feel the wave returning.
3. Make a bulletin board or poster display of marine foods. Look for pictures advertising seafoods in magazines or store flyers at home, and bring them to school. You could also bring recipes calling for seafood ingredients. Give your display a title such as "Small Fish That Feed Thousands."

3. b
4. evaporation
5. false
6. The Gulf Stream carries warm water from the Gulf of Mexico along the east coast. This current produces a warm climate.
7. The water in the Labrador Current comes from northern Greenland and other Arctic regions.
8. c

Review

9. God formed volcanoes of materials that erupted from within the earth.
   He formed other mountains by the crumpling of the earth's crust, perhaps at the Flood.
10. head—the distance that the water falls;
    flow—the amount of water running over the falls
11. to generate hydroelectricity
12. The glaciers slide slowly down into the valleys and melt as they reach lower altitudes.

Activities

1. The deepest known point is Challenger Deep in the Mariana Trench, located between Japan and Australia in the Pacific Ocean. Its depth is about 36,200 feet or 6¾ miles (10.9 km).
3. To make the display, staple the pictures and recipes to a medium blue background representing water. Use construction paper of various colors to make borders for the pictures, to cover the cut or torn edges. A narrow strip of light blue along the top could represent the sky. A small strip of dark green treetops between the sky and water could simulate a fish-eye view of a neighboring forest. Light brown paper liberally sprinkled with gold glitter would simulate a sandy bottom for the display. Place cutouts of several crustaceans on the bottom, and cutouts of fish in various locations throughout the display. You might add a real fishline baited with an artificial lure or a gummy worm, and rig a dark hairnet to look like a fishnet. A dark blue scalloped border will add a finishing touch to the display.

Lesson 15

Raging Storms

Vocabulary

hail, small, rounded pieces of ice that form high in a thunderhead and fall to the earth.

hurricane (hėr′·i·kān′), a huge, rotating tropical storm with winds of at least 75 miles per hour (120 km/h).

lightning, a powerful discharge of static electricity between clouds or between a cloud and the ground.

thunderstorm, a common storm accompanied by lightning, rainfall, and sometimes hail.

tornado (tôr·nā′·dō), a violent, spinning funnel of air extending downward from a storm cloud.

waterspout, a tornado occurring over a body of water.

The breeze picks up and begins whistling in the maples. Boiling clouds roll overhead. A streak of lightning flashes in the west, followed by a threatening rumble. The rushing sound of heavy rainfall on leaves reaches us from the wooded ridge. Then the rain sweeps across the cornfield, drenching the world in seconds. Another summer storm has arrived.

Probably most of us have experienced thunderstorms such as the one described above. More violent storms include tornadoes and hurricanes. In this lesson we will study all three.

What causes storms? Which storms are the most violent? What is a hurricane like? You should be able to answer these questions after studying this lesson.

God controls the raging tempest. He sends stormy weather to water the earth and to show us His great power. When Pharaoh did not cooperate with God by allowing the children of Israel to leave Egypt, one of the plagues God sent was a terrible storm of hail and fire that destroyed everything in the fields. It killed plants, animals, and even the servants that were outdoors. The Bible tells us that "there was hail, and fire mingled with the hail, very grievous, such as there was none like it in all the land of Egypt since it became a nation" (Exodus 9:24). But this grievous storm did not hurt the

Lesson 15

Lesson Concepts

1. A thunderstorm is produced by strong updrafts that form a thunderhead and produce rain, high winds, and static electricity.
2. As rain is blown upward into the cold part of a thunderhead, it may freeze and form hail.
3. A hurricane is a huge rotating mass of air that forms over a warm body of water.
4. Tornadoes and waterspouts are twisters produced by thunderheads.
5. Storms are used by God to teach us His power and control.

Teaching the Lesson

Many children fear storms, particularly thunderstorms. Storms should be respected, but we will keep them in perspective if we portray them as manifestations of God's power. God has designed the storm, and every storm is well within His control.

If a thunderstorm occurs during the school day, stop what you are doing and discuss thunderstorms with your students. The children probably will not be able to concentrate well on other things anyway. Show your appreciation for the storm and your interest in it. If you are uneasy during a storm, your fear will probably affect your students and cause them to miss the profound beauty of the elemental violence.

Make this lesson as practical as you can. Every area has some type of storm at least occasionally. What stories can your students tell about storms? How about their parents? Help your students appreciate

Israelites. God controlled the storm and kept His people safe while He punished the Egyptians.

Thunderstorms

"The voice of thy thunder was in the heaven: the lightnings lightened the world: the earth trembled and shook" (Psalm 77:18).

A ***thunderstorm*** develops when the sun-warmed earth sends strong updrafts of warm, moist air high into the sky. As the rising air cools, its moisture condenses into a tall, thick cloud. More and more air rises, and violent updrafts moving as fast as 60 miles per hour (100 km/h) build the cloud higher and higher. Finally it is a towering, storm-producing thunderhead. As rain forms in this cloud and begins to fall, it creates cold downdrafts. This gives a thunderhead both updrafts and downdrafts, making it a very turbulent cloud.

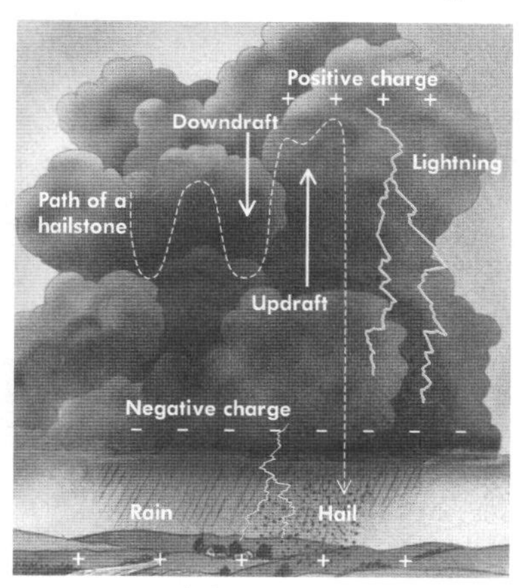

A thunderhead

Updrafts in a thunderhead are strong enough to catch falling raindrops and carry them high into the cloud, where the temperature is below freezing. These raindrops freeze into ice pellets known as ***hail.*** The pellets may be repeatedly tossed up and down, passing sometimes through rain and sometimes through freezing air. As this happens, layers of ice build up until the hailstones are too heavy for the updrafts to carry anymore. Then they fall to the ground in great numbers, sometimes causing severe damage to crops.

The winds in a thunderhead generate static electricity. As the air whirls wildly about, an electrical charge builds up in the same way as when you rub your feet on a carpet in winter. The strength of the charge

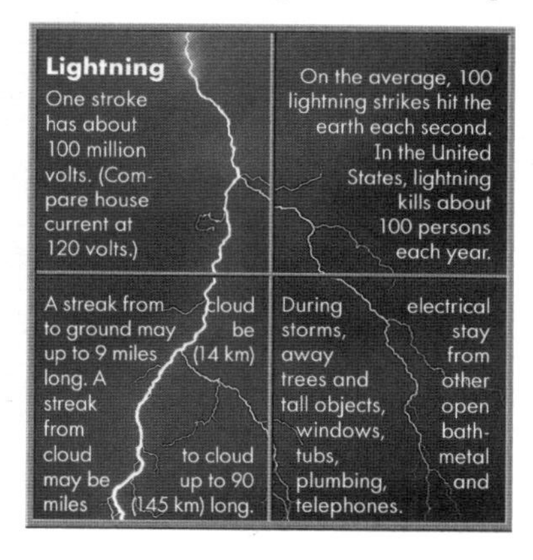

the storms that surround them. Help them to understand how and why storms occur. Weather is a complex subject, but a few basics will help the children grasp the idea that storms are products of weather conditions and are predictable to some extent. But leave the prediction part for the next lesson.

Discussion Questions

1. What storm in the Bible carried a man to heaven?
   the whirlwind that took Elijah to heaven (2 Kings 2:11)
2. What storm in the Bible struck the four corners of a house and killed all the children in one family?
   the storm that destroyed the house where Job's children were gathered (Job 1:19)
3. What parable did Jesus tell in which He referred to a storm?
   parable of the wise man and the foolish man (Matthew 7:24–27)
4. Which would you rather experience—a tornado? or a hurricane? Why?
   This should be a discussion comparing the two. A tornado is usually more destructive than a hurricane in the small area it affects. But a hurricane affects a much larger area and probably does greater damage overall.

increases until it is strong enough to arc across to another cloud or to the ground. Then the static electricity discharges with the brilliant flash that we call ***lightning.*** A single sharp clap of thunder follows the lightning strike. The familiar drawn-out rumble is the echo from other clouds or from nearby hills or buildings.

Tornadoes

Sometimes a severe thunderstorm develops into a ***tornado.*** In such a thunderstorm, the updrafts are so strong that the air rushing in starts to twist. The mass of air spins faster and faster until a whirling, funnel-shaped cloud extends below the thunderhead. This tornado cloud makes a sucking, roaring sound because of its powerful, high-speed winds.

Tornadoes are usually several hundred yards wide and move 20 to 40 miles per hour (30 to 65 km/h). The whirling winds within a tornado blow from 150 to 300 miles per hour (240 to 480 km/h), with some possibly greater than 500 miles per hour (800 km/h). These powerful, twisting winds can uproot trees and almost completely destroy everything in their path.

Besides the twisting winds, another destructive force of tornadoes is low air pressure. In the center of the whirling funnel, the air pressure is so low that it causes buildings to explode. When this low pressure combines with the high-speed, twisting winds and

A wall cloud sometimes forms underneath a severe thunderhead. If the wall cloud rotates persistently for 15 minutes, it may trigger a tornado.

strong updraft of a tornado, great destruction results. Houses, barns, trees, billboards—nearly everything in the path of a tornado is torn to pieces, and the rubble is scattered all around. Sometimes a tornado sucks up creatures in ponds, and people some distance away see fish and frogs raining from the sky!

The deadliest tornado in United States history struck in 1925. It left a path of destruction 200 miles long (320 km) across the states of Missouri, Illinois, and Indiana. It was also wider and faster than most other tornadoes. The funnel was one-half to one mile wide (1 to 2 km), and it moved along the ground at 60 miles per hour (97 km/h). It caused the deaths of 689 people.

A ***waterspout*** is a tornado that occurs over water. It is not as violent as a tornado over land. A waterspout is generally too weak to lift water from the sea. The funnel usually churns up a small cloud of spray where it touches the surface, but it does little damage.

Hurricanes

A ***hurricane*** is a huge tropical storm with a rotating mass of air. It begins over a warm section of the ocean where warm, moist air rises and starts to move in a circular path. This storm is usually 200 to 300 miles across (320 to 480 km)—much larger than a tornado. But the winds in a hurricane are not nearly as strong as tornado winds. They usually blow between 75 and 100 miles per hour (120 to 160 km/h), but sometimes they reach 150 miles per hour (240 km/h).

Hurricane winds cause severe damage when they reach land. Trees are uprooted and roofs are blown from

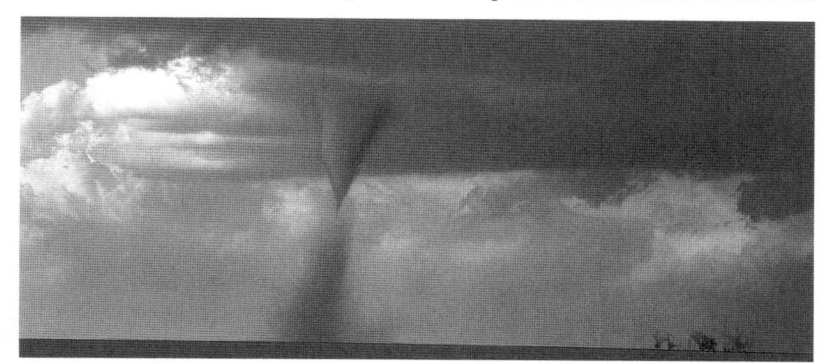

A tornado cloud is extending downward. Surface winds often swirl before the funnel reaches the ground.

Severe hurricane wind and rain

Hurricane flooding filled these West Indies houses with debris and washed roofs away.

houses. Giant waves pound the shore, smashing boat docks and buildings into useless wreckage. However, a hurricane begins losing strength as soon as it moves over land, for heat and moisture from the ocean are the fuel that drives its winds.

Hurricanes produce torrents of rainfall that can cause severe flooding. The high waves and floodwaters may work together to wash away houses and other buildings. Often many people drown in these floods.

The most intense hurricane ever recorded in the Western Hemisphere was Hurricane Gilbert. It swept over the West Indies and Mexico in 1988, killing about 300 people. Hurricane Mitch, a storm that struck Central America in 1998, was especially destructive. About 11,000 people lost their lives during that storm.

Storms show us the power of God. He is in control, and He steers the storms as He sees best. No wonder people refer to storms as "acts of God." Only He can make and control something so powerful.

Exercises

1. The storms in this lesson all start with (warm, cold) air that (rises, sinks).
2. Thunderstorms begin with a strong (updraft, downdraft) of (dry, moist) air.
3. Write *updraft, cold, heat,* or *rain* to tell what each sentence describes.
   a. It causes air to rise.
   b. It causes moisture to condense into clouds.
   c. It builds tall thunderhead clouds.
   d. It causes downdrafts in a thunderstorm.
4. Describe how a raindrop becomes a hailstone.

Lesson 15 Answers

Exercises

1. warm, rises
2. updraft, moist
3. a. heat
   b. cold
   c. updraft
   d. rain
4. A strong updraft carries a raindrop high into the thunderhead, where it freezes into an ice pellet. The pellet is tossed up and down, passing sometimes through rain and sometimes through freezing air. Layers of ice build up until the hailstone is so heavy that it falls to the ground.

5. In a thunderstorm, static electricity discharges in bright flashes called ———.
6. Severe thunderstorms sometimes develop twisting, funnel-shaped whirlwinds called ———.
7. What are two things that can make a tornado very destructive?
8. A tornado that occurs over a body of water is called a ———.
9. Tornadoes and hurricanes are similar in the ——— motion of their winds.
10. What are two differences between tornadoes and hurricanes?
11. In addition to strong winds, what other two forces cause destruction in hurricanes?
12. Why do people refer to storms as "acts of God"?

Review

13. The temperature at the center of the earth is much (higher, lower) than on the surface of the earth.
14. How is the motion of waves different from the motion of currents?
15. What percent of the earth's surface is covered by water?
16. Copy one wrong word in each of these statements, and write the correct word beside it.
    a. Cracks in the earth's crust where rock layers shift are called earthquakes.
    b. When an earthquake occurs under the ocean, it causes a powerful wave called a typhoon.

Activities

1. Take careful notice of a thunderstorm sometime. Which do you observe first, the thunder or the lightning? How many seconds pass between a lightning flash and the thunder that follows? If you divide the number of seconds by 5, the answer will show how many miles away the lightning struck. To find the distance in kilometers, divide the number of seconds by 3.
2. Have any notable storms ever struck your home area? Perhaps your parents have photographs or newspaper clippings showing the storm or the damage it caused.

   Many older persons have had interesting experiences with lightning, tornadoes, or hurricanes. Ask your parents or grandparents to tell you stories of such experiences. You will enjoy them, and so will your classmates.
3. Describe the safest places to go if you see a tornado coming.
4. The extremely high winds of tornadoes sometimes do strange things. Describe at least one curious thing that a tornado has done.

5. lightning
6. tornadoes
7. (Any two.)
   the high winds of the funnel;
   the low pressure in the center of the funnel;
   the strong updraft in the center of the funnel
8. waterspout
9. rotating (*or* circular *or* twisting)
10. (Any two.)
    Most tornadoes are only several hundred yards in diameter. Hurricanes measure several hundred miles in diameter.
    Tornadoes have winds measuring 150–300 miles per hour (240–480 km/h). Hurricane winds usually blow at 75–100 miles per hour (120–160 km/h).
    Hurricanes produce much more rain than tornadoes do.
11. waves, flooding from heavy rains
12. Man has no power over storms. Only God can control them.

Review

13. higher
14. Waves cause only up-and-down motion in water.
    Currents are great rivers in the ocean that move water from one place to another.
15. about 70 percent
16. a. earthquakes, faults
    b. typhoon, tsunami

Activities

3. A storm cellar is probably the safest place to go when a tornado comes. If you are outdoors, lying in a low ditch, under a low concrete bridge, or in a drainpipe under a road will give some protection. Indoors, the southwest corner of the basement is probably the safest place. A narrow space between a sturdy appliance and a basement wall offers protection from falling debris. If you have no basement, a small interior room is probably safest.
4. (Sample answers.)
   picking people up and setting them down some distance away, unharmed;
   driving straw into trees;
   emptying ponds of their water;
   stripping feathers from chickens;
   demolishing most of a house but leaving a calendar on the wall

Extra Activity

Merge science and English composition classes sometime by having your students describe a storm they vividly remember. It could be a tornado, hurricane, or thunderstorm, or even a blizzard. Briefly discuss with them what storms are like. Stimulate enthusiasm for the subject before you tell them to write anything. "Get their minds in gear before engaging their pencils."

Lesson 16

Wonders of Weather Patterns

Vocabulary

barometer (bə·rom′·i·tər), an instrument for measuring air pressure.
climate (klī′·mit), the usual weather for an area over a period of many years.
front, the boundary between two different air masses, often bringing rain or snow.
jet stream, a stream of high winds occurring 10 to 15 miles (15 to 25 km) above the earth.
weather, the condition of the atmosphere at a certain time and place, involving things like temperature, cloudiness, wind, and rain.
weather forecasting, an attempt to predict the weather that an area will have at a certain time in the future.

Weather and Climate

What is the difference between weather and climate? ***Weather*** is the day-to-day condition of the atmosphere around us. We may describe the weather as being sunny, windy, rainy, warm, cold, and so on.

Climate is the usual weather conditions of a certain place. Climate is described by such things as average temperature and rainfall over a period of one year. The general pattern of weather that an area has is its climate.

The climate we live in determines the kind of houses we build, the clothes we wear, the crops we plant, and the livestock we raise. People who live in cold climates build tight houses and insulate them well. If the climate is

What contrasting climates do you see here?

Lesson 16

Lesson Concepts

1. Weather is the complex product of many factors that follow God's laws.
2. The changing weather of North America is caused by warm and cold air masses moving across the land.
3. The climate of an area helps to determine what plants, animals, food, clothes, and houses are found in that area.
4. To forecast the weather, it is helpful to observe the clouds, air pressure, and wind direction.

Teaching the Lesson

Weather involves so many factors that the human mind cannot fully comprehend all their effects. But from the things that we do understand, we observe that weather does follow patterns. Included in the folklore of almost any area are weather sayings and tips for predicting the weather. Some of these are valid, and some are not. Nevertheless, they show that people of bygone days took special interest in the local weather.

Today's lesson should help to clear up some of the mysteries of weather for your students. A number of aspects will remain beyond their ability to understand, but teach them what you can. Be sure to point out parts of the lesson that pertain directly to your area. If you can identify weather elements from today's lesson in today's sky, take the children outside and show them. Some students may not understand what is meant by cottony clouds or high, wispy clouds. Be sure to point them out in the sky.

warm, farmers may raise corn, but if the growing season is too short for corn, they may raise barley instead. Hereford cattle do well in cool climates but not in warmer places. Rice needs much water and must be grown in a wet climate; wheat grows better in a dry climate.

What Makes the Weather?

People sometimes say that everybody talks about the weather but nobody does anything about it. Why not? Men cannot control the weather. Only God can do that. In the plagues upon Egypt, God showed His power to Pharaoh by sending lightning and hail. But His daily gifts of sunshine and rain also display His power. Since God controls the weather, we should not complain about it.

Weather includes four basic things: air temperature, air pressure, wind, and moisture. These four things work together in complex ways to make all kinds of weather. Because they are so complex, man finds it hard to understand and predict the weather. However, it is clear that the weather follows laws that are established by God. And while no scientist fully understands the weather, even fifth grade students can understand some of the basic principles.

The sun warms the air. Weather begins with the sun. When the sun warms the land or ocean, some of that heat warms the air. Air in the tropical zone around the equator is usually quite warm, while air around the North and South Poles is usually cold. In the areas between the equator and the poles, the air may be warm or cold depending on the season.

Cold air sinks, and warm air rises. Do you know how it feels to hold your hand several feet above a hot stove? You can feel the warm air rising. This happens because warm air is lighter than cold air. The heavier cold air sinks and forces the lighter warm air to move upward.

Warm air tends to rise.

There are no stoves in the atmosphere, of course, but some places are warmer than the areas around them. For example, an island is usually warmer than the sea surrounding it. A city may be warmer than the surrounding countryside. A plowed field is warmer than a forest on a sunny day. "Hot spots" such as these heat the air over them and cause it to rise. As

With the weather signs given in the lesson, people of many localities in the United States should be able to make short-term weather predictions with reasonable accuracy. Why not have your pupils take turns predicting tomorrow's weather at dismissal time? You will need a way to determine wind direction, and you will also need a barometer to tell what the air pressure is doing. Teach your pupils how to use the barometer. For predicting weather, it is more significant what direction the needle is moving than what the present reading is. To determine this, give the barometer one gentle tap. If the needle jumps upward, air pressure is rising; and if it jumps downward, air pressure is falling.

A demonstration with oil and water can help to clarify the concept of cold air sinking and forcing warm air upward. If you pour water into a clear container that has some oil in it, you will see the heavier water sink to the bottom and force the lighter oil above it.

Discussion Questions

1. What is the difference between weather and climate?
   Weather is the condition of the atmosphere at a certain time.
   Climate is the average weather over a period of time.
2. If there are high, wispy clouds in the sky today, what does that tell us about the coming weather? Why?
   High, wispy clouds are a sign of approaching rain. These clouds precede a warm front by about two days. Therefore, it is likely to rain several days after such clouds appear.

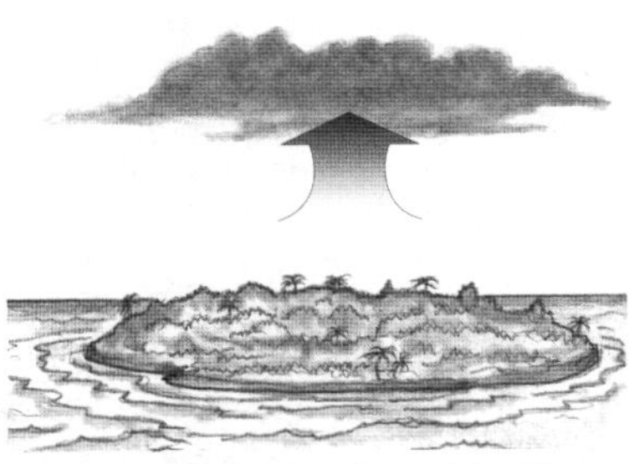

Clouds can tell sailors that an island is near even before they see it.

the warm air moves upward, it cools and clouds begin to form.

Cold, high-pressure air pushes warm, low-pressure air. Air of the same temperature tends to stay together in a body called an air mass. One air mass may cover hundreds of square miles. A cold air mass has high pressure, and a warm air mass has low pressure. Since cold air masses are heavier, they push the lighter, warm air masses ahead of them.

Huge, rotating air masses sweep across North America from west to east. Some of these are high-pressure air masses (often called highs), and others are low-pressure air masses (often called lows). A single high or low may be large enough to cover half of the United States. One after another, these air masses move from west to east across the continent.

A ***barometer*** shows whether air pressure is high or low. Rising barometric pressure tells us that a high-pressure air mass is moving in. High pressure usually brings clear skies and pleasant weather. A falling barometer indicates the coming of a low-pressure air mass, which often brings rain or snow.

A barometer and thermometer

Weather changes occur along the boundaries between different air masses. The highs and lows marching across North America bring weather changes as they come. Most weather changes take place along the boundaries between warm air masses and cold air masses. Such a boundary is called a ***front,*** and it may cause the weather to change from mild to cold, from dry to rainy, or from warm and humid to cool and comfortable.

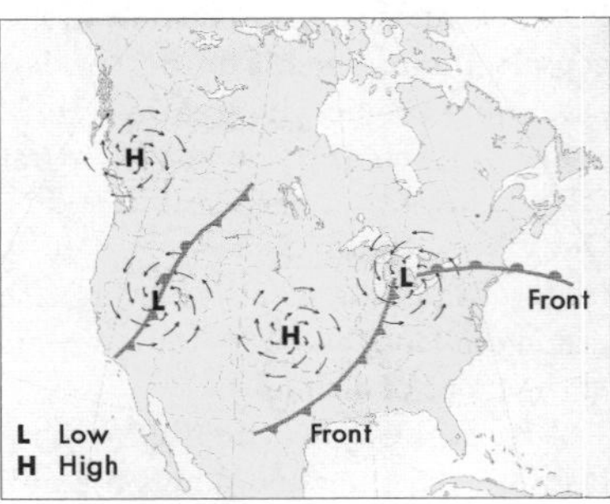

A weather map

High-altitude, high-speed winds influence our weather. Miles up in the sky, a river of air known as the ***jet stream*** flows from west to east across North America. Its speed is perhaps 250 miles per hour (400 km/h), and its path is not always the same. Sometimes the jet stream loops to the north or south, and this brings changes in weather patterns. Dry areas may be flooded, and wet areas may experience drought.

With all these things working together, it is no surprise that the weather is continually changing. Neither is it surprising that man finds it difficult to predict the weather for tomorrow or next week.

Weather Forecasting

"He answered and said unto them, When it is evening, ye say, It will be fair weather: for the sky is red. And in the morning, It will be foul weather to day: for the sky is red and lowring. O ye hypocrites, ye can discern the face of the sky; but can ye not discern the signs of the times?" (Matthew 16:2, 3). People have been interested in ***weather forecasting*** for hundreds of years. And since weather follows certain patterns, we can often make reasonably accurate predictions.

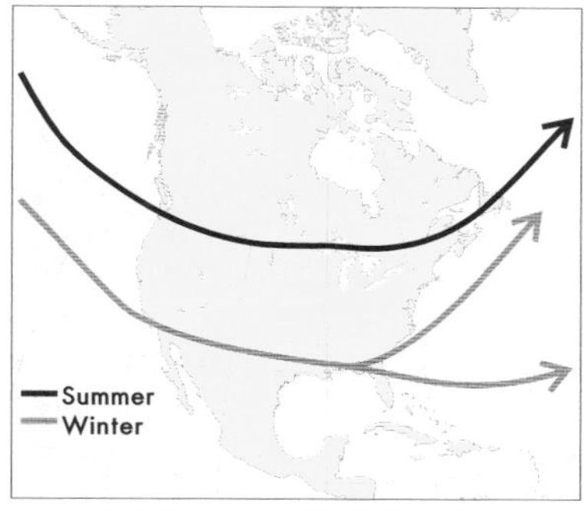

Jet stream over North America

Weather signs tell us what weather is coming. Here are some simple weather signs to help you predict the weather in North America.

1. "Red in the morning,
   Sailors take warning.

   Red in the night,
   Sailors' delight."
2. High, wispy clouds, called cirrus (sir′·əs) clouds, usually mean that rain is coming in a day or two, followed by warmer weather.
3. Fluffy, cottony clouds, called cumulus (kyüm′·yə·ləs) clouds, mean fair weather.
4. A ring around the moon means rain in a day or two.
5. Winds from the west and north bring fair weather.
6. Winds from the south and east bring rainy weather.
7. A rising barometer means fair weather ahead.
8. A falling barometer means rainy weather ahead.

God made the weather so that it follows patterns. This allows us to tell what weather is coming in the near future. But since we cannot observe all the things that affect our weather, our predictions are not always accurate.

Which is a sign of fair weather? Which is a sign of rain in a few days?

Lesson 16 Answers

Exercises

1. a. weather
   b. weather
   c. climate
   d. weather
   e. climate
   f. climate
   g. climate
2. air temperature, air pressure, wind, moisture
3. climate
4. clouds
5. low pressure
6. cold, warm
7. True
8. barometer

Exercises

1. Write whether each sentence describes something related to *weather* or to *climate.*
   a. A thunderstorm is approaching from the west.
   b. Jerry noticed that the temperature was –5°F when he went out to do the chores.
   c. New Mexico receives about 13 inches of rain each year.
   d. The day we went fishing, the sky was clear and the temperature was mild.
   e. Barrow, Alaska, has an average temperature of 9°F.
   f. The average temperature of North Carolina in January is 40°F.
   g. Every summer, many tornadoes occur in the central United States.
2. List the four main things that make up the weather.
3. The (weather, climate) of an area determines the kind of crops that will grow in that area.
4. As warm air moves upward into the sky, it becomes cooler and ——— begin to form.
5. A warm air mass is an area of (high pressure, low pressure).
6. When a warm air mass meets a cold air mass, the heavier ——— air mass pushes the lighter ——— air mass ahead of it.
7. True or false? Air masses move from west to east across North America.
8. A ——— is an instrument that shows changes in air pressure.

9. When a high-pressure area moves in,
   a. rain will come in two or three days.
   b. a steady rain sets in for several days.
   c. the barometer rises and skies become clear.
   d. the weather becomes cold and cloudy for several days.
10. When a low-pressure area moves in,
    a. the barometer rises.
    b. the barometer falls and the skies become cloudy or stormy.
    c. sunny skies will follow for a week or so.
    d. there is a thick layer of clouds, but rain seldom falls.
11. The boundary between a cold air mass and a warm air mass is called a ———.
12. When the ——— ——— changes its path, an area that usually receives much rain may have a drought, and a dry area may receive heavy rains.
13. Study the weather signs given in the lesson. Then write *rainy* or *fair* to tell what kind of weather may be expected after each weather sign below.
    a. a drop in air pressure
    b. a south or an east wind
    c. a rise in air pressure
    d. clouds like cotton balls
    e. a north or west wind
    f. high, wispy clouds

Review

14. What is probably the most common kind of storm?
15. List some of the characteristics of a thunderstorm.
16. What does a tornado look like?
17. Most tornadoes develop from ———.
18. Where do all hurricanes form?

Activities

1. Record the daily weather on a calendar. Write down such things as the temperature at noon and the amount of any rainfall, as well as whether the sky was sunny, cloudy, or partly cloudy. Such a record can be interesting if kept from one year to the next and compared with the weather of another year.
2. Watch for high, wispy cirrus clouds that announce the coming of a warm front. How many days does it take for the rain to come?
3. Use the weather signs in the lesson to predict tomorrow's weather. At dismissal time, check the wind direction, the movement of the barometer, and the clouds. Record this information on a chart. Then write your prediction for the weather tomorrow.

9. c
10. b
11. front
12. jet stream
13. a. rainy
    b. rainy
    c. fair
    d. fair
    e. fair
    f. rainy

Review

14. thunderstorm
15. (Sample answers.) lightning and thunder, occasional hail, heavy rainfall, strong updrafts, tall thunderheads
16. a funnel-shaped cloud
17. thunderstorms
18. over a warm part of the ocean

Activities

1. Encourage your students to record the daily weather on a weather calendar. If you have a record from last year, it would be interesting to compare last year's weather with this year's.

Lesson 17

Wonders Above the Earth

Vocabulary

binary star (bī′·nə·rē), two stars that revolve around each other, usually appearing so close together that they look like a single star.

galaxy (gal′·ək·sē), a large group of stars, often numbering in the billions.

light-year, the distance that light travels in one year; almost 6 trillion miles (9½ trillion km).

magnitude (mag′·ni·tüd′), the brightness of a star as seen from the earth.

nebula (neb′·yə·lə), a hazy spot in the night sky, usually a cloud of glowing gas and dust.

nova (nō′·və), a star whose brightness suddenly increases by thousands or millions of times and then gradually returns to normal.

star cluster, a group of stars numbering from ten to several million, found within a galaxy.

telescope (tel′·i·skōp′), an instrument that makes distant objects appear larger and nearer.

variable star, a star whose brightness changes from time to time.

"O LORD our Lord, how excellent is thy name in all the earth! who hast set thy glory above the heavens.... When I consider thy heavens, the work of thy fingers, the moon and the stars, which thou hast ordained; what is man, that thou art mindful of him? and the son of man, that thou visitest him?" (Psalm 8:1, 3, 4).

You have probably observed the starry sky many times. Perhaps you even tried to count the stars. How many are there? Quite likely, you lost count at some point and gave up. Truly, our God is a great Creator!

How Many Stars Can We See?

"He telleth [counts] the number of the stars; he calleth them all by their names" (Psalm 147:4). Even if we cannot count the stars, God can. According to the psalmist, He has named each of them.

About 6,000 stars are visible without a telescope, but we can see only about half of them at one time.

Lesson 17

Lesson Concepts

1. The heavens are full of testimony to the wonder of God's creative power.
2. About 6,000 stars are visible to the unaided eye, though only about half can be seen at any one time.
3. The telescope helps us to see faint objects by using a large lens or mirror to collect their light, thus making them appear brighter.
4. The nearest star (after the sun) is $4^1/_3$ light-years away.
5. Variable stars vary in apparent brightness, or magnitude.
6. Nebulae are huge clouds of glowing gas and dust in space.
7. Some stars are grouped together to form star clusters.
8. The universe contains millions or billions of galaxies, each containing billions of individual stars.

Teaching the Lesson

When people stand on top of a tall mountain on a clear day, they often comment on the view. "You can see for miles and miles!" If distance is so impressive, go outside and look at the stars on some clear, dark night. Stars are trillions and trillions of miles away. Little pricks of light they are, scattered all over the sky. You will see the same stars that Job and Abraham and Daniel observed. With great precision, God arranged the stars; and with great carefulness, He holds them in place.

The wonder becomes still greater with the use of optical instruments. Even if you have only a pair of binoculars, by all means show your students some of the marvels in the sky. Plan for an evening of stargazing. Encourage your students to study the stars on

(The other half can be seen from the opposite side of the earth.) However, these 6,000 stars are a very tiny part of all the stars God created. There are billions upon billions of stars not visible to our eyes. Some parts of the sky look milky because of the millions of distant stars that fill the area. The stars in those areas are too dim for us to see them individually, but all of them together form a whitish path across the night sky that we call the Milky Way.

Telescopes help us observe the stars by making them appear brighter. Many stars are so far away that they are too dim to see with the unaided eye. By using lenses or mirrors to gather light from a large area, telescopes can make faint stars appear much brighter. The Hale Telescope in California has a mirror with a diameter of 200 inches (500 cm). It can make faint stars appear one million times as bright as they appear to the unaided eye.

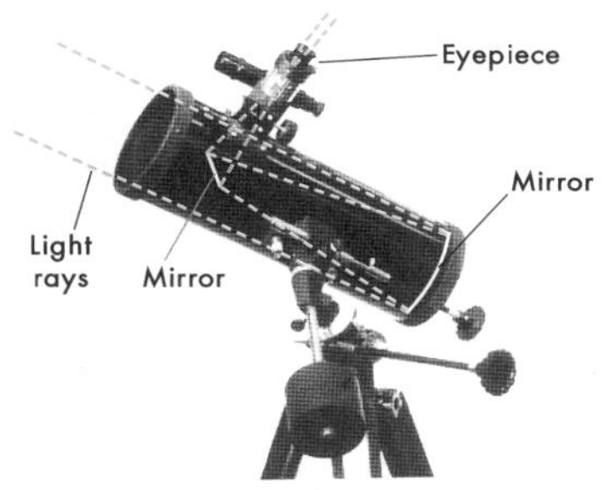

A reflecting telescope makes faint stars appear much brighter.

As larger and more powerful telescopes are built, men are able to see more and more stars. A very powerful telescope makes it possible to see billions of stars that could not be seen otherwise; but it will never reveal all the stars, because God's universe is too large. The stars show us God's greatness and unlimited power.

How Far Away Are the Stars?

The nearest star is our sun, about 93 million miles (150 million km) from the earth. Traveling at the speed of a jet plane, it would take about twenty-one years to reach the sun. Even sunlight takes more than eight minutes to travel to the earth. As you can see, the sun is very far away.

Compared with other stars, however, the sun is a very near neighbor. If the sun were as large as the period at the end of this sentence, the distance to the next nearest star would still be 10 miles (16 km).

Because of the great distances in space, scientists need a larger unit of measure than we use on earth. That is why they use ***light-years.*** Light travels 186,282 miles in one second. A light-year is the distance that light travels in a year, or about 6 trillion miles (9½ trillion km). The nearest star besides the sun is about 4⅓ light-years away. That means even if you could travel at the speed of light, it would take 4⅓ years to reach the nearest

their own as well. Watching for meteors and studying the planets with a telescope are also worthwhile. They will need close supervision when handling a telescope, but arousing their interest never hurts.

Stars may seem unrelated to practical life for most of us, but really they are not. Anything that inspires praise for our God is worth noticing. Read Psalm 8. David's sense of wonder and humility may also be ours. The majesty and infinity of God are mirrored by the celestial realm. Pass to your students an enthusiasm that culminates in praise and adoration for our all-powerful Creator.

Discussion Questions

1. God not only knows how many stars there are, but He also knows all their names. What does this tell us about God?
   Our God is great and all-powerful. His knowledge and understanding are unlimited.
2. Imagine a night sky full of stars all having the same brightness. Why was it wise for God to create the stars with varying magnitudes?
   If all the stars were as dim as some stars, we would not be able to see any stars at all.
   If all the stars were as bright as some stars, there would hardly be any darkness, for the whole sky would glow with light.
   With varying magnitudes, it is also easier to tell which star is which than if they were all of the same brightness.

Distance to the Nearest Star

←4⅓ light-years→

Pluto's orbit around the sun (7 billion miles across)

Nearest star (25,000 billion miles away)

For this diagram to be completely accurate, the orbit of Pluto would need to be about one-thousandth of an inch across!

star! If you traveled only at the speed of an ordinary jet, the trip would take 5¾ million years. And this trip would be to one of the nearest stars! Many other stars are millions of times farther away than that. Can you imagine the size of the whole universe?

How Bright Are the Stars?

"There is one glory of the sun, and another glory of the moon, and another glory of the stars: for one star differeth from another star in glory" (1 Corinthians 15:41). God made the stars in such a way that some are brighter than others. The brightness of a star is its ***magnitude.*** The brighter a star is, the lower its magnitude number will be. Polaris, the North Star, has a magnitude of 2.04. Rigel (rī′·jəl), the star marking the left knee of Orion, has a magnitude of 0.08. This means that Rigel is much brighter than Polaris.

Name	Magnitude
Arcturus	0.00
Betelgeuse	0.41 (variable)
Capella	0.05
Deneb	1.26
Mizar/Alcor	2.27/3.95
Polaris	2.04
Procyon	0.34
Regulus	1.35
Rigel	0.08
Sirius	−1.50
Spica	1.12
Vega	0.04

Magnitudes of some familiar stars

Some stars change in brightness and are called ***variable stars.*** In one kind, the two stars of a ***binary star*** revolve around each other, and one star comes in front of the other from time to time. This causes the brightness of the binary star to vary. An example is Algol in the constellation Perseus. Most binary stars have a constant brightness because both stars are visible at once. Mizar and Alcor are a good illustration of this. Together they form the binary star in the bend of the Big Dipper's handle.

A ***nova*** is also a variable star. It is an ordinary star that explodes to form

A binary star is actually two stars revolving around each other.

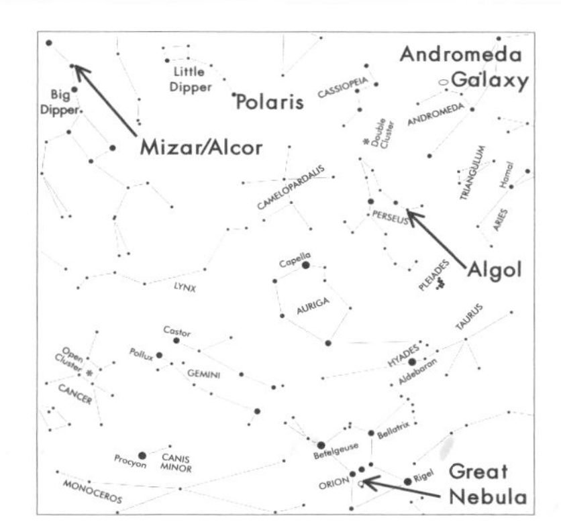

Great Nebula in Orion

a very bright spot in the sky. Most novae increase in brightness very rapidly and then diminish more slowly. They often take from several weeks to a few years to return to normal brightness.

Sometimes when a nova explodes, it produces a great cloud of glowing gas and dust that is called a ***nebula.*** Many nebulae are brilliantly colored clouds. The easiest nebula to find in the sky is the Great Nebula in the sword of Orion. This nebula is impressive even when seen through a binocular.

Star Clusters

A group of stars relatively close together and moving together is called a ***star cluster.*** Globe-shaped clusters generally contain thousands of stars.

A telescope view of two galaxies. The galaxy on the left is the Andromeda Galaxy.

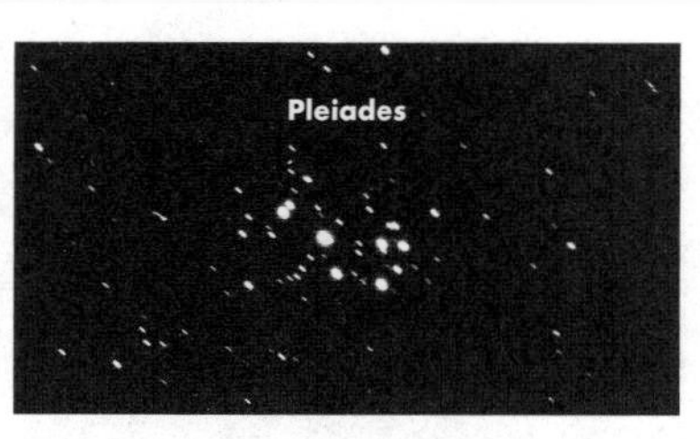

Smaller, odd-shaped star clusters, called open clusters, contain only from ten to several hundred stars. Pleiades is a good example of an open cluster. Seven stars may be seen with the unaided eye, though counting all seven is a good test of eyesight. Through a binocular, you can see many more stars in this cluster.

Galaxies

A ***galaxy*** is one of the vast star groups in the universe. The sun is a star in the Milky Way Galaxy. In fact, all the stars, novae, clusters, and nebulae that have been named are in this same galaxy. The galaxy nearest to ours is about 150,000 light-years away. The Andromeda (an·drom′·i·də) Galaxy, the closest one visible in North America, is 2,000,000 light-years away. The Andromeda Galaxy may be seen as a faint spot with the unaided eye, but it looks like a glowing cloud through a binocular.

Scientists estimate that up to one billion galaxies can be photographed with large telescopes. Some believe that up to 100 billion galaxies may exist. If each of the 100 billion galaxies contains an average of 100 billion stars, the universe contains an estimated total of 10 sextillion stars. (That is *10* followed by 21 zeroes.) Of course, this number is just a guess. No man really knows how many stars there are, but we do know that the number is beyond imagination.

Such immense numbers and distances make us feel very small. But we know that God made everything in the universe. He knows the number of the stars, and He calls each one by its name. We must conclude with the psalmist, "O LORD our Lord, how excellent is thy name in all the earth!" (Psalm 8:9).

Exercises

1. About how many stars can we see on a clear, dark night?
2. What instrument helps us to see the stars and planets better?
3. Since the distance to the stars is so great, scientists use the ——— as a unit of measure.
4. True or false? With the powerful telescopes available today, scientists can now see all the stars.
5. What is a light-year?
6. A telescope is useful because it makes the stars appear ———.

Lesson 17 Answers

Exercises

1. 3,000 stars
2. the telescope
3. light-year
4. false
5. the distance that light travels in one year (about 6 trillion miles or 9½ trillion km)
6. brighter

7. True or false? A bright star will have a low magnitude number.
8. Stars whose brightness changes are called ——— stars. Two kinds of such stars are ——— stars and ———.
9. When a star explodes to become a nova, it sometimes forms a great cloud of gas and dust called a ———.
10. Why does a binary star appear brighter at some times than at others?
11. A group containing from ten to several thousand stars is called a ——— ———.
12. True or false? The universe contains only a few galaxies besides our own.
13. Who deserves the honor and glory for making the stars?

Review

14. What kind of weather does a low-pressure area generally bring?
15. What kind of weather does a high-pressure area generally bring?
16. How are weather and climate different from each other?
17. What are the four things that make up the weather?
18. True or false? Mountains may greatly affect the weather.

Activities

1. Go out on a clear, dark night, and observe the stars. Try to find an open area away from lights, where your view is not blocked by trees or buildings. Can you find the North Star? How about several constellations? A star guide will help you to locate various stars and constellations.
2. Which of the heavenly wonders mentioned in this lesson have you observed in the sky? Most of them can be seen with the unaided eye. The Great Nebula in Orion and the Andromeda Galaxy are impressive when seen through a binocular. Pleiades and the Milky Way are also interesting to see through a binocular.

   See if you can find all the heavenly bodies listed below. Perhaps your teacher or parents can help you if you have trouble.

   a. Milky Way Galaxy
   b. Betelgeuse
   c. Orion the Hunter
   d. Polaris (North Star)
   e. Big Dipper
   f. Mizar and Alcor
   g. Great Nebula in Orion
   h. Pleiades
   i. Andromeda Galaxy
3. Sometimes we hear of falling stars. Do stars really fall out of the sky? Use an encyclopedia to find out.

7. True
8. variable, binary, novas
9. nebula
10. When both stars in the pair are visible, the binary star appears brighter. When one star hides the light of the other, the binary star appears less bright.
11. star cluster
12. false
13. God the Creator

Review

14. rain or snow
15. clear and pleasant
16. Weather is the condition of the atmosphere at a certain time. Climate is the average weather over a period of time.
17. air temperature, air pressure, wind, moisture
18. True

Activities

1–2. Can you, the teacher, locate objects in the night sky? Can you identify the heavenly bodies listed in the pupil's text? If not, get a star map and use it on several clear, dark nights to become acquainted with the heavens. You cannot expect to learn in just a few nights where all the major constellations are. But you should at least view the heavenly bodies mentioned in the lesson. If you can encourage your students by example to study the magnificent night sky and glorify the Creator, you will have been successful with today's lesson.

3. Falling stars are not stars but meteors. These chunks of rock enter the earth's atmosphere at such high speeds that they become glowing hot because of friction with the air. Sometimes they fall all the way to the earth, and then they are called meteorites.

Lesson 18

Unit 3 Review

Review of Vocabulary

earthquake	food chain	lava	seismograph
earth's crust	glacier	magma	tremor
energy	head	ocean current	tsunami
fault	hydroelectricity	plankton	volcano
flow	kinetic energy	potential energy	wave

The earth has many mountains on its surface. A __1__ is a mountain that forms when materials erupt through a crack in the __2__. One of the materials is molten rock, which is called __3__ before it erupts and __4__ after it reaches the surface. Other mountains were possibly formed when the earth's crust folded and wrinkled as the Flood waters receded. A high mountain may have a huge river of ice called a __5__ sliding slowly down into a valley.

When stress builds up along a __6__, rock layers may shift with a sudden jolt. Such a shift is what causes an __7__. An instrument called a __8__ can detect even a faint __9__ that shakes the ground. When an earthquake occurs under the ocean, a __10__ often forms and causes great destruction when it strikes the shore.

Waterfalls provide a good source of __11__, or power to do work. The amount of power is determined by the __12__ (height) and the __13__ (amount of water) at a waterfall. Today waterpower is used mostly to produce __14__. When no waterfall is available, men build dams to raise the water level. Water stored behind a dam has __15__ because it is stored and waiting to be used. Water flowing through a turbine has __16__ because it is moving and actually doing work.

Two kinds of water movement occur in the ocean. A __17__ simply moves the water up and down, forming crests and valleys on the ocean surface. An __18__ is a large, slowly moving river in the ocean that moves water from place to place. The __19__ in the ocean starts with tiny plants that make food. Tiny sea animals are the next link. These tiny plants and animals are called __20__. This provides food for larger animals, which are eaten by even larger animals, and some of these may provide food for man.

Lesson 18 Answers

Review of Vocabulary

1. volcano
2. earth's crust
3. magma
4. lava
5. glacier
6. fault
7. earthquake
8. seismograph
9. tremor
10. tsunami
11. energy
12. head
13. flow
14. hydroelectricity
15. potential energy
16. kinetic energy
17. wave
18. ocean current
19. food chain
20. plankton

Lesson 18

One of the objectives of this unit was to inspire wonder in the student. As you review, highlight the wonders of God's creation and show how they glorify Him.

barometer	hurricane	nebula	tornado
binary star	jet stream	nova	variable star
climate	lightning	star clusters	waterspout
front	light-year	telescope	weather
galaxy	magnitude	thunderstorm	weather forecasting
hail			

The __21__ is the most common kind of storm. It includes __22__, thunder, heavy rainfall, and occasionally little balls of ice called __23__. Sometimes a severe thunderstorm develops into a violent whirlwind called a __24__. If the whirlwind occurs over water, it is called a __25__ and seldom does much damage. The __26__ is a large, circular storm that forms over a warm ocean and brings winds of 75 to 150 miles per hour (120 to 240 km/h).

__27__ is the condition of the atmosphere at a certain time. __28__ is the average of that condition over a period of time. __29__ is the study of weather patterns in an attempt to predict the weather at a certain time in the future. The boundary between a cold air mass and a warm air mass is called a __30__, and it brings changes in weather. A __31__ tells us whether the air pressure is high or low. The __32__ is a channel of high-altitude, high-speed winds that bring changes in weather patterns.

A __33__ helps us to observe the many wonders that God created in the night sky. The __34__ of a star is the brightness of the star. A __35__ changes in brightness from one time to another. One kind is the __36__, a pair of stars revolving around each other and sometimes hiding each other's light. Another kind is the __37__, a star that explodes and suddenly becomes many times brighter than before. The exploding star may produce a giant cloud of glowing gas and dust called a __38__. A __39__ is a huge group of stars often numbering in the billions. Within it are smaller groups called __40__. The universe is so large that scientists do not use miles or kilometers to measure distances between stars. Instead, they use the __41__, which is the distance that light travels in a year.

Multiple Choice

1. Mountains (choose two)
   a. were formed by slow changes in the earth's crust over millions of years.
   b. were probably formed when the earth's crust folded after the Flood.
   c. are still being formed today by volcanic eruptions.
   d. have not changed since the Creation.

21. thunderstorm
22. lightning
23. hail
24. tornado
25. waterspout
26. hurricane
27. Weather
28. Climate
29. Weather forecasting
30. front
31. barometer
32. jet stream
33. telescope
34. magnitude
35. variable star
36. binary star
37. nova
38. nebula
39. galaxy
40. star clusters
41. light-year

Multiple Choice

1. b, c

2. c

2. Mountains affect the weather by
   a. warming the air that flows up over them.
   b. putting moisture into the air.
   c. cooling the air and making rain fall.
   d. changing the direction of winds.

3. d

3. The shaking we call an earthquake is caused by
   a. two sections of the earth's crust bumping into each other.
   b. tsunamis and crumbling buildings.
   c. the earth's crust folding to form mountains.
   d. two sections of the earth's crust shifting at a fault.

4. c

4. All the following are results of an earthquake **except**
   a. crumbling buildings.
   b. shaking ground.
   c. heavy rainfall.
   d. fire.

5. d

5. Which of the following determine the amount of energy in a waterfall?
   a. head and depth
   b. flow and depth
   c. flow and speed
   d. head and flow

6. c

6. The head of a waterfall is
   a. the depth of the river above the falls.
   b. the width of the falls.
   c. the height of the falls.
   d. the amount of water flowing over the falls.

7. c

7. Oceans affect the weather by
   a. causing cold winds to blow over land in the winter.
   b. causing warm winds to blow over land in the summer.
   c. providing moisture to fall on land as rain.
   d. producing jet streams.

8. a

8. Thunderstorms are known for bringing
   a. lightning, wind, and rain.
   b. lightning, hail, and a destructive funnel-shaped cloud.
   c. thunder and lightning but little rain.
   d. lightning, hail, and winds of 75 to 100 miles per hour.

9. b

9. Tornadoes are storms that usually
   a. develop out over the ocean.
   b. develop from severe thunderstorms.
   c. develop from the jet stream.
   d. develop from hurricanes.

10. Hurricanes are like tornadoes in that they
    a. are often hundreds of miles wide.
    b. have winds that rotate around the center of the storm.
    c. form over water and then move toward land.
    d. cause giant waves that destroy much property.

10. b

11. The four things that work together to make our weather are
    a. air temperature, wind, rain, and moisture.
    b. air pressure, wind, snow, and rain.
    c. air temperature, air pressure, wind, and moisture.
    d. snow, air temperature, wind, and moisture.

11. c

12. Most of the following sentences refer to climate. Which one refers only to weather?
    a. The state of Vermont receives 39 inches of rain per year.
    b. Hawaii has an average temperature of 75°F in July.
    c. In March 1925, a terrible tornado swept through Missouri, Illinois, and Indiana.
    d. Northwestern Pennsylvania receives about 7 feet of snow each winter.

12. c

13. We can forecast weather because
    a. weather follows patterns that God has established.
    b. weather is actually very simple to understand.
    c. scientists have learned all about weather patterns.
    d. weather seldom changes.

13. a

14. Which weather signs point to rainy weather? (Choose all the correct answers.)
    a. falling barometer   c. high, wispy clouds
    b. north wind   d. ring around the moon

14. a, c, d

15. On a clear, dark night, we can see ——— stars with the unaided eye.
    a. about 6,000   c. over 1,000,000,000
    b. about 2,000,000   d. about 3,000

15. d

16. The magnitude of a star is
    a. how far away the star is.   c. how bright the star appears.
    b. how large the star is.   d. how fast the star is moving.

16. c

17. Which one of the following contains the other three?
    a. galaxy   c. star cluster
    b. nebula   d. nova

17. a

18. About how much of the earth's surface is covered by water?
    a. 20% (2 tenths)   c. 50% (half)
    b. 70% (7 tenths)   d. 80% (8 tenths)

18. b

Unit 4

Title Page Photo

A busy carpenter crew uses a host of simple machines. Hammers, nails, saws, and numerous other things can be classified as simple machines or a combination of simple machines. Ask your students to identify some of the simple machines they see in the picture.

Introduction to Unit 4

We live in an age when we have machines that help us do many things easily and quickly. A thorough understanding of the laws of motion and machines is not necessary before a person can use a machine or the laws of motion to his advantage. However, understanding these laws helps us to use machines more efficiently. It also serves to satisfy the curiosity that fills many children to know how and why a machine works.

Make this study as practical as you can. Bring in the machines that are familiar to them. A textbook cannot be specifically tailored to every type of community. But you, the teacher, must adapt it to the interests and surroundings of your students.

Unit 4

Wonderful Laws of Motion and Machines

"Thou hast a mighty arm: strong is thy hand, and high is thy right hand" (Psalm 89:13).

Our God is great and mighty. He does not need tools and powerful machines to do His work. But we are not like that. Without tools and machines, we are almost helpless.

Suppose you had no knife or scissors. How would you cut paper? Suppose your wagon had no wheels. Could you bring in a load of firewood with it? If you lost your hammer, how would you pull a nail out of a board? Without a pump, how would you get water out of a well three hundred feet deep? And how hard it would be to loosen a tight bolt with your fingers! You would surely be glad to use a wrench.

Many things would be difficult or impossible to do if we had no tools or machines. But God has created us with the intelligence to make machines and use them to do our work better. In this unit you will learn a number of things about motion and machines.

Story for Unit 4

Caleb's Machines

"Father," Caleb said one evening as the Shetler family sat around the supper table, "Brother Mark said we must make a list of fifteen examples of simple machines that we often use. I think I use only five or six. I can't see how I'll ever find fifteen!"

"When do you need to have the list finished?" Father asked.

"It has to be done by tomorrow's science class," Caleb said.

"Well, do you know what the six different kinds of simple machines are? Can you list them for me?" Father asked.

"Let's see. I think I can list them. Lever, pulley, wheel and axle, inclined plane, screw, and . . . and . . . I can't think of the last one," Caleb said.

"Wedge," said Steven, a ninth grader.

"Oh, yes! It's a wedge." Caleb remembered now.

"Well, are there any of the simple machines around you here at the table?" Father asked. "Let's see if we can find any."

"You could start with the machines on either side of your plate," Steven suggested.

"Machines? You mean my knife, fork, and spoon?" Caleb asked.

"Surely," Father answered. "That's a good suggestion. Your spoon and fork are levers, and your knife is a wedge."

"I never thought of that. But that is only three. I need fifteen," Caleb said.

"Look over in the sink. Did Mother use any machines when she made supper?" Father asked.

Everyone had finished eating and was excused from the table, so Caleb walked over to the sink. "Oh, is this a machine?" He held up a can opener for Father to see.

"Yes. That has both a lever and a wheel and axle,"

Lesson 19

God's Laws for Moving Things

Vocabulary

foot-pound, a unit used to measure work; the amount of work done when one pound of force is used to move something one foot.

force, a power that causes an object to move, change direction, or stop.

friction, a kind of resistance that results when two things rub together.

inertia (i·nėr′·shə), the tendency of a resting object to stay at rest, and of a moving object to stay in motion.

molecular attraction (mə·lek′·yə·lər), a kind of resistance that holds together the tiny particles (molecules) of which things are made.

resistance, a power that hinders motion.

work, the moving of an object by applying a force.

All around us, things are moving. Cars travel down the road. The beaters of Mother's mixer spin around inside the mixing bowl. Marlin uses his wagon to pull a load of firewood to the house. Lynette slides down the hill on her sled. But things do not move by themselves. What makes them go?

Force

Suppose your wagon is standing on a level driveway. What would happen to it if nothing pushed or pulled it? Since the wagon cannot move by itself, it would just sit there. It needs a push or a pull to make it move. This push or pull is called a ***force.*** All moving things must have a force that is making them move. Things never move without a force being applied.

Force moves things. The force of wind makes the curtains flutter. The force of an engine pushes a car down the road. When you throw a ball, your arm supplies the force needed to push the ball through the air.

God has often used force to carry out His will. One example is the story of the Israelites at the Red Sea. "He divided the sea, and caused them to pass through; and he made the waters to stand as an heap" (Psalm 78:13). God used the force of the wind to divide the sea. Wind normally cannot do that. But God is not limited by impossible things. He can do anything because He is God.

Father said. "That's good. What else can you find?"

Caleb pushed aside a kettle. "Oh, here is a carrot shredder. That would be like a knife, wouldn't it?"

"Yes, it's a wedge just as a knife is," Steven said.

"Look at the eggbeater. What kind of machine might that be? Is it a wedge too?" Caleb carried it over to where Father was sitting.

"I'm not sure what you'd call that. But I see a wheel and axle in it. See the crank and gear? I think you could list that as a wheel and axle," Father said.

"What about that empty mayonnaise jar sitting there? Can you see a machine on it?" Steven asked.

"Oh, yes. There is a screw where the lid fits on." Caleb was enjoying the machine hunt now. "There are machines everywhere, aren't there," he said.

"Let's go out to the garage now," suggested Father. "There will be more machines out there."

"I see some pulleys on the garage door," Caleb said. "Those are the first pulleys we have found."

He walked over to the workbench. "Here's a screwdriver. That must be a lever. Right, Father?"

"Yes, a screwdriver can be used as a lever or as a wheel and axle," Father answered.

"And here is a hammer. And here is a wrench. They are both levers. Let's see, what else would be a machine?"

"What is hanging beside the hammer?" Steven asked.

"Oh, another kind of knife. Let's see . . . a utility knife would be a wedge, just like a table knife."

"That's right," Father encouraged. "Now what is in the top left drawer under the workbench?"

Caleb opened the drawer. "Bolts. I hadn't thought of them as being machines, but I guess they are screws. Now I only need two more."

"I know of two you use a lot. They are both found on the inner side of a car door. Can you find them?"

Caleb opened the car door. "Of course, the door latch handle and the window crank. Would they both be levers?"

"No, I think the window crank would be a wheel and axle," Father said.

"Good, now I'm finished. That must be about all of them anyhow," Caleb said happily.

"About all of them!" Steven exclaimed. "I should say not. What about a shovel, a rake, a hoe, a sledgehammer, a handsaw, or a jack?" he said, pointing them out to Caleb. "And don't forget your bicycle and Abner's tricycle. Besides, we haven't even started in the barn yet."

"Sure enough," Caleb answered. "There are still lots of machines. Maybe I should keep hunting for more. I think I'll try to find thirty machines. That is twice as many as I need." And he did.

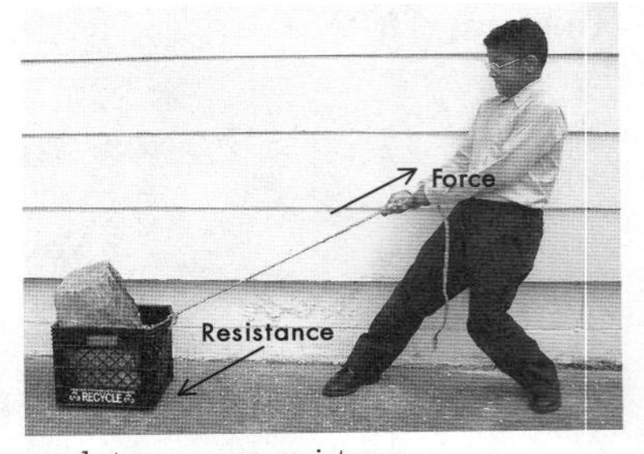

To push or move something, a force needs to overcome resistance.

Four Kinds of Resistance to Motion

Whenever something moves, there is ***resistance*** that hinders the movement. You will learn about four kinds of resistance. They are gravity, inertia, friction, and molecular attraction.

1. Gravity. This resistance is a hindrance to many forces. If you have ever pedaled a bicycle up a hill, you have felt the resistance of gravity. You needed to apply strong force to overcome this downward pull. There are many machines that work against gravity. Elevators in tall buildings must overcome gravity to lift people to higher floors. Airplanes, cranes, and forklifts also must overcome the resistance of gravity.

Strong force is needed to overcome gravity.

The heavier an object is, the greater will be the resistance of gravity. That is why a loaded truck must shift to a lower gear to climb a steep hill. The same truck has little difficulty when it is empty because the resistance of gravity is not nearly as strong.

2. Inertia. The resistance called ***inertia*** keeps a stopped object from starting to move. Have you ever been standing on a hay wagon that was pulled forward unexpectedly? Why did you fall (or almost fall) instead of moving with the wagon? Inertia was keeping you from starting to move.

Like gravity, the resistance of inertia becomes greater as an object increases in weight. That is why a loaded truck cannot start off as quickly

Lesson 19

Lesson Concepts

1. Objects of themselves do not have the ability to change their state of rest or motion.
2. A force is required to make something move or to stop it.
3. A force is a push or a pull.
4. The four kinds of resistance that must be overcome are gravity, inertia, friction, and molecular attraction.
5. Work is done when a force overcomes resistance and moves something.

Teaching the Lesson

God created laws that govern how things move. These orderly laws of motion are manifest all about us. We are well acquainted with a host of physical forces: falling water, running engines, raging wind, and many others. The things that resist motion are also familiar. Who has not felt the resistance of gravity and inertia, or of friction and molecular attraction?

Help your students to identify the forces and resistances in their own world. Show them that God designed even moving things to behave in orderly ways. They may be too young to grasp the idea of perpetual motion, but this is a good time to begin impressing on them the fact that all moving things need a force to make them go. Provide them with the concepts that will lead them to reject as preposterous any story of perpetual motion that they might encounter.

Inertia

Things at rest tend to stay at rest.

Things in motion tend to stay in motion.

as an empty one. The loaded truck must overcome the greater resistance of inertia before it can move.

Inertia also keeps a moving object from stopping. The loaded truck that needs such a powerful engine to get it moving also needs powerful brakes to stop it.

3. Friction. The resistance of ***friction*** occurs when two things rub against each other. When you slide a box of apples across the floor, there is friction between the box and the floor. Sometimes wagon wheels squeak as they turn. That is because of friction between the wheels and the axles. We reduce this friction when we can. We haul the apples on a cart with wheels to reduce the friction, and we oil the wagon wheels to reduce the friction there.

4. Molecular attraction. All materials are made of tiny particles called molecules, and these are held together by a force called ***molecular attraction.*** When you split firewood,

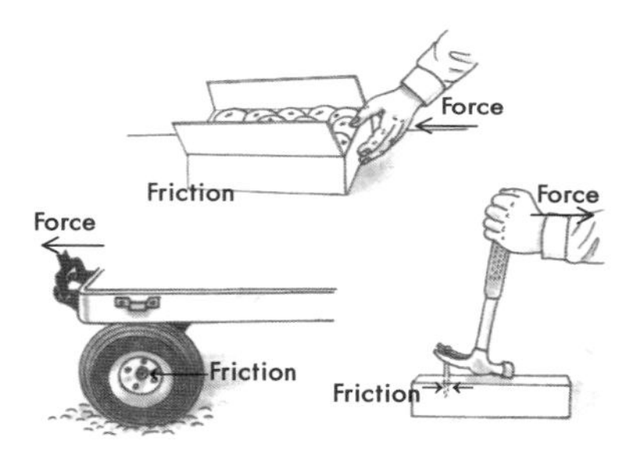

Friction occurs when two things rub against each other.

Molecular attraction keeps wood from splitting easily.

Discussion Questions

1. List a few things that move, and determine the forces that make them go.
   (Individual answers.)
2. Identify some of the resistances that affect the moving things you listed in number 1.
   (Individual answers.)

molecular attraction binds the wood fibers together and makes it hard for you to divide the wood in pieces. Cutting cloth, tearing paper, breaking sticks, and sawing wood are all activities hindered by the resistance of molecular attraction.

Sometimes more than one kind of resistance affects a machine. A heavily loaded truck starting off uphill is hindered by both gravity and inertia. In order to push a box of apples across the floor, you must overcome the resistances of inertia and friction.

Defining Work

When you do a job for your parents, you say that you have worked—and you have. However, the scientific definition of work is quite different from the definition we normally use.

In science, ***work*** is done whenever a force causes something to move. When you throw a softball, you are doing work even though you are playing. When a loaded truck moves uphill, work is being done. The force of the engine has overcome the resistances of inertia, gravity, and friction, and the truck moves.

If the engine stalls and fails to move its load, no work is done. The force is not strong enough to overcome the resistances. Simply applying force is not working. In order for work to be done, the force needs to overcome the resistance and move something.

Work is done only when a force makes something move.

Measuring Work

It is sometimes useful to measure the amount of work performed. This is simple to do. As stated in the previous paragraphs, work is done when a force moves an object. To measure work, you need to know how much force is exerted to move the object. This force may be measured in pounds. Suppose a bag of flour weighs 10 pounds. To lift the bag, you must pull upward with a force of at least 10 pounds.

To measure work, you must also know how far the object is moved. If you lift the bag of flour 1 foot, you have done a certain amount of work. If you lift the bag 2 feet, you have done twice as much work.

Since work involves both force and distance, the unit used to measure work includes both force and distance. If a force of 1 pound moves an object 1 foot, one ***foot-pound*** of work has been done.

How much work did you do in lifting the bag of flour? To find out, multiply the pounds of force exerted by the number of feet the bag moved. In other words, **Work = Force × Distance.** When you lifted the 10-pound bag 1 foot off the ground, you did 10 foot-pounds of work (10 pounds × 1 foot = 10 foot-pounds). When you lifted the bag 2 feet, you did 20 foot-pounds of work.

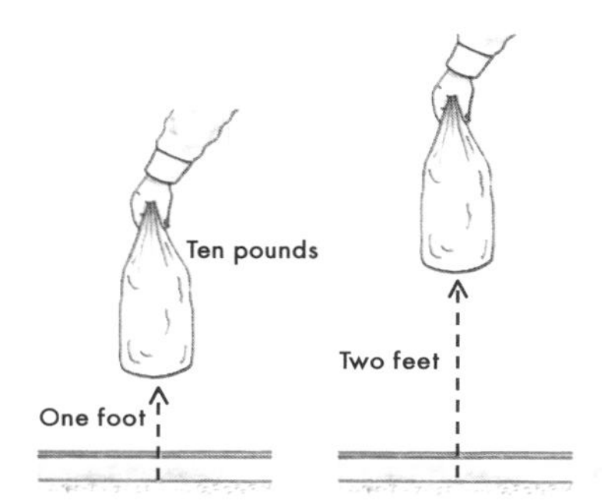

10 pounds × 2 feet = 20 foot-pounds

70 pounds × 9 feet = 630 foot-pounds

Suppose the upstairs floor in your house is 9 feet above the downstairs floor. How much work do you do when you climb the stairs to go to bed? Every time you go upstairs, you lift your weight 9 feet to the second floor. If you weigh 70 pounds, you do 630 foot-pounds of work just climbing the stairs.

Suppose your little sister weighing 40 pounds is sitting in a wagon. You push her a distance of 5 feet, using a force of 10 pounds. Would you find the amount of work by multiplying like this: 40 × 5 = 200 foot-pounds? Or would you multiply like this: 10 × 5 = 50 foot-pounds? Remember: to figure work, you multiply the pounds of force by the number of feet. If you push with a force of 10 pounds for a distance of 5 feet, you do 50 foot-pounds of work, no matter how much your sister weighs.

10 pounds × 5 feet = 50 foot-pounds

Lesson 19 Answers

Exercises

1. a. work
   b. inertia
   c. friction
   d. force
   e. resistance
2. a. false; No object can begin to move all by itself.
   b. true
   c. false; Gravity pulls things downward.
   d. true
   e. true
3. a. inertia
   b. gravity
   c. molecular attraction
   d. friction
4. d

Review

5. Some were formed of materials that erupted from within the earth.
   Others were formed by the crumpling of the earth's crust.
6. head—the height of the waterfall; flow—the amount of water running over the falls
7. Waves cause only up-and-down motion in water.
   Currents are great rivers in the ocean that move water from one place to another.
8. (Any three.) lightning, thunder, heavy rainfall, occasional hail, strong updrafts, tall thunderheads

Exercises

1. Choose the correct words from the list to match the descriptions.

   force — inertia — work
   friction — resistance

   a. Force overcoming resistance and moving an object.
   b. Kind of resistance that keeps stopped objects from moving.
   c. Kind of resistance that occurs when two things rub together.
   d. A power that causes an object to move; a push or pull.
   e. Gravity, inertia, friction, or molecular attraction.
2. Write *true* or *false* for each sentence. If a sentence is false, correct it by changing one or two words.
   a. An object may begin to move all by itself.
   b. A force can be either a push or a pull.
   c. Friction pulls things downward.
   d. Gravity is the resistance that makes it hard for trucks to climb mountains.
   e. If you try with all your might to lift a farm tractor, but you do not budge it at all, you have done no work.
3. For each example, write whether the main resistance is *gravity, inertia, friction,* or *molecular attraction.*
   a. Jerry had to pull hard to get the empty hay wagon moving.
   b. Loren is pulling a wagon loaded with wood up the hill to the house.
   c. Ruby found it hard to cut a piece of heavy cardboard with her scissors.
   d. Margaret is pushing boxes around in the attic to make room for some other things.
4. Suppose your little sister is in a wagon, and you want to push her up a hill. To move the wagon, you must push with a force of 25 pounds. The hill is 125 feet long. How many foot-pounds of work will you do?
   a. 25 foot-pounds
   b. 125 foot-pounds
   c. 150 foot-pounds
   d. 3,125 foot-pounds

Review

5. In what two main ways has God formed mountains since the Creation?
6. Name and describe the two things that determine how much energy a waterfall has.
7. How is the motion of waves different from the motion of currents?
8. List three of the things that identify a thunderstorm.

Activities

1. The four kinds of resistance are important to us. The world would be very different without them. Write sentences describing something that would happen in a world
   a. without friction.
   b. without gravity.
   c. without molecular attraction.
2. Load a small wagon with stones. Try these experiments to see if you can identify resistance.
   a. Pull the loaded wagon up a small hill. Do you feel gravity pulling against you? Empty the wagon, and try again. Why does it go more easily now?
   b. Try pulling the loaded wagon rapidly. Did you need to start slowly? What held you back? Now try it again with an empty wagon. Why could you start off more rapidly this time?
   c. Now fill a box with stones. Try pushing it across the lawn. Is it harder to push the box than the loaded wagon? Why?
   d. Pull your box of stones up a hill. What two kinds of resistance are hindering you? How could you reduce the one kind?
3. Stack half a dozen Carrom rings on top of each other. With another Carrom ring, shoot away the bottom ring of the stack. Why doesn't the stack fall over? Which ring does inertia try to keep moving? Which rings does inertia try to keep from moving? What work was done?

4. Set a cup half full of water on a sheet of heavy paper. Pull the paper slowly, and both paper and cup move. You have overcome the inertia of both paper and cup.

   Now jerk the paper away very quickly. The paper moves, but the cup stays where it was. You have overcome the inertia of the paper, but not of the cup.

Activities

1. (Sample answers.)
   a. Without friction, we would be unable to walk or ride a bicycle. No brakes would work.
   b. Without gravity, things would not stay on the earth. We would be in danger of floating off into space. Without gravity, we could not plow the soil. We would not have the traction to pull the plow, and the loosened soil would float away.
   c. Without molecular attraction, all material things would disintegrate into individual molecules. The whole world would dissolve like a lump of sugar in water.

2–4. Help your students with some of these demonstrations.

Lesson 20

Friction, a Help and a Hindrance to Work

Vocabulary

bearing, a machine part in which another part turns, often with balls or rollers to reduce friction.

brake, a device used to slow or stop a machine, usually by friction.

lubricant (lü′bri·kənt), a slippery substance that reduces friction between two moving objects.

traction (trak′·shən), friction that keeps an object (such as a tire) from slipping as it pulls across a surface.

What Is Friction?

One of the most common kinds of resistance is friction. It is the greatest resistance to be overcome in doing most work. Every machine is affected by friction. Have you ever tried to ride a bicycle with a wheel rubbing against the frame? Such a bicycle is hard to pedal because friction hinders you. Friction also makes it hard to slide a box of books across the floor. Whenever two things rub against each other, friction is produced.

Two things affect the amount of friction between objects. The first thing is pressure, or the amount of force that pushes the objects together. It is much harder to push a twenty-pound box across the floor than a five-pound box because the heavier box presses much harder on the floor. The greater the pressure, the greater the friction.

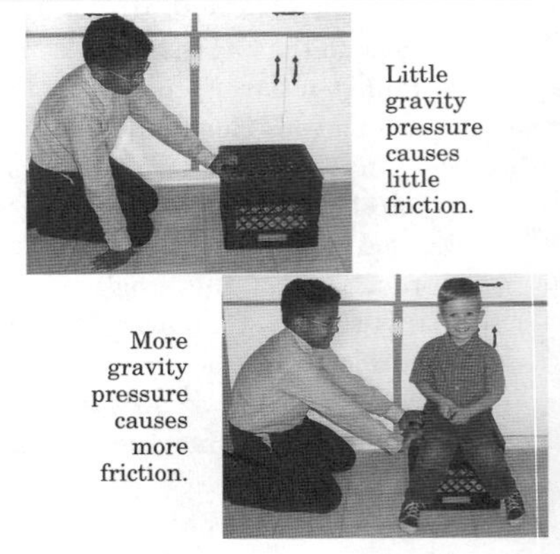

Little gravity pressure causes little friction.

More gravity pressure causes more friction.

The second thing that affects friction is the roughness of the surfaces that rub together. Which has more friction when rubbed over wood—notebook paper or sandpaper? The sandpaper has much more friction because a rough surface has more friction than a smooth one.

Lesson 20

Lesson Concepts

1. Friction is the greatest resistance to be overcome in doing most work.
2. The amount of friction is affected by pressure and by the roughness of the rubbing surfaces.
3. Friction generates heat and causes the wearing of surfaces.
4. Friction can be reduced by using a lubricant, by using wheels, by smoothing the rubbing surfaces, and by separating the rubbing surfaces.
5. Friction is useful for providing traction, for braking, and (in some cases) for producing heat.

Teaching the Lesson

Friction is a broad subject—too broad, in fact, to cover completely in one lesson. Friction is a great hindrance at some times but indispensable at other times. Show both sides of friction: its usefulness and its hindrances.

Be a friction sleuth. Prove that your students are surrounded by friction. Point out how friction helps in walking. Identify friction in the classroom window that is difficult to open and close. Show the importance of friction in grasping a pencil. Discuss the possibilities for reducing the unwanted friction in the classroom—such as trying a little candle wax on that sticky window.

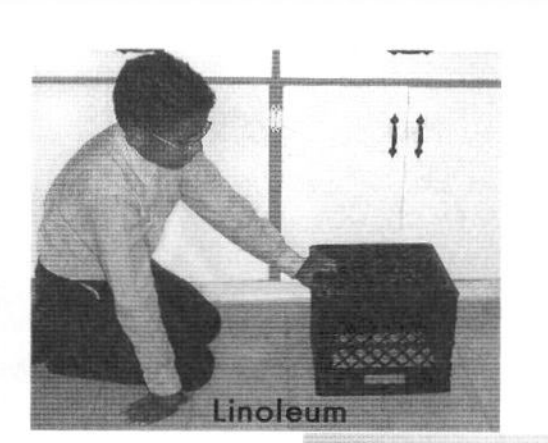

A smooth surface causes little friction.

A rough surface causes more friction.

Friction as a Hindrance

When your bicycle wheel rubs against the frame of the bicycle, friction is a hindrance. It takes much force to overcome that friction. Other machines are hindered by friction in the same way. When there is friction between the moving parts in a machine, much force must be used to overcome the resistance.

A rubbing wheel produces unwanted friction.

Friction causes wear on the rubbing objects. Take a close look at the brake pedal on your father's truck. Notice how the rubber pad is worn smooth on the right side. That is where the driver's foot rubs over the pedal when he slides his foot over to the accelerator. In an older vehicle, sometimes the rubber pad must be replaced because it is worn right through to the metal.

On a bicycle with hand brakes, you can see the wear that friction causes on the brake pads. As you keep using the bicycle, the pads keep wearing down until they must be replaced. Friction causes other things to wear out in the same way.

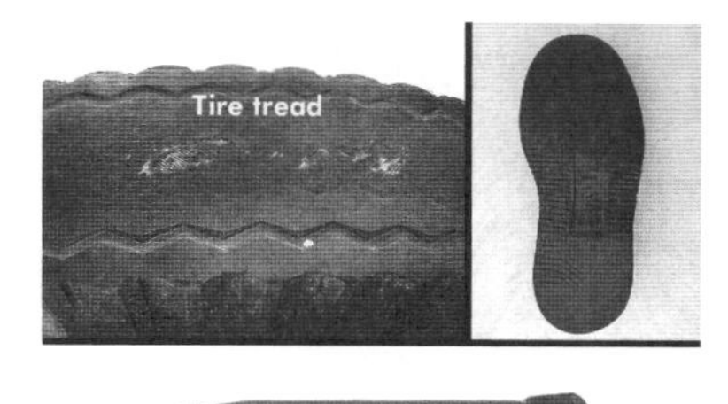

Friction causes wear.

Did you ever feel your hand get warm when you slid it down a stair railing too fast? Perhaps you even received a brush burn. Why does this happen? It is friction that causes the heat.

Discussion Questions

1. What are five places in our classroom where friction occurs?
   (Sample answers.)
   squeaking desk lids;
   sticking windows;
   door hinges (note the wear on the pins);
   between shoes and floor;
   between chalk and chalkboard (Would oiling the chalkboard be a good idea?)
2. For each place in number 1 that friction is a hindrance, how might we reduce the friction?
   by using a lubricant (in most cases)
3. Name some common lubricants, and tell where they are used.
   (Sample answers.)
   oil—hinges, engines, wheelbarrows;
   grease—wheels, joints;
   water—ice skating, grindstones;
   paraffin—sled runners;
   graphite—pencil lead on a sticky zipper

In an engine, heat from friction can become so great that it ruins the engine. To keep an engine from getting too hot and wearing out because of friction, we use motor oil between the moving parts.

Ways to Reduce Friction

1. Use a lubricant. A ***lubricant*** is a slippery substance that reduces friction between two surfaces. One example is the oil in your lawn mower engine. All the moving parts

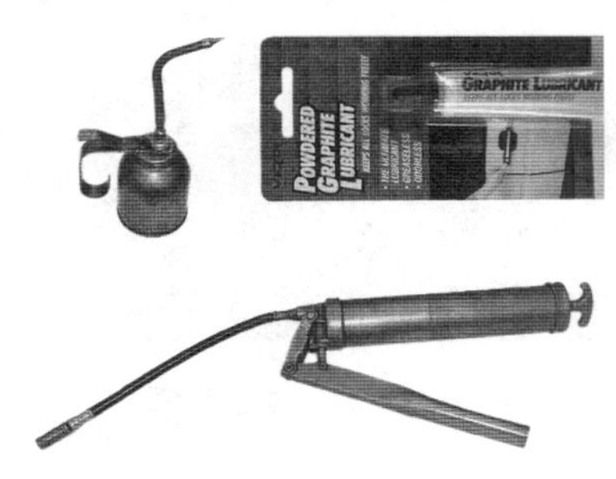

Common lubricants for reducing friction

are covered with a thin layer of oil, which greatly reduces the friction between them.

Sometimes a lubricant causes problems. A car will slide much more easily on a wet road than on a dry road. A wet surface has less friction than a dry one because water is a lubricant that reduces the friction between the tires and the road.

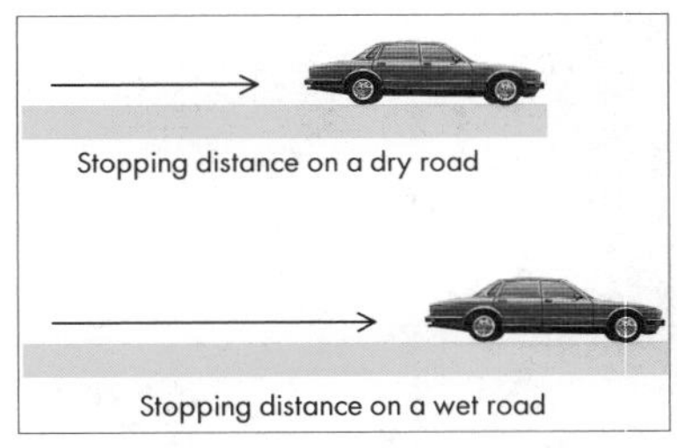

A motorist on a wet road needs to allow an additional 20% braking distance to stop safely.

2. Use wheels or rollers. If you need to bring a large crate of apples from the orchard to the cellar, how can you reduce the friction between the crate and the ground? Oil and grease would not help very much, so you need to reduce the friction in some other way. A cart or wagon will solve the problem nicely because it has wheels. Using wheels is a very useful way to reduce friction.

Wheels help reduce friction.

Many machines have special ***bearings*** that reduce friction. These bearings hold turning shafts and let them spin freely. Many bearings are made to hold balls or rollers between the shaft and the machine. The weight of the shaft rests on the balls or rollers; and as the shaft turns, the balls or rollers turn too, thus reducing friction. Such bearings also need a lubricant like grease or oil to keep them turning smoothly.

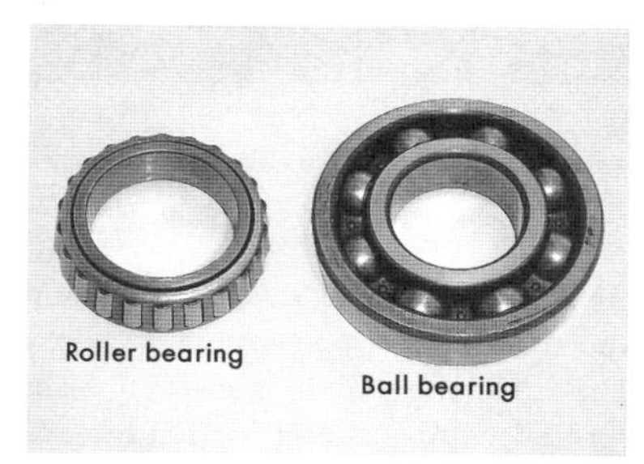

3. Smooth the surfaces. Another method of reducing friction is to smooth the surfaces of the rubbing objects. Smooth surfaces do not cause nearly as much friction as rough surfaces. For that reason, a mechanic tries to keep rust off his new nuts and bolts. If he keeps them shiny and smooth, they will fasten parts much more easily. A smooth, clean surface has less friction than a rough, rusty, or dirty one.

4. Stop the rubbing. Sometimes we can reduce friction by separating the rubbing surfaces. If a fan belt is rubbing on a guard, we can move the guard and get rid of the friction altogether.

How is this boy overcoming unwanted friction?

Friction as a Help

Friction is often very useful. Let us consider three ways that friction is a help to us.

1. Traction. Friction keeps your feet from slipping as you walk. When you walk on ice, there is very little friction between your feet and the ice. This allows you to slide when you play on the ice.

Both cinders and tire chains increase traction.

To apply friction, the brake pads squeeze tightly against both sides of the rotating wheel or disk.

A truck needs friction so that it can move. Friction between the tires and the road gives it ***traction.*** Sometimes mud, snow, or ice will lubricate the road so much that the truck wheels lose their traction and begin to spin. That is why road workers put cinders on snowy roads in winter. The cinders are gritty and

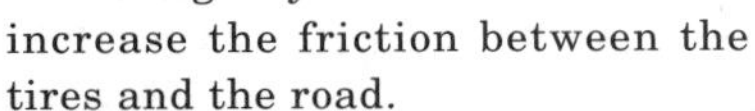

increase the friction between the tires and the road.

2. Braking. Another helpful use of friction is for stopping things. A bicycle with hand ***brakes*** is a common machine that uses friction to stop. When you squeeze the handle, the brakes force the brake pads tightly against the rim of the wheel. Friction between the pad and the rim slows the bicycle. The harder you squeeze the handle, the harder the brake will press against the rim. Do you remember why? As the pressure increases, so does the friction.

3. Heat. You have already seen that the heat from friction is often harmful. But it is helpful when you rub your hands together to make them warm. Heat from friction is also useful when you strike a match on a rough surface. The tip of the match is covered with special chemicals, and friction makes them so hot that they burst into flames. We could also use friction to start fires in other ways, but matches have made the job very simple.

Before matches were invented, people used friction to start fires. The bow has its string looped around the vertical stick. By grasping the stick loosely and sliding the bow back and forth rapidly, the stick spins fast enough to kindle bits of shredded paper on the board.

Exercises

1. Choose the correct words from the list to match the descriptions.
   bearing brake lubricant traction
   a. A substance such as oil that reduces friction between moving parts.
   b. A device that reduces the friction of a turning shaft, sometimes with balls or rollers.
   c. Friction that keeps car tires from slipping.
   d. A device that uses friction to stop something.
2. What are the two things that affect the amount of friction between surfaces?
3. True or false? Most machines are affected by friction.
4. True or false? Friction is always a hindrance.
5. True or false? A layer of oil between moving parts helps to keep the parts from wearing out.
6. What are the four ways of reducing friction?
7. Tell which of the four ways of reducing friction would be best in each of the following cases.
   a. Conrad noticed a rusty spot on the sliding board.
   b. Little Wesley was having a hard time pushing his scooter. Caleb saw that the back wheel was rubbing against the frame.
   c. Timothy wanted to use his sled to pull a load of wood to the house, but most of the snow had melted.
   d. Philip wanted to use the wheelbarrow to haul sawdust to the garden, but the wheel was squeaky and hard to turn.
8. The heat that lights a match is caused by ———.
9. Two benefits of friction are b——— and t———.

Review

10. Before an object can move, a ——— must be applied.
11. Choose the kind of resistance that each phrase illustrates.
    a. A loaded truck trying to go uphill. — friction
    b. A loaded truck trying to start moving. — gravity
    c. A bicycle wheel rubbing against the bicycle frame. — inertia
    d. Wood that is hard to split. — molecular attraction
12. Work is accomplished if the ——— is strong enough to overcome the ——— and move something.
13. If a force of 15 pounds pulls an object for 2 feet, how much work has been done?
14. The tendency of a resting object to stay at rest is called ———.

Lesson 20 Answers

Exercises

1. a. lubricant
   b. bearing
   c. traction
   d. brake
2. pressure between the surfaces; roughness of the surfaces
3. True
4. false
5. True
6. using a lubricant; using wheels or rollers; smoothing the surfaces; stopping the rubbing
7. a. smoothing the surface
   b. stopping the rubbing
   c. using wheels or rollers
   d. using a lubricant
8. friction
9. braking, traction

Review

10. force
11. a. gravity
    b. inertia
    c. friction
    d. molecular attraction
12. force, resistance
13. 30 foot-pounds
14. inertia

Activities

1. Yes, all friction produces heat. The brakes of a car become hot when they are used; moreover, if they overheat, brakes become useless. (This is why truck drivers sometimes lose control of their vehicles when coming down a mountain.) As another example, a lawn mower blade becomes hot from friction when it is sharpened with a bench grinder.

2–4. Try these activities yourself. Can you explain to the class how they work? Demonstrating the concepts will help your pupils to understand and remember them.

Activities

1. You have learned that friction produces heat. Tell whether this is true of all friction. For example, do the brakes of a car heat up when they stop the car? Give other examples if you can.
2. You have seen that lubrication helps to reduce friction. See for yourself how it works. Get two blocks of wood, and rub them together. Then put some petroleum jelly such as Vaseline between them, and rub again. Is it any easier to rub them together now? Why?
3. To show that wheels reduce friction, get a thin rubber band, a piece of board for a ramp, and a small toy farm wagon. Prop one end of the ramp on four or five books, and hook the rubber band onto the hitch of the toy wagon. Pull the wagon up the ramp with the rubber band. Measure how much the rubber band stretches as you pull. Then flip the wagon upside down. Pull it up the ramp that way, and measure the rubber band again. Does the rubber band stretch any farther when you turn the wagon upside down? Do you know why?

4. To see how smoothness or roughness affects friction, set up a ramp like the one in number 3. You will also need a piece of sandpaper, a rubber band, several tacks, and a piece of wood measuring about 1½ by 1½ by 6 inches. Fasten the rubber band to the piece of wood with a tack. Pull the block up the ramp with the rubber band. Measure how far the rubber band stretches. Now tack a piece of sandpaper to the ramp, and pull the block of wood up over it. Measure how far the rubber band stretches this time. Does it stretch any farther? What does this tell you?

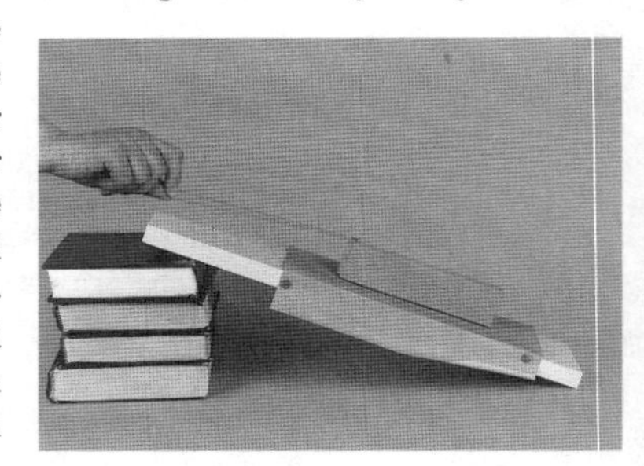

Lesson 21

Using Machines to Perform Work

Vocabulary

effort, force exerted to perform work.

machine, any device that changes the amount, the speed, or the direction of an effort.

mechanical advantage, the number of times that a machine multiplies the effort put into it.

What Is a Machine?

You may think that a ***machine*** is something large and complicated, like a farm tractor, an automobile, or a jet airplane. These machines are large and complex, but many machines are quite simple and common. In fact, we use machines every day. David uses a screwdriver to pry the lid off a paint can. Mary uses a knife to peel potatoes. Mother uses a can opener or an eggbeater to prepare dinner. The whole family eats dinner by using knives, forks, and spoons—all machines. After dinner, we may sweep the kitchen floor with a broom, another machine. At school, we sharpen our pencils with a machine, the pencil sharpener. At recess, we use a machine called a baseball bat to send a softball sailing into the distance.

All these machines and many others are alike in one way—they multiply our effort so that we can do tasks with greater speed and force.

How Machines Help Us

All machines, whether simple or complex, help us in at least one of three ways. They may change the direction of a force, the amount of a force, or the speed and distance of a force.

1. Some machines help us by changing the direction of a force. With a pulley, a worker can make a bucket of bricks move *up* the side of a building by pulling *down* on a rope. That is easier than climbing to the top of the building and pulling the bucket up. To lift a large rock, you can put

A pulley

Lesson 21

Lesson Concepts

1. A machine helps us do work by changing the amount, speed, or direction of an effort.
2. The number of times the effort is multiplied is called mechanical advantage.

Teaching the Lesson

We are surrounded by machines. Some are simple and some are complex, but all are machines. They are so abundant that we often do not even recognize them as machines. Point out the machines around you. Pencil sharpeners, scissors, wrenches, brooms, hole punches, staplers, and a host of other simple machines are right in your classroom. If you have not already used the story at the beginning of Unit 4 in the teacher's manual, you may wish to read it aloud as an introduction to this lesson.

Discussion Questions

1. List five machines found in the classroom.
   (Individual answers.)
2. What is the purpose of finding the mechanical advantage of a machine?
   so that we can compare machines and use the one with greater mechanical advantage

one end of a long bar under it and push down on the other end. You push your bicycle pedals down to move the bicycle forward. Each of these machines changes the direction of a force.

The early American settlers used the force of moving water to grind their grain. They did this by using a machine called a water wheel to change the direction of the water's force. As the water rippled down over the water wheel, it made the wheel

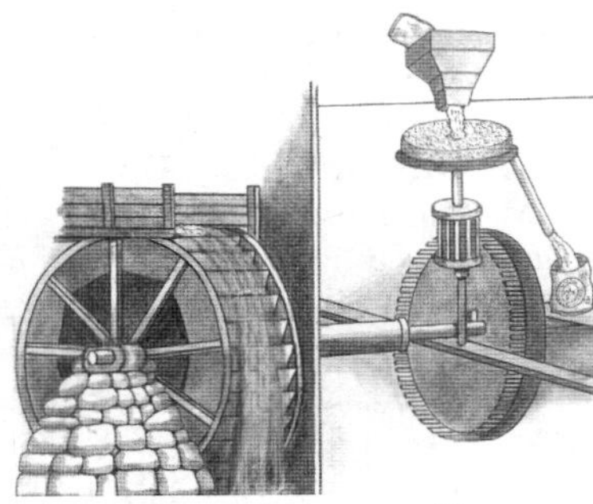
A water wheel turning a millstone

turn. The water wheel was fastened to a shaft that went into the mill and turned the millstone. In this way, the settlers changed the downward force of falling water into a rotating force that did work for them.

2. Some machines help us by increasing the amount of a force. Without machines, you may be unable to overcome powerful resistances such as friction or gravity. If you try to loosen a rusty bolt with your fingers,

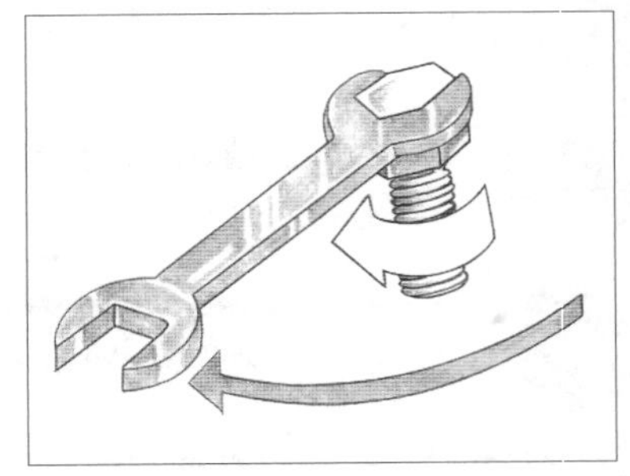
A wrench increases turning force.

you soon find that friction makes it impossible. But if you get a long wrench, you can turn the bolt easily. The wrench is a machine that makes your work easier by increasing the amount of the force.

You could hardly pull a nail out of a board with your fingers. But with a familiar machine, the hammer, you can pull nails with little trouble because it greatly increases the force you use. Of course, the hammer also changes the direction of the force. You pull the hammer *backward* to pull the nail *up*.

A hammer increases nail-pulling force.

But you use a hammer to pull nails mainly because it increases the amount of the force.

3. Some machines help us by increasing the speed and distance of a force. A bicycle not only changes the direction of the force you apply to the pedals. It also helps you go faster. As you pedal, the wheels go around much faster than the pedals do. This lets you ride swiftly without needing to pedal very fast. A bicycle changes the speed and distance of a force.

If you pedal this bicycle 50 revolutions per minute, your foot would travel 164 feet per minute. But the wheels would turn 120 revolutions per minute and travel 657 feet per minute.

A baseball bat also increases the speed and distance of a force. As you swing the bat, you move the handle only a short distance. But the heavy end of the bat moves much farther and faster, sending the ball off at a great speed.

Because a broom increases the speed and distance of a force, you can sweep an entire room without moving

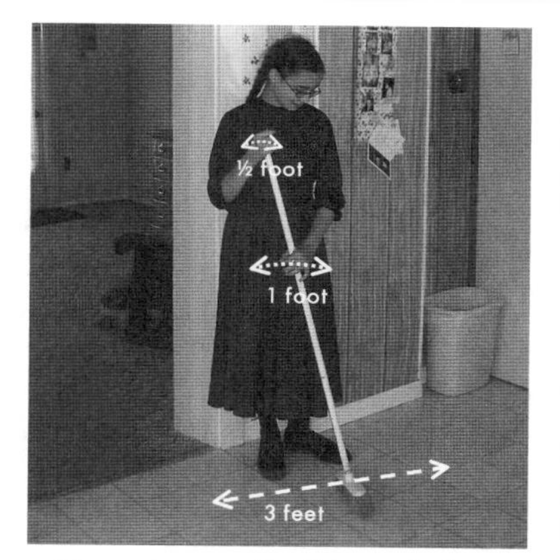

A broom increases the distance of a force.

your arms very much. Because a fly swatter increases the speed and distance of a force, you can swat a fly even though it is faster than you are.

Mechanical Advantage

A machine does not create force. It simply takes our ***effort,*** the force we exert, and changes it in some way to make work easier for us. As you have seen, one of the most important ways a machine changes a force is to multiply the force (effort) that we exert. This enables us to overcome resistances that we are not strong enough to overcome without machines.

Some machines increase our effort much more than others do. To describe how much a machine increases our effort, we use a special term—***mechanical advantage.*** The mechanical advantage of a machine is the number of times the machine multiplies the effort.

Suppose you are using a pole to lift a 60-pound rock. If you must push down with a force of 60 pounds to lift the rock, there is no mechanical advantage. But if you place the pole correctly, you may need to push down with only 10 pounds to lift the rock. Your machine, the pole, has multiplied your 10-pound effort 6 times to overcome a 60-pound resistance. The mechanical advantage of this machine is 6.

To find the mechanical advantage of a machine, divide the resistance by the effort. In other words, **Mechanical advantage = Resistance ÷ Effort.** If a resistance of 80 pounds is overcome by an effort of 10 pounds, the mechanical advantage is 8 because 80 ÷ 10 = 8.

For another example, suppose a man uses a chain hoist to lift a heavy stove off a pickup truck. If the stove weighs 250 pounds and the man pulls on the chain with an effort of 25 pounds, you can find the mechanical advantage like this: 250 ÷ 25 = 10. The machine multiplies the effort 10 times.

Mechanical advantage: 60 ÷ 60 = 1
(no mechanical advantage)
60 lb. 60 lb.

Mechanical advantage: 60 ÷ 10 = 6
10 lb. 60 lb.

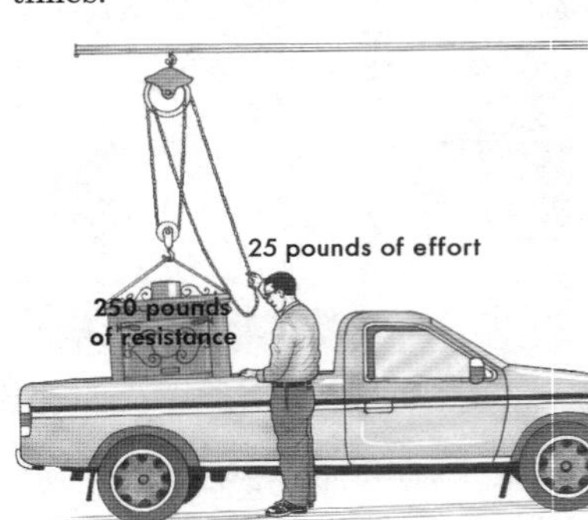

Exercises

1. What are the three ways that machines help us do work?
2. The most important reason a hammer is used to pull nails is because it changes the (direction, amount, speed) of a force.
3. A wrench is used to tighten bolts because it changes the (direction, amount, speed) of a force.
4. A bicycle is used for travel mostly because it changes the (direction, amount, speed) of a force.

Lesson 21 Answers

Exercises

1. by changing the direction of a force;
   by increasing the amount of a force;
   by increasing the speed and distance of a force
2. amount
3. amount
4. speed

Note on 2 and 4: A hammer also changes the direction and speed of the force. A bicycle increases the speed of the force, decreases the amount of the force, and changes the direction of the force as well. However, we use hammers to pull nails mainly to increase force, and we ride bicycles to travel faster than we could by walking.

5. The amount of force put into a machine is called (effort, resistance, mechanical advantage).
6. Mechanical advantage tells the number of times the effort is ———.
7. Some machines change a force in more than one way. Tell how each of the machines below changes a force.
   a. bicycle (two ways)
   b. hammer (two ways)

Review

8. In order for work to be done, the (effort, resistance) must be strong enough to overcome the (effort, resistance).
9. The four kinds of resistance are ———, ———, ———, and ———.
10. What two things affect the amount of friction?
11. What are three ways of reducing friction?
12. If you lift a 21-pound sack 3 feet from the floor onto a table, how much work have you done?
    a. 7 foot-pounds
    b. 18 foot-pounds
    c. 24 foot-pounds
    d. 63 foot-pounds

5. effort
6. multiplied
7. a. changes speed and distance of a force
   b. changes amount and direction of a force

Review

8. effort, resistance
9. gravity, inertia, friction, molecular attraction
10. pressure on the surfaces; roughness of the surfaces
11. (Any three.)
    using a lubricant;
    using wheels or rollers;
    smoothing the surfaces;
    separating the rubbing surfaces
12. d

Lesson 22

Simple Machines: The Lever and the Inclined Plane

Vocabulary

fulcrum (fůl′·krəm), the point on which a lever pivots.

inclined plane (in·klīnd′), a sloping surface.

lever (lev′·ər, lē′·vər), a simple machine made of a stiff bar pivoting on a fixed point, used to increase lifting or pushing effort.

Many machines are very complex. For example, automobiles and lawn mowers have many parts. But even complex machines are made of six kinds of simple machines put together in various ways. In this lesson and the next, you will study the six simple machines: the lever, the inclined plane, the screw, the wedge, the wheel and axle, and the pulley.

The Lever

One of the most common and useful of the simple machines is the ***lever.*** You have been using levers every day for years. Spoons and forks are levers. So are brooms and baseball bats and seesaws and wheelbarrows. Every time you use a scissors or pliers, you are using levers. Even your arms act as levers.

Suppose that you and your father are riding a seesaw. If both of you are normal in size, you will not balance well at all. What can be done to even things up? If your father moves the plank to have the center closer to himself, your weight will lift his, even though he is much heavier than you.

The seesaw is a simple machine, a lever. The lever multiplies your weight so that you can lift someone much heavier than yourself.

Lesson 22

Lesson Concepts

1. The mechanical advantage of a lever is found by dividing the length of the effort arm by the length of the resistance arm.
2. An inclined plane with a gradual slope has a much greater mechanical advantage than one with a steep slope.

Teaching the Lesson

This lesson introduces the lever, perhaps the most basic of the simple machines. Make it real to the pupils by doing some of the activities. Children use levers every day—their arms if no others. But if they examine levers closely and feel for themselves how the effort changes as the fulcrum moves, they will most likely find the lesson interesting and impressive. If your playground has a seesaw, use it to demonstrate some of the ideas in the book. A pry bar used to raise a heavy weight will be more impressive than a yardstick and some books.

All levers have three parts: (1) the ***fulcrum,*** or pivot point, (2) the effort arm, and (3) the resistance arm. The effort arm of a lever always extends from the fulcrum to the point where the effort is applied. Likewise, the resistance arm of a lever always extends from the fulcrum to the resistance. Notice the three parts of a lever in the pictures at the right.

Now look at another example of a lever—a pair of scissors. Actually, the scissors are two levers, each working against the other. The screw or rivet joining the two levers is the fulcrum.

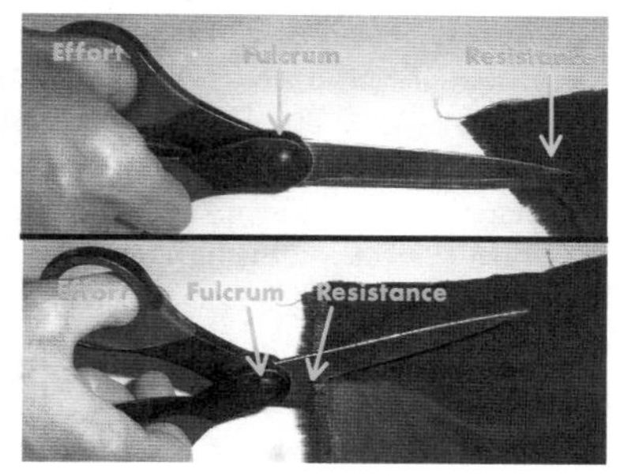

Mechanical advantage is greater if the resistance arm is shorter.

Have you ever used a screwdriver to open a paint can? If so, you were using a lever. The fulcrum is the edge of the can, the effort arm is the part you push down on, and the resistance arm is the tip of the screwdriver prying the lid up.

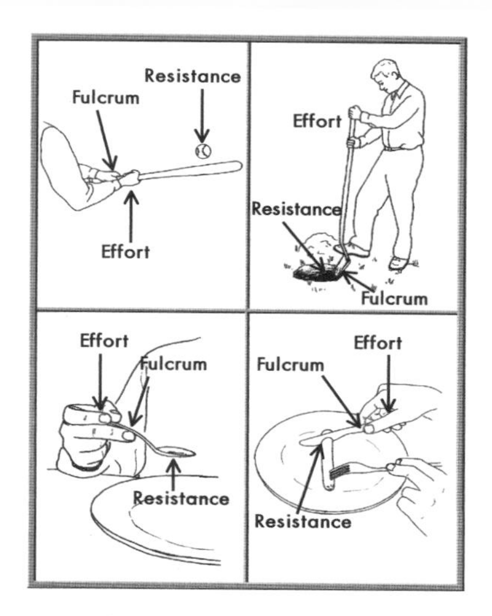

A wheelbarrow is a different kind of lever. In all the levers pictured above, the fulcrum is found between the effort arm and the resistance arm. But a wheelbarrow has the fulcrum at the end of the lever, where the wheel is. The effort arm extends all the way from the fulcrum to the handlebars where

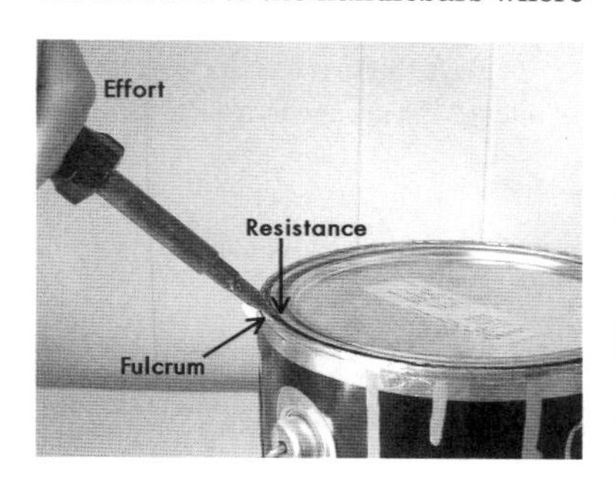

As presented in today's lesson, the method for finding the mechanical advantage of a lever is simplified because it does not take friction into account. It yields only the theoretical mechanical advantage. But your goal is to introduce mechanical advantage as simply and clearly as possible, so that technicality can be disregarded. The pupils can learn about actual and theoretical mechanical advantage later.

Though the formula for the mechanical advantage of an inclined plane is not taught in the lesson, the more observant students will figure it out if you do Activity 4. Allow your children to feel the difference in each case.

Discussion Questions

1. List five levers found in everyday life, starting in the classroom.
   (Individual answers.)
2. Why do the builders of roads and railroads pay much attention to the steepness of a slope?
   If the slope is too steep, the locomotives and trucks will be unable to pull their heavy loads to the top.
   Coming down, the brakes of a truck may overheat if the slope is too great.
3. How steep an incline is too steep for vehicles?
   Allow the children to use the 4-foot board to show the greatest slope they think is permissible. Point out that when the slope becomes steep, warning signs may be posted at the top of the hill. They may warn of a 6% slope, or 8%, or even 11%. An 11% slope is quite steep. To show an 11% slope with your 4-foot board, raise one end 5¼ inches.

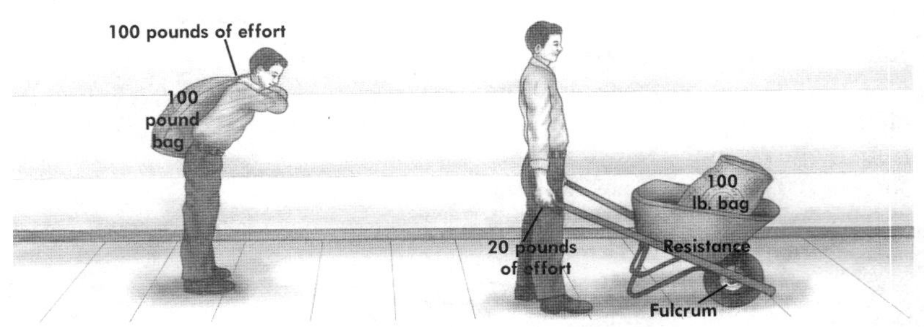

Which method is the easier way to carry 100 pounds?

you lift. The resistance arm is part of the effort arm. It extends from the fulcrum to the weight on the wheelbarrow.

Mechanical Advantage of a Lever

As you have seen, the lever is a simple machine that multiplies effort. We put forth much less effort if we move 100 pounds of soil with a wheelbarrow than if we carry it on our backs. Opening a paint can is much easier if we use a screwdriver. And with a lever, a 150-pound man can be lifted by a boy who weighs only 50 pounds.

To find the mechanical advantage of a lever, divide the length of the effort arm by the length of the resistance arm. In other words, **Mechanical advantage = Length of effort arm ÷ Length of resistance arm.** Look at the lever below. The boy's weight is pushing down 6 feet from the fulcrum. Therefore the effort arm is 6 feet long. Because the father's weight is 2 feet from the fulcrum, the resistance arm is 2 feet long. So we find the mechanical advantage of the lever by dividing like this: $6 \div 2 = 3$. The boy's effort of 50 pounds is multiplied 3 times to lift the father's weight of 150 pounds (3×50 pounds = 150 pounds).

A hammer is a lever that multiplies our effort so much that we can pull

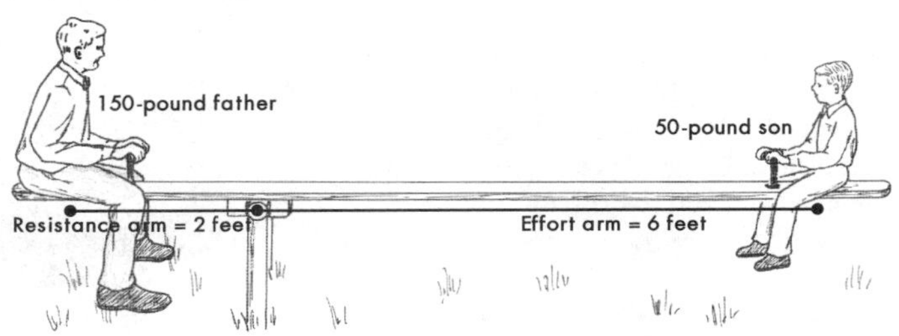

nails out of solid wood. How can we calculate the mechanical advantage of a hammer? Look at the hammer below. The effort arm is 12 inches long. The resistance arm is 2 inches long. So the mechanical advantage of the hammer is 12 ÷ 2, or 6. This means that if we pull on the hammer handle with a force of 20 pounds, the force exerted on the nail is 6 × 20 pounds, or 120 pounds.

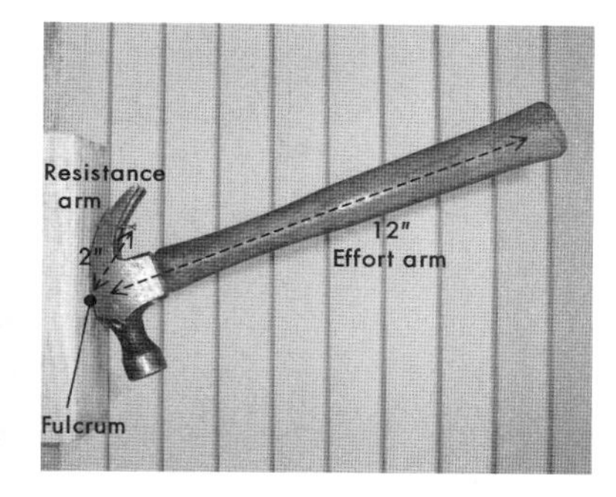

The Inclined Plane

David was helping his father load apple crates onto a trailer. He hauled them over to the trailer on his wagon. But the bed of the trailer was so high that David could not lift the crates onto the trailer by himself. So his father used several planks to make a ramp. Then David could easily pull the wagon up the ramp and set the crates on the trailer. We call this simple machine an ***inclined plane.***

An inclined plane will help David.

The inclined plane is the oldest and simplest of the six simple machines. When God created the first hill, He made the first inclined plane. Every time you ride your bicycle up a hill, you are using an inclined plane to raise your body weight to a higher level.

An inclined plane with a gradual slope has a much greater mechanical advantage than one with a steep slope. You find it much easier to ride your bicycle up a gentle slope than a

Which slope is easier to go up?

steep slope. When engineers design highways and railroads, they carefully plan the slopes of inclined planes so that the mechanical advantage is great enough to let heavy trucks and trains travel over the hills.

We use an inclined plane to make it easier to pull a hay wagon into the upper story of the barn. A hay elevator is an inclined plane that helps to raise hay bales into the haymow. Even a stairway is a kind of inclined plane, which makes it easier for us to go upstairs.

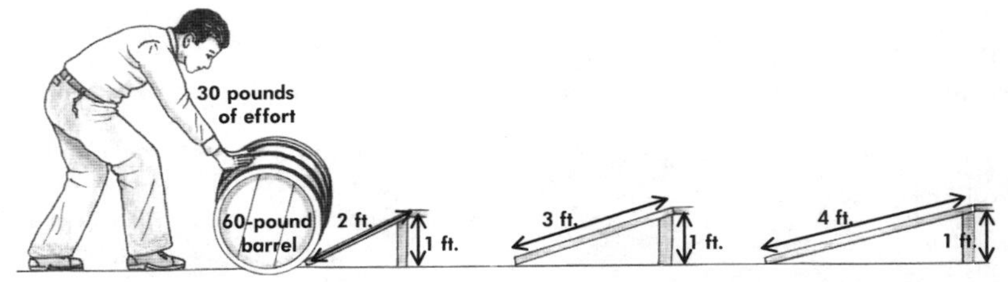

A man rolls a 60-pound barrel up three different inclined planes. The first plane is 2 feet long to raise the barrel 1 foot and takes 30 pounds of effort. The second plane is 3 feet long for 1 foot of height and takes 20 pounds of effort. The third plane is 4 feet long for 1 foot of height and takes 15 pounds of effort.

Lesson 22 Answers

Exercises

1. lever, inclined plane, screw, wedge, wheel and axle, pulley
2. fulcrum
3. fulcrum
4. fulcrum
5. effort, resistance
6. effort
7. d

Exercises

1. List the six simple machines.
2. The ——— of a lever is the point on which the lever pivots.
3. The effort arm of a lever is the distance from the effort to the ———.
4. The resistance arm of a lever is the distance from the resistance to the ———.
5. The mechanical advantage of a lever is found by dividing the length of the ——— arm by the length of the ——— arm.
6. A lever with a mechanical advantage of 9 multiplies the (effort, resistance) by 9.
7. If you apply an effort of 25 pounds to the effort arm of a lever with a mechanical advantage of 5, how much will you be able to lift with the resistance arm?
   a. 5 pounds
   b. 50 pounds
   c. 75 pounds
   d. 125 pounds

8. Answer the following questions about the lever shown here.
   a. How long is the effort arm?
   b. How long is the resistance arm?
   c. What is the mechanical advantage?

Effort 6 feet — Resistance 2 feet

9. Which of the levers described below would lift a heavy weight most easily?
   a. effort arm—6 feet; resistance arm—2 feet
   b. effort arm—10 feet; resistance arm—5 feet
   c. effort arm—4 feet; resistance arm—1 foot

Review

10. The number of times that a machine multiplies effort is called the ——— of the machine.
11. Friction is useful in at least two ways: it provides t———, and it makes b——— possible.
12. True or false? Work is done when the resistance matches the effort so exactly that nothing moves.
13. True or false? Wheels help to reduce inertia.
14. If Jane pushes a lever 1 foot with an effort of 20 pounds to lift a 100-pound rock, how much work has she done?
    a. 20 foot-pounds
    b. 80 foot-pounds
    c. 100 foot-pounds
    d. 2,000 foot-pounds

Activities

1. Answer the following questions about the lever in this picture.
   a. How long is the effort arm? (Remember to figure from the fulcrum.)
   b. How long is the resistance arm? (Remember to figure from the fulcrum.)
   c. What is the mechanical advantage?
   d. With an effort of 8 pounds, this lever could lift how many pounds?

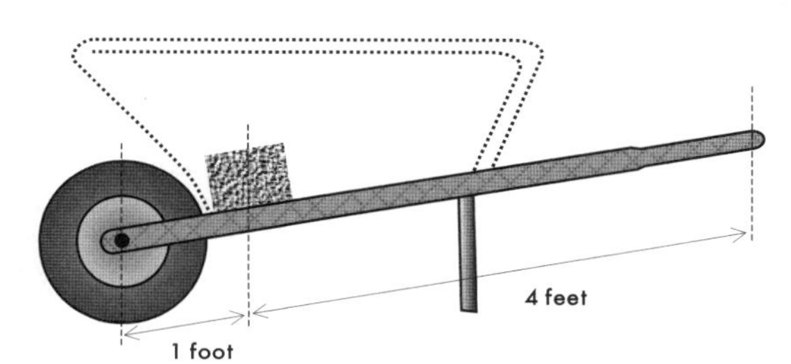

8. a. 6 feet
   b. 2 feet
   c. 3

9. c

Review

10. mechanical advantage
11. traction, braking
12. false
13. false (Wheels reduce friction.)
14. a

Activities

1. a. 5 feet
   b. 1 foot
   c. 5
   d. 40 pounds

2–4. Do some of these activities with your students. Hands-on experience helps them to understand the principles in focus.

2. Try out a lever. Get a yardstick, a bar eraser, a ruler, and a thin rubber band. Use the yardstick as a lever by placing the eraser under it for a fulcrum. Let the effort arm of the lever extend out over the edge of the table. Place a small book on the resistance arm, and hook the rubber band around the end of the effort arm. Pull down on the rubber band until it lifts the book on the resistance arm. Measure how far it stretches the rubber band.

   Next, move the fulcrum closer to the rubber band. Pull down on the rubber band again and measure it. Did it stretch any farther? Move the fulcrum up close to the book. How far does the rubber band stretch now?

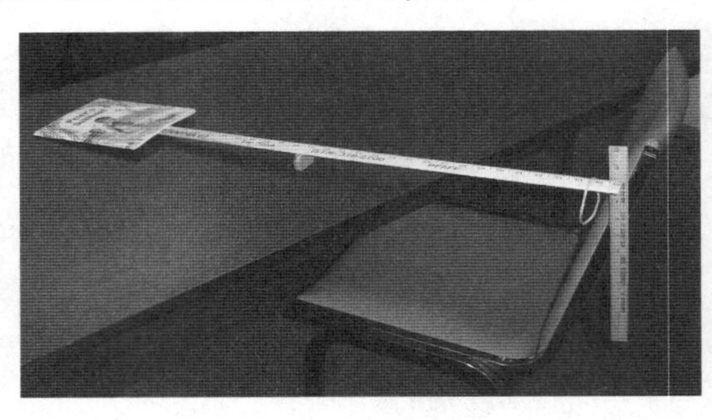

   Try moving the rubber band back and forth. What difference does that make on the effort needed to lift the book?

3. If you have a seesaw at your school, try some balancing experiments like the one described in the lesson. If you do not have a seesaw, a two-by-ten plank about 10 feet long placed on a fence rail could serve the same purpose.

4. Use a board 4 feet long to learn about the mechanical advantage of an inclined plane. Block one end of the board up 1 foot from the floor. If your board is 4 feet long, this forms an inclined plane with a mechanical advantage of 4. Fasten a thin rubber band to a toy wagon with a load of blocks. Pull the wagon up the ramp with the rubber band. Measure the length of the rubber band as you pull.

   Now increase the height of the inclined plane to 2 feet. This decreases the mechanical advantage to 2. Pull the wagon up this new inclined plane, and measure the rubber band. Which inclined plane needs a stronger force to pull the wagon up? How can you tell?

   Dropping the height of the 4-foot inclined plane to 2 inches produces a mechanical advantage of 24. Notice how little force is required to pull the wagon when the mechanical advantage is this high.

Lesson 23

Four Other Simple Machines

Vocabulary

pulley, a wheel with a grooved rim in which a rope runs, used to change the direction of its pulling effort.

screw, a simple machine made of an inclined plane wrapped around a cylinder.

wedge, a simple machine made of wood or metal, which tapers from a thick edge at one end to a thin edge at the other end.

wheel and axle, a simple machine made of a wheel and an axle turning together, used to increase the pulling effort of a rope wound around the axle.

Remember that the six simple machines are the lever, the inclined plane, the screw, the wedge, the wheel and axle, and the pulley. In Lesson 22 you studied the lever and the inclined plane. In this lesson you will study the remaining four machines.

The Screw

The ***screw*** is another powerful machine. You cannot push a nail into a piece of solid lumber with a screwdriver. But try a screw. With a little pressure and a few twists of the screwdriver, the screw sinks in. Why is it possible to turn a screw into wood but impossible to push in a nail? It is because of the screw's mechanical advantage. The screw has threads that are like an inclined plane wrapped around it.

The spiral-shaped inclined plane gives the screw such a high mechanical advantage that we can turn it into a board. As we turn the screw, it slowly digs in and pulls down into the wood.

We use screws in many ways. Threaded bolts have such a great mechanical advantage that we can fasten things tightly. With a jackscrew, we can lift a car off the ground and change a flat tire. Augers allow us to

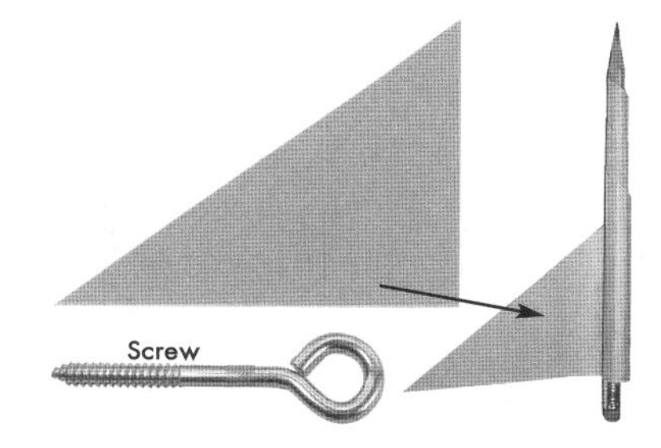

A screw is a spiral-shaped inclined plane.

Lesson 23

Lesson Concepts

1. The screw and the wedge can give a high mechanical advantage. Both of these are forms of the inclined plane.
2. The wheel and axle multiplies our effort by using a large wheel to turn a small axle. (This is actually a form of the lever.)
3. Movable pulleys provide a mechanical advantage because pulling a greater amount of rope causes a smaller amount of lift. (In the lesson example, the force of pulling 4 feet of rope is concentrated into 1 foot of lift.)

Teaching the Lesson

This lesson deals with the other four simple machines: the screw, the wedge, the wheel and axle, and the pulley. Some preparation beforehand can make this an interesting class. To illustrate a screw, wrap a paper triangle around a pencil as shown in the text. This should be helpful for those who may not have understood the illustration in the textbook.

Be sure to have a screwdriver available, and have the students try loosening a screw first by holding the shaft and then by using the handle.

Do you have several small pulleys? If not, you can find some at the hardware store. Rig up a fixed pulley, a movable pulley, and a block and tackle. Hang a weighted bucket or paint can on the resistance end, and let your pupils feel the difference in effort required for each of the three setups.

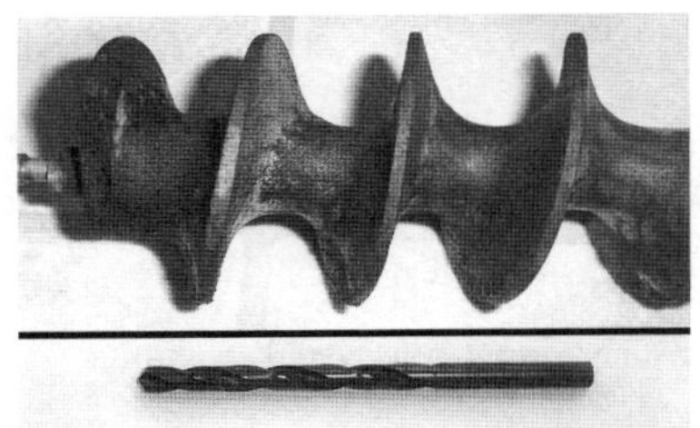
Several kinds of screws

lift feed into bins. The screw on a drill bit bites into the wood and cleans the shavings out of the hole we are drilling.

The Wedge

Imagine trying to split apart a solid chunk of firewood with your bare hands. Molecular attraction holds the wood fibers tightly together, making it impossible for you to force them apart. What you need is a machine to multiply your effort. The simple machine called a ***wedge*** is just the thing.

A wedge is actually a double inclined plane. It is a powerful tool. If we use a sledgehammer to pound a wedge into a log, we can split the whole log apart.

A wedge is a double inclined plane.

The longer and thinner a wedge is, the more mechanical advantage it has. The wedge will work faster if it is thicker, but we will need to use more force because it has less mechanical advantage.

Which wedge will split the log more easily?

We use wedges for many things besides splitting wood. The sharp edge on your mother's knife is a wedge that helps her to cut carrots easily. Your wedge-shaped teeth make it simple for you to bite an apple. Axes and plows also make use of wedges.

The Wheel and Axle

A simple machine that you use every day is the ***wheel and axle.*** The wheel and axle consists of a wheel

Discussion Questions

1. How many of the simple machines can you locate in your classroom?
   (Individual answers.)
2. List the simple machines, and give one or two examples of each.
   (Individual answers.)

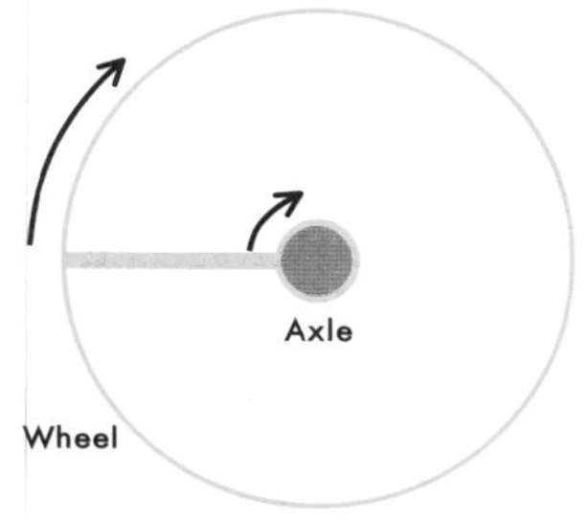

A simple wheel and axle

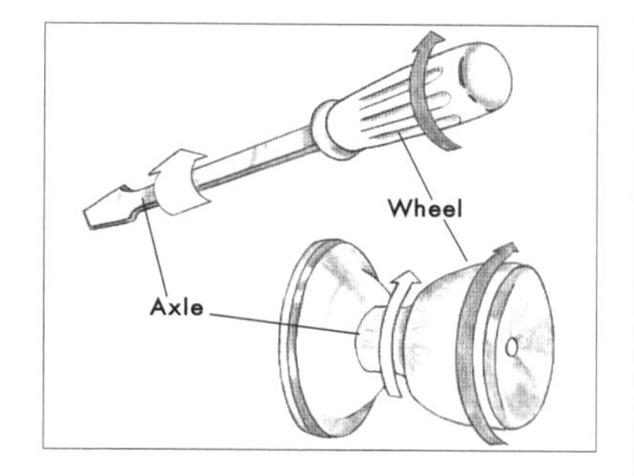

fastened to an axle. A little force applied to the wheel is multiplied to become a large force at the axle.

Did you ever try to open a door when the doorknob was missing? The doorknob with its rod is a wheel and axle that makes it much easier to move the door latch. Imagine turning a screw into a board without a screwdriver. The screwdriver is a wheel and axle that multiplies your effort so that you can easily turn the screw. The handle of the screwdriver is the wheel, and the shaft is the axle. The steering wheels used to guide tractors and automobiles are other wheel and axles.

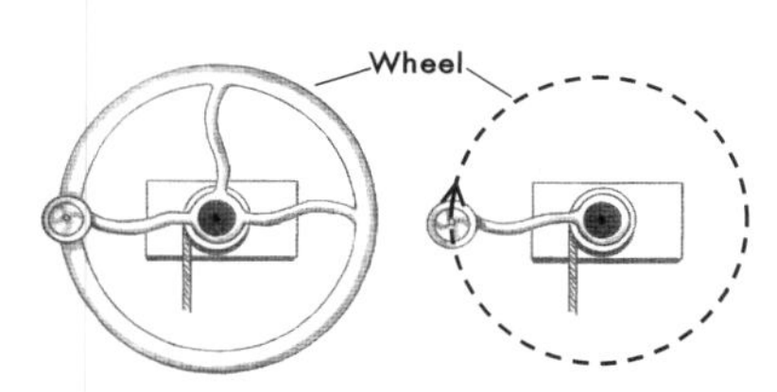

A crank serves as a wheel even if the whole wheel is not there.

Some wheel and axles have a crank instead of a large wheel. The pencil sharpener in your classroom is an example. Could you sharpen your pencils if the pencil sharpener had an axle but no crank? Probably not. The crank greatly multiplies your effort.

Pictured below is a wheel and axle called a windlass. Long ago, the windlass was commonly used to draw buckets of water from an open well. The windlass made it easier to lift a heavy bucket because a large crank was used to wind up the rope on a small shaft.

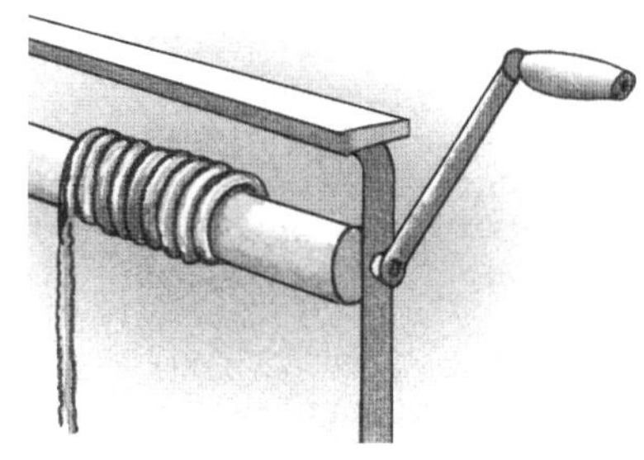

The Pulley

Another simple machine that uses a wheel is the ***pulley.*** A pulley is a grooved wheel with a rope running around it. Pulleys may be rigged in several ways to help us do our work.

A fixed pulley is one that is fastened so that it does not move up or down as the rope is pulled. Fixed pulleys are useful for changing the direction of an effort. For instance, to raise venetian blinds at windows, it is more convenient to pull downward on a string than to pull upward at the top of a window.

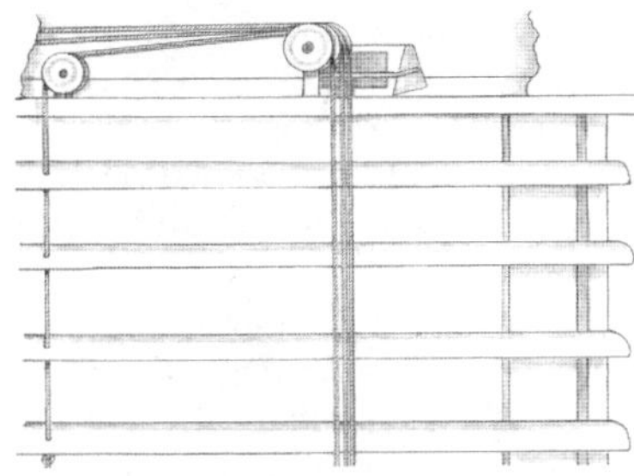

Venetian blinds use pulleys for lifting.

A fixed pulley does not increase effort. If you want to raise a 10-pound weight with a fixed pulley, you must pull on the rope with a force of at least 10 pounds.

Sometimes a pulley is rigged in such a way that the pulley moves as the rope is pulled. This is called a movable pulley. The movable pulley makes our work easier by multiplying the effort by 2. Its mechanical advantage is 2.

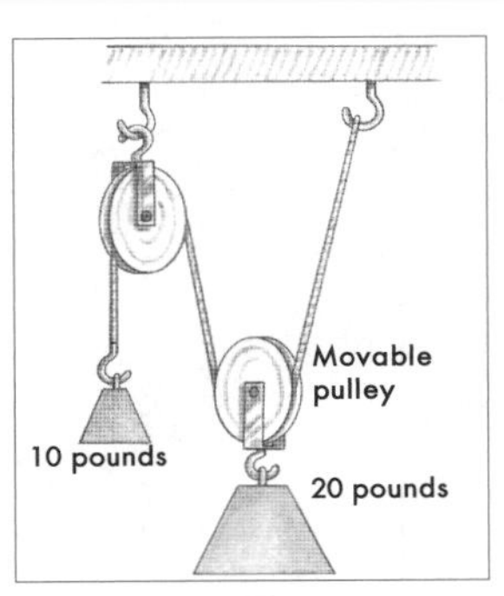

A 10-pound weight balances a 20-pound weight hanging on a movable pulley.

For greater mechanical advantage, pulleys are often joined together in a device called a block and tackle. The more movable pulleys you use together, the greater the mechanical advantage will be.

With the block and tackle pictured here, the force will be multiplied 4 times. You gain this mechanical advantage by pulling 4 feet of rope

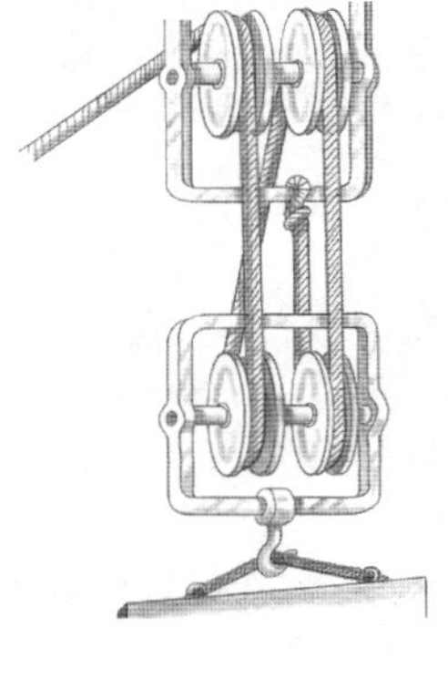

This block and tackle multiplies the force 4 times.

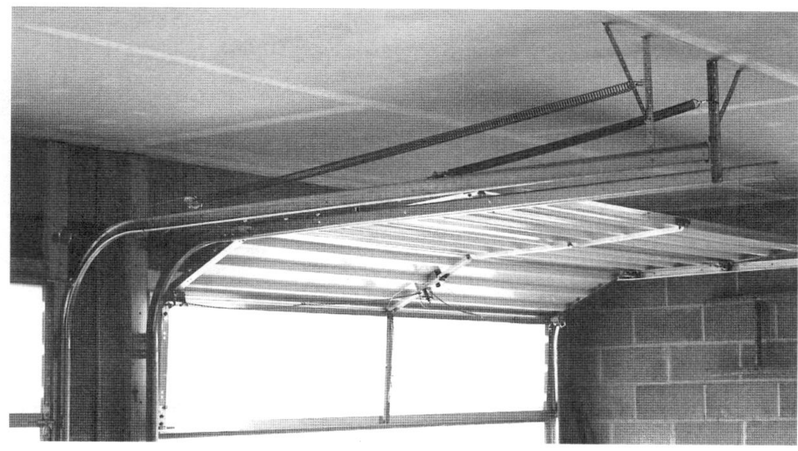

Springs and pulleys help raise garage doors.

for every 1 foot that the weight is lifted.

Can you find places where pulleys make your work easier? Perhaps your windows use pulleys and weights to help you open them. Springs and movable pulleys help us to raise heavy overhead garage doors. And your father may have a block and tackle for lifting heavy weights.

Exercises

1. The screw is a machine that is closely related to the (lever, inclined plane, wheel and axle).
2. To split a tough piece of wood, it is best to use a wedge that is (long and thin, short and fat).
3. True or false? With a wheel and axle, a little force applied to the wheel becomes a large force at the axle.
4. What is the mechanical advantage of a movable pulley?
5. What is the name of a device with several movable pulleys joined together?
6. Identify the simple machine in each of these common things.
   a. crowbar
   b. mountain road
   c. nut and bolt
   d. stairway
   e. canning jar ring
   f. pliers
   g. steering wheel

Lesson 23 Answers

Exercises

1. inclined plane
2. long and thin
3. True
4. 2
5. block and tackle
6. a. lever
   b. inclined plane
   c. screw
   d. inclined plane
   e. screw
   f. lever (two opposing levers)
   g. wheel and axle

Review

7. by changing the direction, increasing the amount, or increasing the speed and distance of a force
8. force (effort), feet
9. True
10. false
11. True
12. lever, inclined plane, screw, wedge, wheel and axle, pulley

Activities

1. a. closer together
   b. one with threads close together
2. a. wheel and axle (crank), screw (spirals on cutters), wedge (cutting edges)
   b. wedge (cutting edge), lever
   c. pulley
3. Most likely, many of your pupils are wearing a block and tackle to class today. A shoe with laces is probably the most common block and tackle. If the eyes in a shoe were good pulleys, what would be the mechanical advantage of your shoe and laces? A shoe with five pairs of eyes has a theoretical mechanical advantage of 10. Except for friction, you would be able to draw your shoe together with a force of 200 pounds by pulling with only 20 pounds on the laces. That explains how we can tie our shoes so tightly with such a relatively light pull. Of course, since the eyelets do not have bearings as a pulley does, the actual mechanical advantage is much lower than 10.

   An older office chair has a screw to raise and lower the seat. A newer chair has a lift cylinder that includes a lever as part of the adjusting apparatus.

Review

7. Give three ways that machines help us do our work.
8. Foot-pounds are calculated by multiplying the pounds of ——— by the number of ———.
9. True or false? If nothing is moved, no work is done.
10. True or false? If a machine uses a 50-pound effort to overcome a 600-pound resistance, the mechanical advantage of the machine is 30,000.
11. True or false? If a lever has a 12-inch effort arm and a 6-inch resistance arm, its mechanical advantage is 2.
12. List the six simple machines.

Activities

1. Study the lesson drawing which shows that a screw can be made by wrapping an inclined plane around a pencil.
   a. If the slope of the inclined plane were more gradual, would the threads of the screw be closer together or farther apart?
   b. Which screw has the greater mechanical advantage, one with threads close together or one with threads far apart?
2. Identify the simple machines in these tools. Some have more than one.
   a. a wall-mounted pencil sharpener (3 machines)
   b. a shovel (2 machines)
   c. your shoelaces
3. Make a list of the simple machines around you. Put levers in one group, screws in another, pulleys in another, and so forth for all six simple machines. Examine common, everyday objects. Are they simple machines? If not, do they have simple machines in them? Start with your teacher's chair. Is it adjustable? If so, what simple machine is used to raise and lower the seat?
4. Fasten a pulley at a high spot, such as the door lintel. Tie a weight to a string, and use the fixed pulley to lift it. Now use a movable pulley to lift the same weight. Can you feel the difference in effort?
5. Study a ten-speed bicycle.
   a. Count how many times the rear wheel goes around each time the pedals go around once. Shift gears, and count again. Can you find the tenth gear? It is the gear that makes the wheel spin the greatest number of times with one turn of the pedals.
   b. Can you find several examples of the wheel and axle on the bicycle? Can you find several levers?

Lesson 24

Pumps to Move Liquids

Vocabulary

centrifugal pump (sen·trif′·yə·gəl), a pump that moves a liquid by slinging it from a rotating impeller.

gear pump, a pump that uses two interlocking gears to move a liquid.

impeller (im·pel′·ər), a wheel with blades on it, used to move a liquid.

piston pump (pis′·tən), a pump that uses pistons and valves to move a liquid.

pressure, the amount of force applied to a liquid.

siphon (sī′·fən), a tube shaped like an upside-down J, which lifts a liquid over a barrier and carries it to a lower place without using a pump.

In ancient times, people used buckets and muscle power to move liquids. If they wanted to get water out of a deep well, they let a pitcher down into the well with a rope. After the pitcher was full, they pulled it up with the rope. When Rebekah drew water for Abraham's servant, she likely used such a rope and pitcher. It was hard work for her to draw water for all his camels. She could not just open a faucet and let the water gush out as we do today when we water our animals.

From High Pressure to Low Pressure

Pumps are machines that move liquids for us. All pumps move liquids by

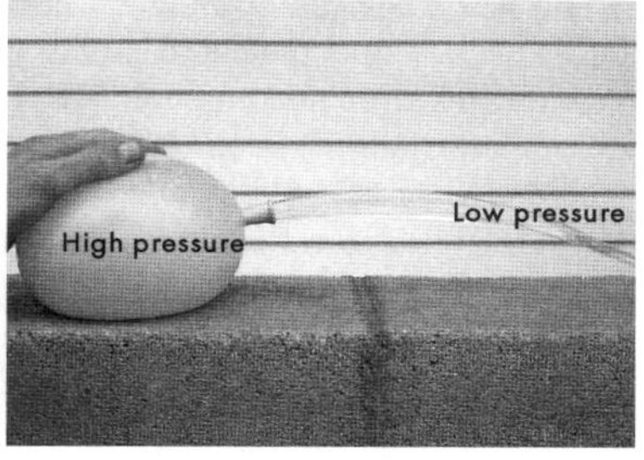

Liquids move from high pressure to low pressure.

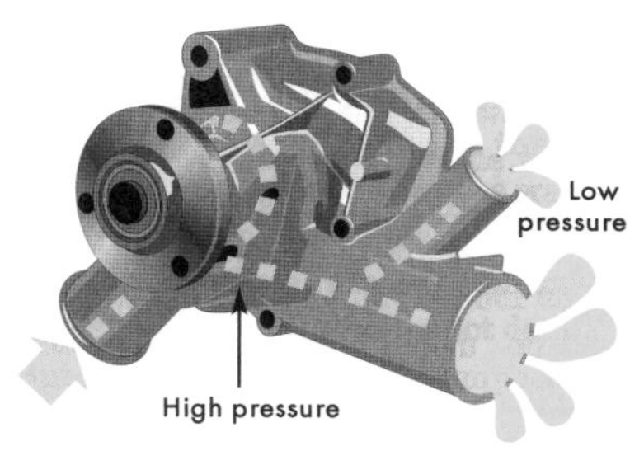

A water pump for a car

Lesson 24

Lesson Concepts

1. Liquids flow when the pressure at one point is greater than the pressure at another point.
2. In a siphon, the greater amount of liquid in one part of the tube outweighs the amount in the other part. The difference in weight (or pressure) causes the liquid to flow to a lower level.
3. The piston pump uses a piston and valves to build up pressure and move liquids.
4. The gear pump uses two interlocking gears to move liquids.
5. The centrifugal pump uses inertia to build up high pressure at the edge of a rotating impeller.

Teaching the Lesson

Pumps are important laborsaving machines. They come in many types and in numerous different sizes. Some of the tiniest pumps can deliver only one pint per minute. Other giant models handle much more. Some delicate pumps produce only a few ounces of pressure. Others achieve pressures of several thousand pounds per square inch. The versatility of pumps makes them one of the most common and useful machines.

How do pumps touch the students' lives? Most of them probably have well pumps at home. Surely they eat foods from the freezer or refrigerator. The compressors in these appliances are pumps. How do the children travel? Automobile engines have gear pumps for changing the ***pressure*** of the liquid. Liquids flow when the pressure is greater at one place than at another. Think of a balloon. If you inflate a balloon by attaching it to a water faucet, the water inside the balloon is under pressure. If the neck of the balloon is opened, the water will squirt out. It is moving from the high-pressure area inside the balloon to the low-pressure area outside the balloon.

That is how pumps make liquids move. The pump changes the pressure of the liquid at one place. Then the liquid moves from an area of higher pressure to an area of lower pressure.

The Siphon

A ***siphon*** is a hose or pipe that moves a liquid, such as water, up over the edge of a container and down to a place lower than it was before. Look at the siphon pictured below. It has two sides, A and B, both full of water. Notice that side A is longer than side B. Since it is longer, it has more water

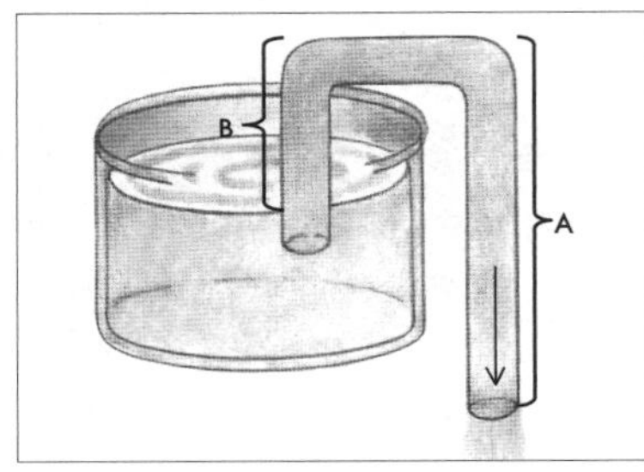

A siphon

in it. This makes the water in side A heavier than the water in side B, so the water in side A goes down. That pulls the water up on side B. The water will keep flowing through the siphon until side B is out of the water or until the water level is the same in the two containers.

Siphons are useful, but they have two disadvantages. First, a siphon usually cannot move a liquid very fast. Second, a siphon can move a liquid only to a lower position. What can you do if you want to move a liquid more quickly or move it to a higher position? You will need to use a pump.

Different Pumps for Different Jobs

Piston pumps. Years ago, people had ***piston pumps*** beside their sinks. A piston pump is a simple device with a piston and two valves. How does it work? Refer to the diagram of a piston pump as you read the explanation below.

First, as the piston moves up, it forces the air out of the pump chamber. This creates a vacuum and pulls water up into the chamber. As the piston goes back down, the bottom valve closes and the water in the chamber comes up through the valve in the piston. When the piston moves up again, it draws more water into the chamber and also pushes the water above the piston out through the spout. Every time the

lubrication, and fuel pumps for delivering fuel. Students may pump up their bicycle tires with a tire pump or with Father's air compressor, which is also a pump. Do they spray their garden vegetables? A sprayer is a pump of some kind. Do they live on dairy farms? They may start a vacuum pump twice a day.

Pumps will be a mystery for some students. Endeavor to show them how these machines work. Use diagrams on the chalkboard and demonstrations to make pumps simple. Even though they are quite ordinary, pumps can be very interesting if one takes the time to examine and understand them.

Discussion Questions

1. What is one advantage of using a siphon?
   All you need is a hose or pipe. You need no power source; gravity does the work.
2. The most common kind of pump is the centrifugal pump. What advantages do centrifugal pumps have over other pumps?
   Centrifugal pumps can pump liquids faster than other kinds of pumps.
   They can pump at high pressures.
   They have fewer moving parts, so friction will not wear them out as quickly or hinder their movement.

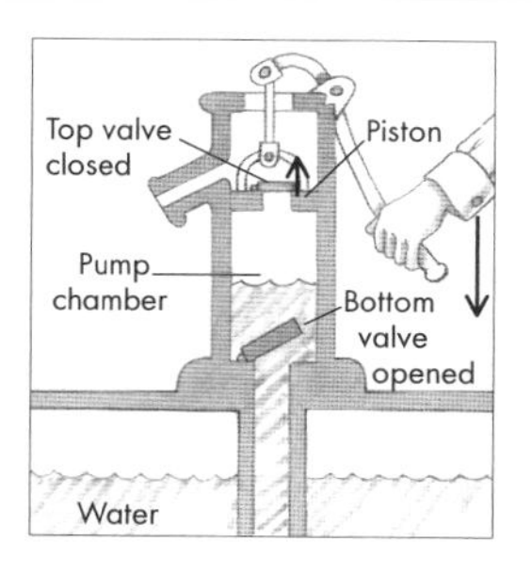

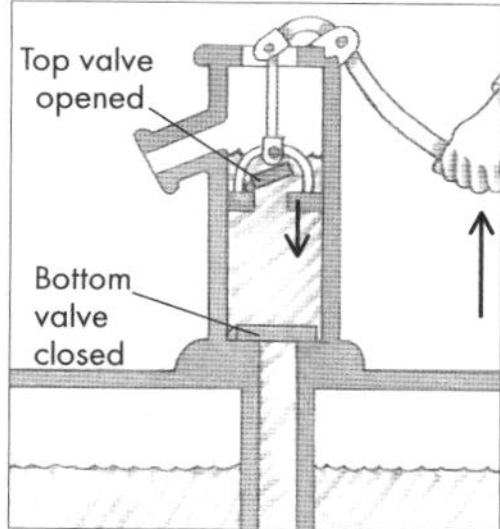

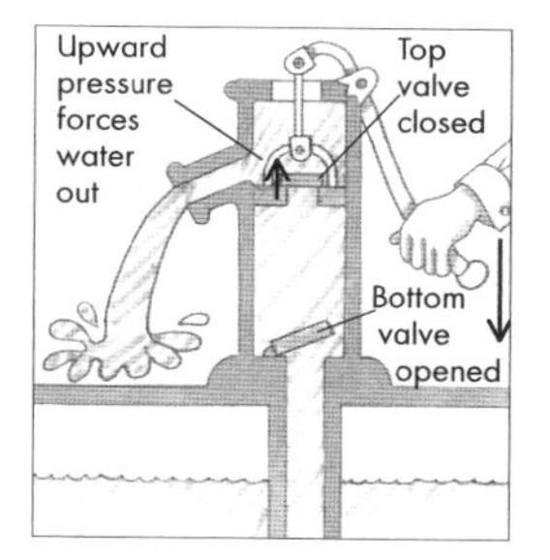

How a piston pump works

piston goes down, the water in the chamber moves above the piston, but no water comes out of the spout. Every time the piston moves up, the water above the piston is forced out through the spout and more water is drawn in below the piston.

Gear pumps. Another common pump is the ***gear pump.*** The oil pumps in many engines are gear pumps. The two gears fit snugly side by side in an oval pump chamber. As they turn, oil is trapped in the spaces between the teeth going around the outside of the chamber. When the teeth come together in the middle, they push all the oil out from between themselves. This puts pressure on the oil and forces it through a pipe to oil the engine.

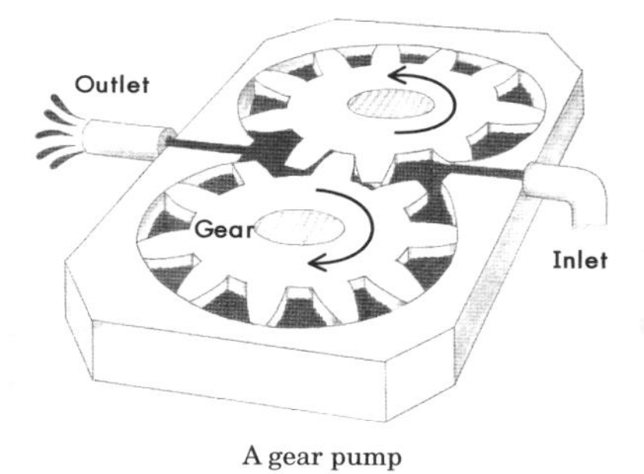

A gear pump

A gear pump is better than a piston pump for moving thick or sticky liquids and for providing a steady, dependable stream. But gear pumps do have some limitations. They usually do not pump as fast as other pumps do. You must also be careful to keep any pieces of dirt from going through a gear pump. Dirt that gets between the gears and the chamber walls can wear out a gear pump very quickly.

Centrifugal pumps. The third common type of pump is the ***centrifugal pump.*** A centrifugal pump has an ***impeller*** spinning inside the pump chamber to raise the pressure of the liquid. To see how this pump works, put some water into a bowl and stir it with a spinning motion. If you make the water spin fast enough, some of it

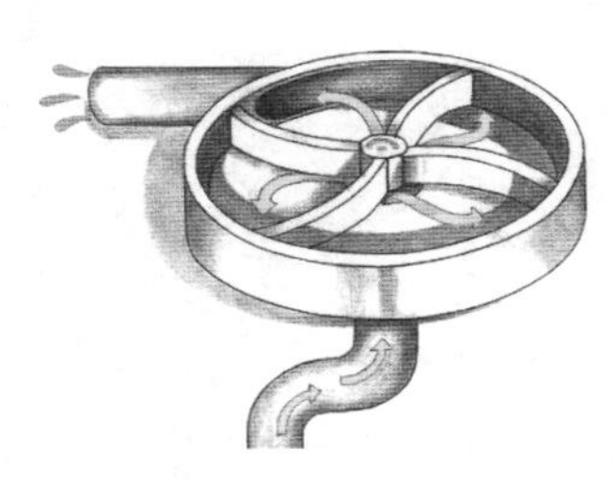

A centrifugal pump

will fly over the edge of the bowl.

What causes this to happen? It is inertia, the force that tends to make moving objects keep moving in a straight line. Inertia puts enough pressure on the water so that it moves upward and flies over the edge of the bowl.

In a similar way, centrifugal pumps use inertia to move liquids. The impeller inside the pump spins very rapidly. When the blades of the impeller catch water, the water slings outward off their edges. The force of the slinging causes pressure that pushes the water through the outlet of the pump. As this water moves away, more water comes into the pump at the center of the impeller.

Fire engines use centrifugal pumps to boost the water pressure in their hoses. The pressure is so great that they can pump as much as 1,500 gallons (5,700 liters) of water per minute. That means you could fill a 55-gallon (208-liter) barrel about every two seconds. Centrifugal pumps are excellent for delivering the great amounts of water needed to fight fires.

Exercises

1. Choose the correct words from the list to match the descriptions.

   centrifugal pump — piston pump
   gear pump — siphon

   a. A pipe that transfers a liquid over the edge of one container and down into another at a lower level.
   b. A pump that uses two interlocking gears to move a liquid.
   c. A pump that uses a piston and valves to move a liquid.
   d. A pump that uses inertia to sling water off the edge of a rotating impeller.
2. A pump causes a liquid to flow by raising the ——— at one place in the liquid.
3. In what two ways are pumps better than siphons?
4. When the piston of a piston pump moves downward, it is
   a. pushing water out through the spout.
   b. pushing water through the valve in the piston.
   c. lifting more water into the pump chamber.

Lesson 24 Answers

Exercises

1. a. siphon
   b. gear pump
   c. piston pump
   d. centrifugal pump
2. pressure
3. Pumps can move liquids faster than siphons can.
   Pumps can lift liquids to higher levels; siphons can move liquids only to lower levels.
4. b

5. What are two advantages of a gear pump?
6. A centrifugal pump has an ——— that uses the force of ——— to put pressure on the liquid.
7. Why are centrifugal pumps excellent for use on fire engines?

Review

8. What are the four kinds of resistance?
9. List the six simple machines, and give an example of each.
10. How can you find the mechanical advantage of a lever?

Activities

1. Pumps are used to move air as well as liquids. Usually air pumps are called compressors. Your father may have an air compressor for inflating tires, or you may have used a simple tire pump to inflate your bicycle tires.

   Look at the diagram to the right. Such a tire pump needs two valves. Can you show where each valve would be located to make the tire pump work properly?
2. Before a siphon will work, it must be filled with water. Filling the siphon is called priming the siphon. The siphon is usually primed by sucking on it or by immersing it.

   Is it possible to make a self-priming siphon? Try it. Make a siphon shaped like the one in the picture. You can make it from three flexible drinking straws. Two of the straws should be full length. Bend the third straw, and shorten the long side by cutting it 1½ inches longer than the short side. Tape the short sections of the whole straws onto either end of the cut straw. Use long pieces of tape to hold the siphon in the correct shape.

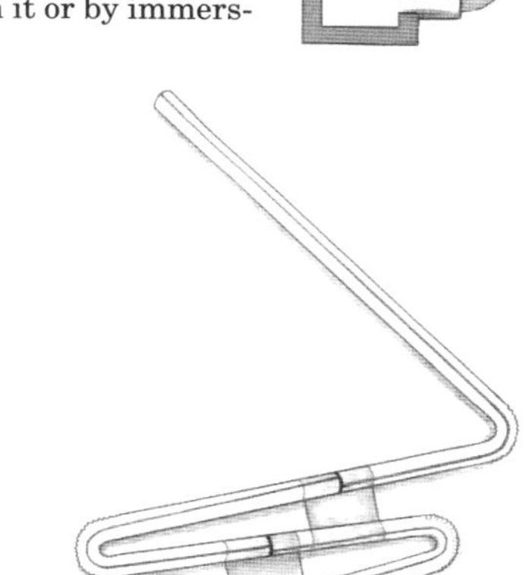

5. It pumps thick liquids well.
   It provides a steady, dependable stream.
6. impeller, inertia
7. They can pump great amounts of water at high pressures.

Review

8. gravity, inertia, friction, molecular attraction
9. (Possible examples are given.)
   lever—scissors;
   inclined plane—barn hill;
   screw—jar lid;
   wedge—knife blade;
   wheel and axle—steering wheel;
   pulley—pulley on garage door
10. Divide the length of the effort arm by the length of the resistance arm.

Activities

1. One valve would be located at the piston, to let air move down past the piston but not up past it. A rubber collar on the edge of the piston, angled downward, may do this job.

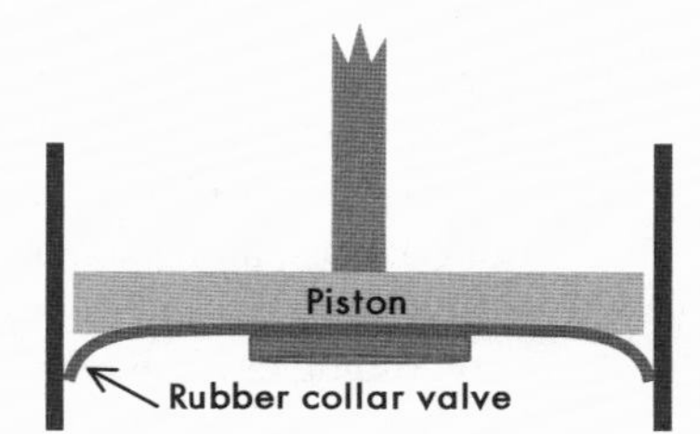

The tire chuck on the end of the hose contains the second valve. This valve allows the air to move out of the hose but not into it. (Students may place the second valve at the other end of the hose. Either location would work.)

Now fill a bucket to the brim with water. Plunge the siphon rapidly down into the water as far as it will go. After several tries, the water will flow from the end of the siphon. If the water does not begin coming through, blow your siphon clear before trying again. You have made a self-priming siphon. As the water rushes through the siphon, it builds up enough speed that inertia moves it up over the last bend in the siphon and down to drain out through the end. Your siphon is now primed.

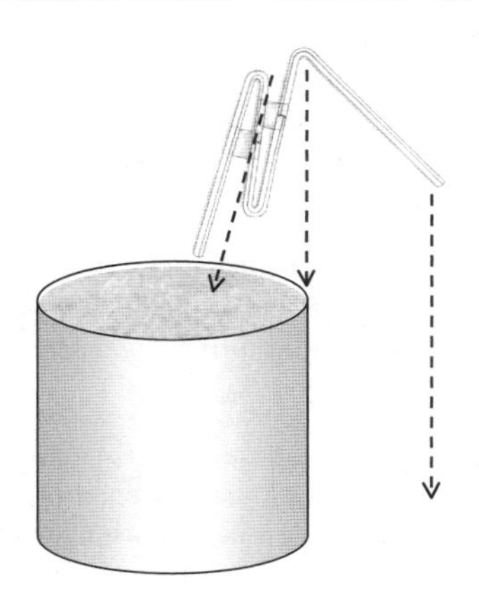

3. You can imitate the action of a centrifugal pump. Fill a five-gallon bucket and a garden hose with water. (Clear, quarter-inch vinyl tubing may be better if you have it.) Place one end of the hose into the bucket, and let the other end hang down lower than the bucket so that the water begins to siphon. Pick up the hose about four feet from the end, and lift it into the air, keeping the end of the hose on the ground so that water is still siphoning out through the end of the hose.

   Now, with one sweeping motion, begin to swing the hose in circles over your head. If you swing it at the right speed, the water will continue to fly from the end of the hose. With a little practice, you should be able to lift all the water from the bucket simply by slinging the other end of the hose around over your head.

Extra Activities

1. Build a simple piston pump. You will need 2 glass marbles, 2 corks with holes, 3 clear plastic syringes without needles (check with a pharmacy or a dairy farmer), about 2 feet of small vinyl tubing of a size that fits on the bottom of the syringes, and 3 small lengths of glass or metal tubing to fit the vinyl tubing and the corks. (A science supply house has most of these items.) The children can actually see the valve action with this apparatus.

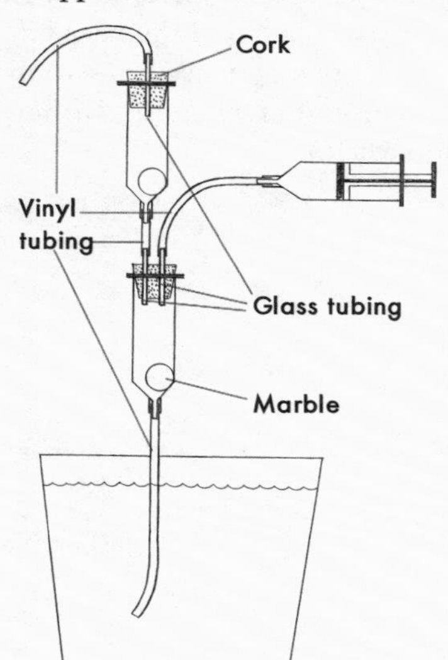

2. Some students might remember seeing a hand-operated piston pump that ejected water on both the upstroke and the downstroke. They might wonder how such a pump works. Illustrate with a diagram on the chalkboard; even a relatively crude drawing will serve the purpose. Show them the path of the water through the pump, and explain how the valves work as the piston moves up and down.

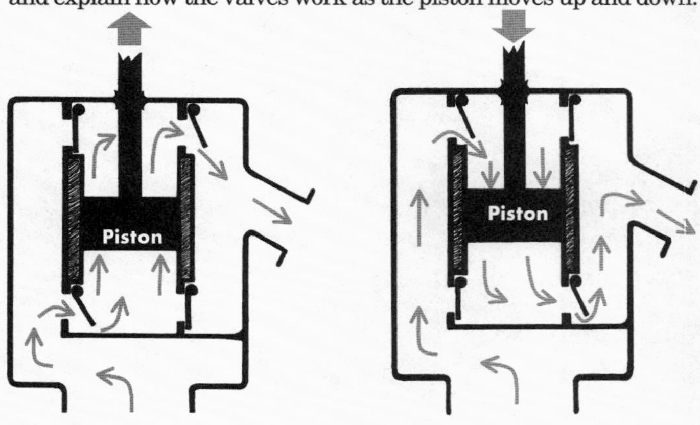

Lesson 25

Unit 4 Review

Review of Vocabulary

bearing	impeller	pressure
brake	inclined plane	pulleys
centrifugal pump	inertia	resistance
effort	lever	screw
foot-pound	lubricant	siphon
force	machines	traction
friction	mechanical advantage	wedge
fulcrum	molecular attraction	wheel and axle
gear pump	piston pump	work

Every time something moves, a __1__ is needed to make it move. When something is made to move, we say that __2__ has been done. It takes one __3__ of work for one pound of force to move something one foot.

There are several kinds of __4__ that hinder movement. Gravity must be overcome when we lift things. __5__ must be overcome when we split firewood. __6__ tries to keep a stopped truck from moving. __7__ occurs when two surfaces rub against each other. To reduce the amount of heat and wear caused by rubbing, we need to use a __8__ between the surfaces. In the case of a turning shaft, we can use a __9__ to support the shaft.

Friction has several important uses. One is to provide __10__ for car tires so that they will not slip. Another is to stop a moving vehicle by using a __11__.

We use six simple __12__ to make our work easier. A __13__ can be used to lift things by placing a plank on a pivot called a __14__. To lift things with ropes, we can set up several __15__ as a block and tackle. A windlass is a good example of a __16__. When we tighten a bolt or a jar lid, we are using the machine called a __17__. The __18__ makes it possible to lift heavy things by moving them up a slope. The knife we use to peel potatoes is an example of a __19__. If we want to know how much easier a machine makes our work, we can find the __20__ of the machine. This tells us how many times the machine is multiplying our __21__.

Pumps move liquids by making the __22__ of a liquid greater at one place than at another. A __23__ is a very simple tool for moving a liquid to a lower level. A __24__ uses

Lesson 25 Answers

Review of Vocabulary

1. force
2. work
3. foot-pound
4. resistance
5. Molecular attraction
6. Inertia
7. Friction
8. lubricant
9. bearing
10. traction
11. brake
12. machines
13. lever
14. fulcrum
15. pulleys
16. wheel and axle
17. screw
18. inclined plane
19. wedge
20. mechanical advantage
21. effort
22. pressure
23. siphon
24. piston pump

Lesson 25

Glance back through your teacher's manual, and notice the sections titled "Lesson Concepts." Also take a brief look at the unit test. Have you stressed the main concepts well enough so that your pupils know them well? Evaluate their performance in this review lesson, and give extra drill in areas where they are weak.

25. centrifugal pump
26. impeller
27. gear pump

a piston and valves to pump water. A __25__ is excellent for a fire engine because it can pump large amounts of water at a high pressure. It pumps water by slinging it from a spinning __26__. The oil pump in an engine is usually a __27__ because it can pump a thick liquid in a steady stream.

Multiple Choice

1. b
2. d
3. b
4. b
5. c
6. d
7. c

Multiple Choice

1. An object will move if
   a. the resistance is greater than the force.
   b. the force is greater than the resistance.
   c. the force and the resistance are equal.
   d. there is enough resistance.
2. The four kinds of resistance are
   a. gravity, inertia, force, and friction.
   b. inertia, friction, molecular attraction, and force.
   c. friction, molecular attraction, gravity, and force.
   d. gravity, inertia, friction, and molecular attraction.
3. Work is done
   a. every time force is applied.
   b. every time something is moved.
   c. every time the resistance is great enough.
   d. every time friction is applied.
4. Which things affect the amount of friction between rubbing surfaces?
   a. gravity and pressure
   b. pressure and surface roughness
   c. surface smoothness and temperature
   d. pressure and temperature
5. Which one of the following problems is **not** caused by friction?
   a. bearings getting hot
   b. moving parts wearing out
   c. metal rusting
   d. needing more force to move
6. We **cannot** reduce friction by
   a. lubricating the surfaces.
   b. using wheels or rollers.
   c. smoothing the surfaces.
   d. using stronger motors.
7. Friction is helpful because
   a. it is a good source of heat for our homes.
   b. it makes things wear out.
   c. it can be used to stop vehicles.
   d. it makes things move more easily.

8. Mechanical advantage
   a. has no practical use.
   b. tells how many times a machine multiplies the effort.
   c. tells how many times a machine increases the resistance.
   d. tells how much friction a machine must overcome.
9. To find the mechanical advantage of a lever, you should
   a. divide the length of the resistance arm by the length of the effort arm.
   b. multiply the pounds of force by the number of feet moved.
   c. divide the length of the effort arm by the length of the resistance arm.
   d. divide the length by the height.
10. Liquids move
    a. from a low-pressure area to a high-pressure area.
    b. from a high-pressure area to a low-pressure area.
    c. more swiftly through a siphon than through a pump.
    d. only if we pump them.
11. Siphons are useful because
    a. they can pump liquids very rapidly.
    b. they can lift liquids to higher containers.
    c. they are easy to make and use.
    d. they pump liquids at very high pressures.
12. Piston pumps
    a. use impellers to build up pressure to pump water.
    b. use pistons and valves to pump water.
    c. use high-pressure water to pump water.
    d. use gears to pump water.
13. Gear pumps
    a. are good for pumping thick or sticky liquids.
    b. are good pumps for fire engines.
    c. are good for pumping liquids that may contain sand or gravel.
    d. produce a strong but uneven flow of liquids.
14. The purpose of the impeller in a centrifugal pump is
    a. to reduce the pressure of the liquid being pumped.
    b. to raise the pressure of the water with friction.
    c. to raise the pressure of the water with inertia.
    d. to reduce the friction of the water.
15. If Jacob pushes with an effort of 20 pounds to make a 40-pound object move 10 feet, how much work has he done?
    a. 20 foot-pounds
    b. 200 foot-pounds
    c. 400 foot-pounds
    d. 800 foot-pounds

8. b
9. c
10. b
11. c
12. b
13. a
14. c
15. b

Unit 5

Title Page Photo

These are a few of the household chemicals that we use frequently. We often fail to recognize them as chemicals. Ask your students which two of the chemicals are safe to drink.

Introduction to Unit 5

We live in a chemical world. The food we eat is made of complex chemicals. Life itself depends on multitudes of chemicals and chemical reactions within our bodies. We tend to think of chemicals as toxic materials that come in bottles and barrels labeled "Poison." This chapter portrays chemicals in a different way. Everything is made of combinations of some one hundred basic elements. Everything has a chemical structure. Essentially, every substance is either a chemical or a combination of chemicals.

Crucial to an understanding of our surroundings is a basic knowledge of how things change. Where does wood go when it is burned? Why does fire produce water and carbon dioxide? How is it that food nourishes us? A study of basic chemistry is necessary to find the answers to such questions.

While many chemicals are harmless, we do need to be careful when using potent cleaning agents, petroleum products, fertilizers, acids, and so forth. Wise, conservative use should be stressed. Tell the students that wise use sometimes includes not using a chemical at all if the long-range effect is harmful. Teach them to be good stewards of what God has entrusted to us.

Unit 5

The Wonders of Chemicals

You are surrounded by chemicals. Your clothes are made of chemicals. The food you eat is made of chemicals. Your own body is made of chemicals. The paper in this book is made of several chemicals. In fact, everything is made of chemicals.

In this chapter, the term *chemical* means something different than you probably thought at first. Often we use *chemicals* to refer to poisonous liquids and spray materials that must be handled carefully because they are harmful to the body. This chapter uses *chemicals* to refer to the ingredients or building blocks that make up everything you can see, touch, or feel.

Yes, some chemicals are poisonous. But some chemicals are very safe and are needed for life. The next time you take a drink of water, you will be swallowing a pure chemical. Water is just one of the many chemicals that God has created. Water is so necessary for life that God made a special way to spread this chemical over the land. "Who [God] covereth the heaven with clouds, who prepareth rain for the earth, who maketh grass to grow upon the mountains" (Psalm 147:8). In this unit you will learn about the wonder of water and other chemicals that God has created.

To teach this unit effectively, you should collect the supplies needed for the various demonstrations in the lessons. Some may be a bit difficult to find, and some may need to be purchased from a science supply house. Start now so that you have them when needed. Below is a list of items needed.

- Iron filings
- Test tube
- Bunsen burner or propane torch
- Nails
- Vinegar
- Ammonia
- Hydrogen peroxide
- Candle
- Copper wire
- Baking soda
- Various things to test for acids and bases (see list in Lesson 29)
- Litmus paper or hydrion paper
- Limewater
- Turmeric
- Alcohol
- Purple plants (see Activity 3 in Lesson 29)
- Phenolphthalein
- Labels from chemical containers

Story for Unit 5

Chemical Cake

The Miller family sat down to a farmer's breakfast of sausage and eggs one Saturday morning. Samuel's mouth watered when he smelled the sausage. Would there be any left when the platter came to him? He hoped so. He was hungry.

Lesson 26

How Chemicals Are Grouped

Vocabulary

atom (at′·əm), the smallest possible particle of an element.

chemical formula (fôr′·myə·lə), a set of chemical symbols that tells what elements a compound is made of.

chemical symbol, one or two letters that scientists use to stand for an element.

chemistry (kem′·i·strē), the study of chemicals.

compound, a substance in which two or more kinds of atoms are chemically joined as molecules.

element (el′·ə·mənt), one of the simple substances made of only one kind of atom.

matter, any material that takes up space and can be perceived by a sense, such as sight or touch.

molecule, (mol′·i·kyül′), the smallest possible particle of a compound.

property, a quality or characteristic of a substance, such as hardness or shininess.

Look around your classroom. How many different materials can you see? Wood, metal, paper, glass, fabrics, and probably many more are within sight. Scientists use the word ***matter*** to describe all these. Anything that you can see, touch, or feel is matter.

An automobile, your body, water, and even air are all made of matter. Yes, air is matter even though you cannot see it. You can feel it blow against your face or hand. Such things as light, electricity, and gravity are not matter.

All matter is made up of tiny particles called ***atoms.*** Atoms are so small that only the largest ones can be seen with the most powerful microscopes.

As you look around, there seem to be thousands of different kinds of matter in the world. But really, all these are made of only a few more than one hundred different kinds of atoms. And many of these are very rare. Most of the materials around us are made of only about a dozen different atoms.

"Samuel, would you please pass me the sodium chloride?" asked big brother Henry.

"What's that?" Samuel was puzzled.

"Oh, that's the white chemical in that little bottle with holes in the top," Henry said.

"Salt?" Samuel asked, looking at Father for help. Father's eyes were twinkling. "That must be it," Samuel decided, passing the salt to Henry.

After breakfast, Father and Henry went out to the machinery shed. Samuel stayed to help Mother with the dishes.

"Mother, do we put chemicals on our food?"

"Chemicals?" asked Mother, perplexed.

"Yes. Henry called salt a chemical this morning. I thought chemicals were poisonous," Samuel said.

"We use many things in food that are actually chemicals," Mother explained, "even though we usually don't think of them as chemicals. I'm going to bake a cake for dinner. Do you want to help? I can tell you about some of the chemicals I use when I bake."

"Oh, good!" Samuel replied. He knew that Mother had been a schoolteacher long ago. "What kind of cake are you going to bake?"

"I think we'll call it a chemical cake," Mother said.

When the dishes were finished, Samuel got out a cake pan, measuring cups, and a spatula while Mother hunted for a recipe. "Here's a good one," she said, holding up a white index card.

"First we need 2½ cups of flour," said Mother. "Get it out of the big bag in the pantry."

"Is flour a chemical?" Samuel asked.

"Yes, it is a combination of several chemicals," Mother answered.

Samuel watched as Mother measured sugar into the bowl. "Is that a chemical too?"

"Yes, sugar is a sweet chemical that comes from sugar cane and sugar beets."

Mother measured cocoa into the bowl. Then she said, "Please bring me the baking soda. It is made of sodium and some other chemicals. Soda will be the last of the dry chemicals for our cake."

Now Mother reached for the vinegar. "I know where that comes from," Samuel said. "It comes from apple cider. See, it says that here on the jug. Is vinegar a chemical?"

"Yes, vinegar is a chemical with an acid in it," Mother replied. "Now you may add 1½ cups of cold water while I get the vanilla."

"Water isn't a chemical, is it? You can drink lots of water, and it doesn't do anything. It's just plain clear," Samuel said.

"Yes, water is a chemical," Mother explained. "Actually, everything we use is a single chemical or a combination of chemicals. Have you ever heard someone call water H_2O?" she asked.

"I think I have. But I didn't know what they were talking about," Samuel answered.

"H_2O is the chemical formula for water. You'll probably learn that in school sometime. Now let's put our chemical cake into the oven so it will be ready in time for dinner."

At 12:15, Father and Henry came inside. "M-m-m!

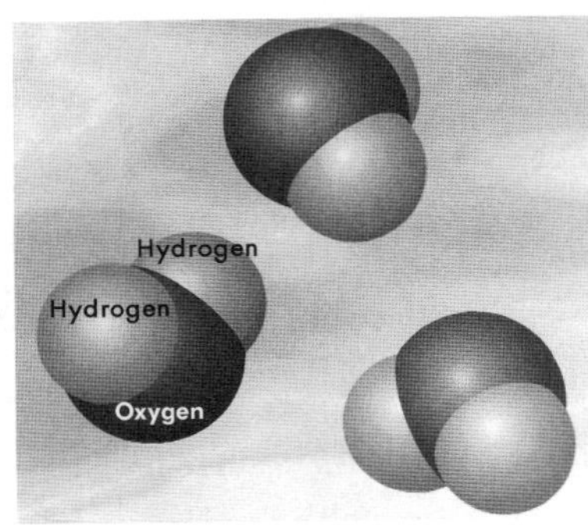

These molecules are scientists' concept of how two hydrogen atoms join with one oxygen atom to make a water molecule.

Atoms sometimes exist alone, but most often they form clusters, or groups, called ***molecules.*** The atoms in a molecule may be all the same kind, or they may be several different kinds.

Compare atoms to students at school. A class may consist of one girl by herself. But usually there are two girls, four boys, or any combination of boys and girls. Include different nationalities of children, and the possible number of combinations of students in one class is almost endless.

In the same way, only a few kinds of atoms can make thousands of different types of molecules. Each different type of molecule is a different material.

If the molecules of a material are made of only one kind of atom, that material is called an ***element.*** Since there are only about one hundred different kinds of atoms, there are only about one hundred different elements. The study of how elements make up different kinds of materials is called ***chemistry.*** In doing this study, all materials are called chemicals.

Iron, gold, copper, tin, oxygen, nitrogen, and hydrogen are a few of the familiar elements. Instead of writing the long name of an element each time they refer to it, scientists have

Six Common Elements

Name	*Symbol*	*Properties*	*Uses*
aluminum	Al	strong, light, does not rust	ladders, airplanes, lawn furniture
copper	Cu	good conductor of electricity	electrical wire
iron	Fe	hard, very strong	train rails, machinery, nails
carbon	C	black solid, burns	fuel
hydrogen	H	very light gas, burns	lighter-than-air balloons, fuel
oxygen	O	gas, supports burning	burning, breathing

Lesson 26

Lesson Concepts

1. Chemistry is the study of chemicals.
2. An element is a simple material made of only one kind of atom.
3. An atom is the smallest possible particle of an element.
4. Each element has a chemical symbol to stand for that element.
5. There are about one hundred different elements.
6. The properties of a chemical are the characteristics it has.
7. A compound is a chemical made of two or more elements that are combined.
8. A molecule is the smallest possible particle of a compound.

Something smells good in here!" Father sniffed the air and smiled at Samuel. "Did you help?"

"Yes, I did. Can you guess what we baked for dessert?" Samuel asked.

"Smells like some kind of cake," Henry said as he came back from washing up.

They sat down to eat. When the main course was finished, Mother said, "Samuel, you may get the cake while I open a jar of peaches."

"It's chemical cake," Samuel said. He passed the cake to Henry.

"Chemical cake? This doesn't look like something made of chemicals. It looks and smells like ordinary cake to me," Henry said.

Mother smiled. "All the ingredients are chemicals, aren't they, Samuel."

"Sure thing, every one of them," replied Samuel.

assigned a ***chemical symbol*** to each element. The symbol is often one or two letters from the beginning of the name. The symbol for hydrogen is H, for oxygen is O, and for copper is Cu. Sometimes the symbol comes from an old name for the element and is very different from our modern name. For example, the symbol for iron is Fe. Notice that the first letter of the chemical symbol is always capitalized. If there is a second letter, it is always a small letter.

God gave each element special ***properties,*** or characteristics. Because of their properties, different chemicals are useful for different things. On page 154 is a list of six common elements. Memorize the names and symbols of these elements.

Remember that the molecules of an element are made of only one kind of atom. If the molecules of a material contain more than one kind of atom, the material is called a ***compound.***

Water is a very familiar compound. Each molecule of water has two atoms of hydrogen (H_2) and one atom of oxygen (O). Therefore the chemical formula for water is H_2O. Each compound has a ***chemical formula*** that tells which elements are in the compound and how many atoms of each are in one molecule.

One molecule is the smallest possible particle of water. If you take the

Six Common Compounds

Name	*Formula*	*Elements in One Molecule*	*Uses*
water	H_2O	2 hydrogen atoms, 1 oxygen atom	drinking, washing, fighting fires
table salt	NaCl	1 sodium (Na) atom, 1 chlorine (Cl) atom	seasoning and preserving food
carbon dioxide	CO_2	1 carbon atom, 2 oxygen atoms	making bread and cakes fluffy, fighting fires
iron oxide (rust)	Fe_2O_3	2 iron atoms, 3 oxygen atoms	not useful; ruins vehicles and other things made of metal
calcium oxide (lime)	CaO	1 calcium (Ca) atom, 1 oxygen atom	making mortar, improving soil
sugar	$C_{12}H_{22}O_{11}$	12 carbon atoms, 22 hydrogen atoms, 11 oxygen atoms	sweetening food

9. The chemical formula of a compound tells which elements are in it and how many atoms of each are in one molecule.

Teaching the Lesson

The study of chemistry can be difficult because of new terms, symbols, and concepts. Your job is to make chemistry understandable and even enjoyable to fifth graders. This will not be accomplished simply with words.

Chemistry is interesting when water boils, steam condenses, and solids melt. But even if your pupils find these things interesting, they have seen them many times before. What really fascinates them is seeing something they have never seen before. Chemistry class is your opportunity to open doors to fascinating new ideas for your pupils.

Of the elements and compounds discussed in the lesson, bring as many to class as possible. Though you cannot bring in hydrogen very easily, you could purchase a helium balloon and point out that hydrogen can be used in the same way (though it is more dangerous). Other items are easy to bring—aluminum (an aluminum pan or aluminum foil), copper (copper wire), mercury (a mercury thermometer), carbon (coal or soot), and so on.

Require the pupils to memorize the basic elements and compounds and their associated symbols and formulas. Flash cards are an effective way to drill chemical symbols and chemical formulas.

molecule apart, you no longer have water. You have just hydrogen and oxygen. A molecule is the smallest possible particle of any compound. But since the molecules of an element have only one kind of atom in them, you could say that an atom is the smallest particle of an element.

Atoms are so tiny that for many years scientists could not actually see them. Finally in 1981, the scanning tunneling microscope enabled scientists to examine the surface of a material and see the little bumps made by each atom or molecule next to each other.

There are thousands of useful compounds. On page 155 is a list of six common ones. Memorize the names and formulas of each compound.

The surface of this gold is magnified by a scanning tunneling microscope. The bumps indicate individual atoms.

Chemistry is a large subject. It is also an interesting subject because there are so many useful chemicals. God was very wise and good in creating the chemicals we need.

Lesson 26 Answers

Exercises

1. chemistry
2. a. atom
   b. element
   c. compound
   d. molecule
   e. chemical symbol
   f. chemical formula
3. a. copper
   b. carbon
   c. aluminum
   d. carbon dioxide

Exercises

1. The study of chemicals is called ———.
2. Write the correct word for each description.
   a. The smallest possible particle of an element.
   b. A chemical with only one kind of atom.
   c. A chemical made of two or more different kinds of atoms.
   d. The smallest possible particle of a compound.
   e. The letter or letters that stand for an element.
   f. The letters and numbers that stand for a compound.
3. Name the correct chemical for each chemical symbol or chemical formula in these paragraphs.

   Mary turned on the electrical switch as she entered the dining room. Immediately electricity flowed through the *(a)* Cu wires. She remembered that far away at a power plant, the *(b)* C in coal was being burned to produce the electricity. In the kitchen, Mother was using an *(c)* Al frying pan to make pancakes. The baking soda in the batter was producing bubbles of *(d)* CO_2, which made the pancakes rise.

Discussion Questions

1. a. Is there any substance not made of elements?
   no
   b. Do all the thousands of chemicals in our world contain only about 100 elements?
   yes
   c. What does that tell us about God?
   He is a wise and wonderful Creator.
2. How do the properties of the compounds in the lesson compare with the properties of the elements making them up?
   (Three examples are given.)
   Water: Both hydrogen and oxygen are gases at room temperature, but water is a liquid. Hydrogen and oxygen explode violently when they are mixed and ignited. But when hydrogen and oxygen combine to make water, we can use it to put out a fire.
   Salt: Sodium is a soft, silver-colored metal that reacts very easily with other chemicals. If sodium touches water, it will react furiously. The hydrogen produced in the reaction sometimes explodes, so great caution must be used. Chlorine is a greenish, poisonous gas, which was used as a poison gas on the battlefields of World War I. But the two combine to form salt, which has many practical uses.
   Carbon dioxide: Carbon is a black, gray, or clear substance. Oxygen is a colorless gas necessary for life. The compound of the two is a gas released in burning and breathing.

Mary used *(e)* $C_{12}H_{22}O_{11}$ to make syrup for the pancakes. Then she put plates, glasses, and *(f)* Fe knives, forks, and spoons on the table. She saw a little *(g)* Fe_2O_3 on one knife because it had not been dried properly the last time it was washed. Mary filled a pitcher with *(h)* H_2O and poured it into the glasses.

While the family was eating, Mary reached for the shaker and sprinkled some *(i)* NaCl on her hard-boiled egg. Everyone around the table was breathing *(j)* O. They were thankful for the many useful chemicals God had made for them.

4. Tell whether each of these refers to an *element* or a *compound.*
   a. Fe
   b. Fe_2O_3
   c. $CuSO_4$
   d. O
   e. Cu
5. The chemical formula for iron carbonate is $FeCO_3$. Describe the elements in one molecule of iron carbonate.
6. Tell what elements are named in the following Scripture verses. They are all metals.
   a. Exodus 15:10
   b. Psalm 119:127
   c. Proverbs 27:17
   d. Ecclesiastes 5:10

Review

7. What two great events did God use to shape the earth as it is today?
8. How many main body parts and pairs of legs do the following creatures have?
   a. arachnids
   b. insects
9. In what two main ways has God formed mountains since the Creation?
10. What two things determine the amount of energy in a waterfall?
11. a. In what way are tornadoes and hurricanes similar?
    b. How are they different in size?
12. What is a light-year?
13. List the six simple machines.

Activity

A dictionary or an encyclopedia will give you a list of all the elements. As a class, see how many of the element names you can find. Look on labels of household chemicals and foods for the names of elements and compounds.

e. sugar
f. iron
g. iron oxide (rust)
h. water
i. salt
j. oxygen

4. a. element
   b. compound
   c. compound
   d. element
   e. element
5. 1 atom of iron, 1 atom of carbon, 3 atoms of oxygen
6. a. lead
   b. gold
   c. iron
   d. silver

Review

7. the Creation and the Flood
8. a. 2 main body parts and 4 pairs of legs
   b. 3 main body parts and 3 pairs of legs
9. from material erupting from within the earth;
   from the crumpling of the earth's crust
10. head and flow
11. a. Both are destructive storms with masses of rotating air.
    b. A hurricane is much larger than a tornado. A tornado may be only a few yards in diameter; a hurricane is hundreds of miles in diameter.
12. the distance light travels in one year
13. lever, inclined plane, screw, wedge, wheel and axle, pulley

Activity

Help the children compile their lists for this activity. Below are some elements you are likely to find in your classroom, though some (such as carbon and iron) will be in mixtures that contain other elements.

aluminum—metal ring that holds the eraser on some pencils
carbon—the "lead" in pencils
copper—wires, pennies
iron—desks, chairs, file cabinets
mercury—in some thermostats
oxygen—in the air, which is about one-fifth oxygen

Lesson 27

Changes in Chemicals

Vocabulary

chemical change, a change in which atoms are regrouped to form new molecules and new materials.

chemical reaction, a process that causes a chemical change.

physical change, a change in the form of a substance but not in the molecules, which does not result in new materials; a change caused by cutting, freezing, dissolving, and so forth.

Physical Changes

A piece of paper can be changed in many ways. You can cut it or tear it. Each little piece is still paper. You can crumple it to make a ball of paper. You can put it in water to make wet paper. You can iron it to make hot paper or put it into a freezer to make cold paper. In each of these changes, you still have paper in the end. Changes in size, form, and temperature are called ***physical changes.***

A physical change does not change the molecules of a material. When paper tears, the molecules separate from each other but the atoms within the molecules do not come apart. When rock crushers make gravel, they break the large clusters of molecules that make up the rock into small clusters of molecules called stones. But the material is the same as before.

In ice, the molecules stay close together and do not move about.

In liquid water, the molecules stay close together and move about freely.

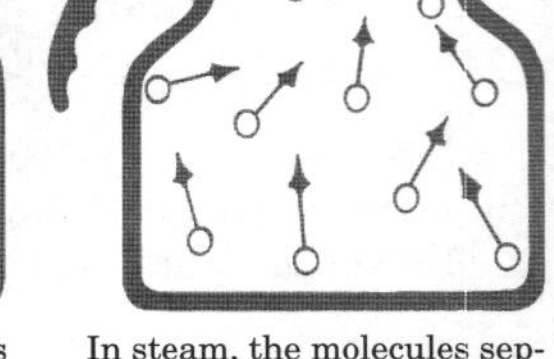

In steam, the molecules separate and fly freely about in the air.

Lesson 27

Lesson Concepts

1. A physical change is a change of state or appearance.
2. A chemical change results in different molecules with different chemical formulas.
3. Chemicals react according to laws that God has established.

Teaching the Lesson

The essential difference between a physical change and a chemical change is that in one the molecules are changed, and in the other they are not. By means of chemical changes, a chemist is able to break apart the molecules of one substance and put the resulting atoms together into new molecules with greatly different properties. This is why George Washington Carver could make three hundred products from the peanut. It also explains why compounds like plastic can be made from many different raw materials.

One of the most important chemical changes is photosynthesis, in which carbon dioxide and water are used to synthesize glucose (sugar) molecules. This

Melting, freezing, and evaporation are other kinds of physical changes. In a solid material such as iron or ice, the molecules do not move around much at all. They stick tightly together one next to another like rows of soldiers standing shoulder to shoulder. This is why ice, wood, and steel are solid, they keep their shape, and they are hard to bend.

When ice melts, the molecules loosen up and slide over and around each other. This melting is a physical change. The molecules are still the same, but in the liquid water they slide freely over each other. This explains why liquid water fills a container whether it is round, square, or any other shape. It also explains why water runs across the floor if you spill it.

When you heat water on a stove, you soon see steam rising from the water. The heat causes the water to evaporate. Evaporation is a physical change because the steam is still water molecules. But the molecules in the steam are no longer staying together and sliding over each other. They have separated and are flying about freely in the air.

If you put salt in water, the salt dissolves. Dissolving is a physical change similar to melting. The molecules of salt let loose from each other and float freely among the water molecules. The salt crystals have melted, but they are still salt and the water is still water. Another physical change has taken place. Physical changes make many materials more useful to us, but the materials themselves stay the same. A different kind of change is needed to make new materials.

Chemical Changes

If you set fire to a piece of paper, some of the paper will turn black. The blackness is caused by carbon atoms. Carbon is no longer paper. Fire produces a ***chemical change*** because it breaks up the molecules of paper and regroups the atoms to form new materials. Some of the carbon in the paper combines with oxygen in the air and forms carbon dioxide. Carbon dioxide is very different from paper.

A chemical change is the result of a ***chemical reaction.*** The following diagram illustrates the chemical reaction that takes place when burning produces carbon dioxide.

(C) plus (O)(O) yields (O)-(C)-(O)

C Carbon plus O oxygen yields CO_2 carbon dioxide

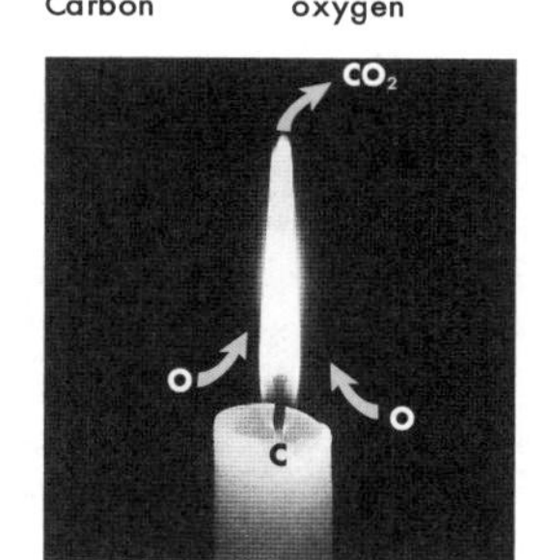

In burning, one atom of carbon combines with two atoms of oxygen to form one molecule of carbon dioxide.

might appear to be a relatively simple chemical change easily duplicated by chemists, but that is not the case. So far, God has withheld this secret from scientists.

A question may arise about the difference between a chemical reaction and a chemical change. One term refers to the process, and the other the effect. A chemical reaction is a process that results in a chemical change.

By all means, take the time to do the activities described in the lesson. This will make the difference between a dull chemistry class and an interesting one. You can get iron filings by cutting a piece of steel with a hacksaw. A propane torch works well for heating a test tube.

Discussion Questions

1. When candle wax melts, is that a chemical change? Explain.
   No. The molecules of wax are not changed. It is still wax.
2. List some common chemical changes.
   burning (combustion), curdling, fading, rusting, photosynthesis, digestion, bubbles rising in a cake, using a chemical to open a clogged drain

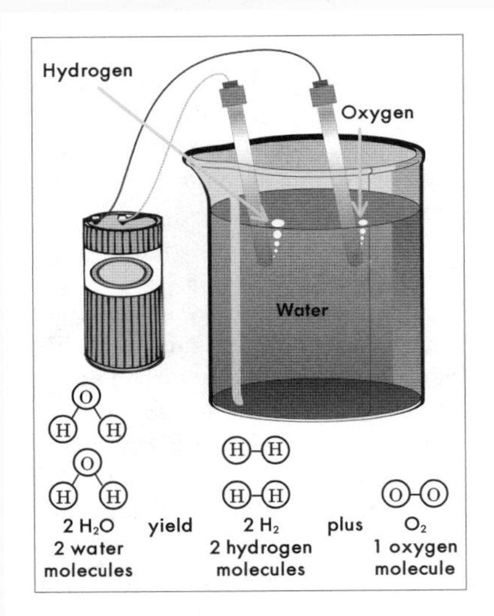

Fire causes some very useful chemical changes. We put gasoline or diesel fuel into an engine, and the fire inside changes the fuel so that it powers the engine and puts out exhaust gases. Fire can also cause chemical changes in baking. If we make a cake batter by mixing flour, sugar, soda, and other ingredients, the heat of a fire can change the batter into a cake.

Electricity can also cause chemical reactions. Remember that water is a compound made of hydrogen and oxygen (H_2O). If electricity is passed through water, it will separate the molecules. The hydrogen atoms come together to form hydrogen molecules, and the oxygen atoms come together to form oxygen molecules. Neither the hydrogen nor the oxygen is a liquid anymore.

Some chemical reactions take place when chemicals come in contact with each other. If you put vinegar in a glass and add some baking soda (sodium bicarbonate), you will see a strong reaction. Many bubbles of gas will form. If you let the reaction continue for a while and then lower a burning match into the glass, the fire will go out. From this, you can tell that the gas produced by the reaction is not oxygen as in air. It is carbon dioxide, which does not allow the fire to burn.

Sometimes water will cause a reaction to take place. If a piece of iron is kept dry, it will not rust. But if the iron is wet, it will slowly combine with the oxygen in the air to make iron oxide, or rust. Rust is not magnetic as iron is. Rust is a different color and is much weaker than iron.

Light causes the color of some materials to fade. This is a chemical reaction. For example, construction paper on a bulletin board tends to lose color over several weeks. Or a farmer may stack nice green bales of hay just inside the barn doors. If the doors are left open, the outside of the stack turns yellow and brown, while the hay inside the stack remains green. Light has faded the hay.

One of the most important chemical changes is called photosynthesis. Green plants take in water from the ground and carbon dioxide from the air. Sunlight causes the water and carbon dioxide to combine into sugar that the plant uses to grow.

Chemical reactions are very important to us. They digest our food and allow us to breathe. They power our machinery when gasoline or diesel fuel burns in engines. A chemical reaction makes concrete hard.

Because all chemical changes follow laws that God established, we can make millions of useful chemicals from the one hundred elements that God made. What a great God we serve! "Thou art the God that doest wonders: thou hast declared thy strength among the people" (Psalm 77:14).

A wood fire produces ash, smoke, steam, and carbon dioxide.

A water and iron reaction produces rust.

Why does a flame go out when it is lowered into a cup containing vinegar and baking soda?

Exercises

1. In a ——— change, a substance is changed to a different form but the molecules stay the same and no new materials are produced.
2. Ice, liquid water, and steam are very different. But all three forms are water because
   a. they are all transparent.
   b. they all weigh the same.
   c. they all contain molecules with two atoms of hydrogen and one atom of oxygen.

Lesson 27 Answers

Exercises

1. physical
2. c

3. chemical

4. a. P
   b. C
   c. C
   d. P
   e. C
   f. P
   g. P
   h. C
5. a. chemical reaction
   b. chemical change

6. a. carbon dioxide
   b. hydrogen and oxygen
   c. fading
   d. sugar
   e. carbon dioxide
   f. iron oxide
7. (Sample answers.) iron rusting, color fading
8. Chemical reactions digest our food and allow us to breathe.
9. vinegar and baking soda reacting

Review

10. True
11. a few more than 100

3. In a ——— change, the atoms in molecules are regrouped and new materials are produced.
4. Write *P* if the sentence describes a physical change, or *C* if it describes a chemical change.
   a. Ice melts to make liquid water.
   b. Electricity causes water to separate into two gases.
   c. Wood burns into smoke and ashes.
   d. Construction paper is cut into tiny pieces.
   e. Pouring vinegar on baking soda makes it fizz and foam.
   f. Sugar dissolves in water.
   g. Direct sunlight makes a tool too hot to hold.
   h. Direct sunlight makes the color fade in a curtain.
5. Write *chemical change* or *chemical reaction* for each description.
   a. It is a process by which new materials are formed.
   b. It is the result of a process by which new materials are formed.
6. For each phrase, name the result of the chemical reaction that takes place. Choose from the list below. You will use one answer twice.

   carbon dioxide — iron oxide
   fading — sugar
   hydrogen and oxygen

   a. Paper burning.
   b. Electricity passing through water.
   c. Sunlight shining on construction paper.
   d. Sunlight shining on green plants.
   e. Vinegar and baking soda being mixed.
   f. Waterdrops standing on an iron surface.
7. Not all chemical reactions work as quickly as when vinegar and baking soda are mixed. Give one example from the lesson to show that some chemical reactions work very slowly.
8. What are two ways that chemical reactions are important to our bodies?
9. What chemical reaction in your lesson seems to be referred to in Proverbs 25:20?

Review

10. True or false? All substances are composed of atoms.
11. About how many elements have been identified?

12. The molecules of an ——— contain just one kind of atom, but the molecules of a ——— contain more than one kind of atom.
13. Write the chemical symbols for these elements. Give them from memory if you can.
    a. hydrogen d. copper
    b. oxygen e. aluminum
    c. carbon f. iron
14. Write the chemical formulas for the compounds below. Give them from memory if you can.
    a. water c. salt
    b. carbon dioxide d. sugar
15. Give the common name for each compound.
    a. iron oxide b. sodium bicarbonate

Activities

1. Observe the effect of water on iron. Wash 9 nails with soapy water to remove any oil on them. Wrap three of them in a dry cloth, wrap three in a wet cloth, and place three in a container of water. Compare the sets of nails after several days.
2. Many chemical reactions result in a change of color. Add a pinch of iron filings to a test tube ¼ full of white vinegar to make ferrous acetate—$Fe(C_2H_3O_2)_2$. Bring the mixture to a boil, and keep it warm for about five minutes. Filter the results through filter paper onto a saucer. Add ¼ test tube of ammonia (NH_4OH). You should see a green color of the compound ferrous hydroxide—$Fe(OH)_2$. Now add ¼ test tube of hydrogen peroxide (H_2O_2). You should see the red-orange color of ferric hydroxide—$Fe(OH)_3$.

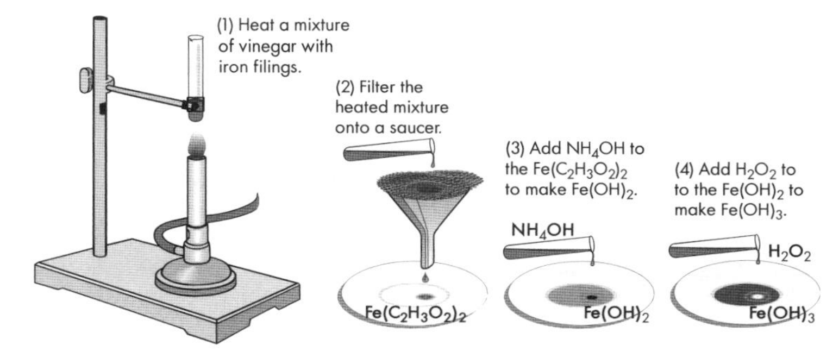

12. element, compound

13. a. H d. Cu
    b. O e. Al
    c. C f. Fe
14. a. H_2O
    b. CO_2
    c. NaCl
    d. $C_{12}H_{22}O_{11}$
15. a. rust
    b. baking soda

Activities

The materials for these activities are easily available. Hydrogen peroxide is a common antiseptic.

2. The formula for steps 1 and 2 is $Fe + 2HC_2H_3O_2 \rightarrow H_2 + Fe(C_2H_3O_2)_2$. The hydrogen bubbles off. The step 3 formula is $Fe(C_2H_3O_2)_2 + 2NH_4OH \rightarrow Fe(OH)_2 + 2NH_4C_2H_3O_2$. The ammonia acetate ($NH_4C_2H_3O_2$) remains in the saucer but is not readily visible. The step 4 formula is $2Fe(OH)_2 + H_2O_2 \rightarrow 2Fe(OH)_3$.

Lesson 28

Fire, an Important Chemical Reaction

Vocabulary

fire extinguisher, a portable device for putting out fires.

fuel, a material that can be burned to produce heat and light.

kindling temperature, the temperature at which a material will begin to burn.

soot, a black powder made mostly of carbon, which forms when a fire burns without enough oxygen.

You hold a match in your hand. There is no fire. You strike the match on a rough surface, and it bursts into flame. The pile of paper and wood in front of you is not burning. You hold the burning match close to some paper, and it begins to burn. Soon the whole pile is on fire, and later nothing is left but some ashes. What is fire?

Fire is a chemical reaction that needs three things. Two of those things are already present before you strike a match. The chemicals in the match, as well as the wood and paper, provide fuel for the fire. ***Fuel*** is a material that will burn. Other fuels include coal, plastic, gasoline, alcohol, natural gas, and propane.

Oxygen is also necessary for fire. There is plenty of oxygen in the air around the match and the pile of fuel. About one-fifth of the air is oxygen.

Besides fuel and oxygen, one more thing is needed to have a fire. A fuel

1. Friction raises the match to its kindling temperature.

2. The burning match raises dry grass to its kindling temperature.

3. The burning grass raises the wood to its kindling temperature.

Lesson 28

Lesson Concepts

1. A fire must have three things: heat, fuel, and oxygen.
2. A fuel must reach its kindling temperature before it will begin to burn.
3. Two products of burning are water and carbon dioxide.
4. Controlled fire can be very helpful.
5. Uncontrolled fire can be very dangerous and destructive.
6. A fire can be extinguished by removing one of the three things it needs.

Teaching the Lesson

Fire fascinates people as few other things do. Be sure that the demonstrations you do today will not tempt the students to play with fire. Your example will speak louder than your cautions and warnings.

Do your students come from homes that are heated with gas, wood, or coal? Are they familiar with gas stoves for cooking? Perhaps they are accustomed to having candles on their birthday cakes. Discover what kind of fire they are most familiar with, and present the lesson from that angle. It will help them make an association between real life and their studies.

Many schools are required by law to have fire drills. Discuss with your students the procedures for

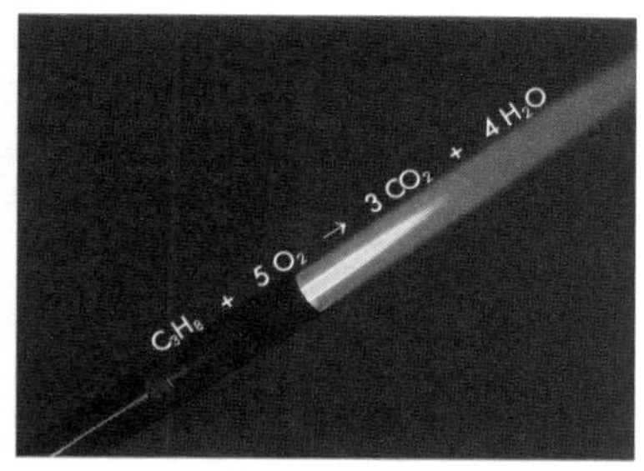

Propane plus oxygen yields carbon dioxide and water.

will not begin burning until it reaches its ***kindling temperature.*** Every fuel has a certain temperature that it must reach before it will burst into flame. Wood must be heated to about 700°F (370°C) before it will catch fire.

The chemicals in the match head have a much lower kindling temperature. They will begin to burn from the heat of friction when you strike the match on a rough place. The burning match will then raise the paper to its kindling temperature, and the burning paper will raise the wood to its kindling temperature.

So three things are needed to have a fire: fuel, oxygen, and heat. If any one of these is missing, there will be no fire.

Fire, a Chemical Reaction

In burning, fuel combines with oxygen to form new materials. For example, your father may have a propane torch. When it burns, the chemical reaction shown above takes place.

Notice that water is one of the products of burning propane. Most fuels have some hydrogen in them, and water forms when it combines with oxygen. Did you ever see water

Cooling a fire is one way to fight it.

escaping from a fire and for extinguishing one. It would be profitable to plan a field trip to your local fire department. Some fire departments will give a live demonstration of how to use fire extinguishers and fire hoses. They usually provide much information about fire hazards too.

Be sure your students understand that fire does not annihilate matter. Fire is a chemical reaction that changes a fuel into one or more other materials. In a barn fire, all those tons of burning wood and hay are chemically changed into an equal mass of smoke, ashes, charcoal, carbon dioxide, carbon monoxide, and water.

Discussion Questions

1. Which of the three things necessary for fire do we usually keep away so that no fire starts?
   Heat. We need oxygen to breathe, so we cannot remove that. Many of the materials we use daily are fuels, so we need them also. The way we prevent fire is by controlling the heat.
2. Discuss some precautions to take in case of fire. Do your pupils know how to file out of a building in a prompt and orderly manner? Have they been told to close doors and windows and to turn off the lights in the event of a fire? Do they know where the fire extinguishers are kept? Have you planned and practiced two escape routes from your classroom?
   This is an excellent time to review your school's fire policy if you have one. If you have none, this would be a good time to establish a policy. Explain why the precautions listed are important.

dripping out of the tailpipe of an automobile on a cold morning? The water forms when the gasoline burns.

Some fuels such as wood also have other elements such as potassium and phosphorus in their molecules. These materials do not burn completely. They are left behind as a white ash, or they rise into the air as smoke. Most of the carbon in a hot fire combines with the oxygen to make the colorless gas carbon dioxide (CO_2).

If a black substance is left behind or rises as black smoke, it is unburned carbon. The black powder made of unburned carbon is called ***soot.*** A fire that produces much black smoke and soot is not getting enough oxygen. If the fire is in a stove, the draft can be opened farther to give the fire more oxygen so that it can burn up the carbon in the fuel.

Fire, a Useful Friend

People have used fire for thousands of years. Its heat and light are helpful in many ways. Many of our houses have stoves or furnaces to keep us warm in cold weather. Some people use gas stoves to cook their food. Factories use fire to manufacture steel and refine oil.

Much of the electricity we use comes from the energy of burning coal. Incinerators use fire to burn trash. The fire of candles and lanterns produces light. Campfires can roast hot dogs, give light, and keep us warm on a cool evening. The burning of gasoline or diesel fuel takes us to and from school, drives large trucks, and powers farm tractors. Fire is very useful and important to us.

Fire, a Dangerous Enemy

People in Bible times knew the destruction of fire. "The fire burneth a wood, and . . . the flame setteth the mountains on fire" (Psalm 83:14). Sometimes whole cities were set on fire. In some cases, God used fire to punish ungodly people, as when He destroyed Sodom and Gomorrah.

Controlled fire is a friend. But when fire gets out of control, it can cause great damage in a short time. It can swiftly destroy buildings, crops, and valuable forests. People and animals are often killed by fires that are out of control. Since fire can quickly become a dangerous enemy, we must be very careful with it.

One way to control fires is to keep them from starting at the wrong time or place. Sometimes a community has a "burn ban," which says that no open fire is to be started because the grass is so dry. Matches must be used very carefully because a little fire, like unkind speech, can quickly get out of control. "Behold, how great a matter a little fire kindleth!" (James 3:5).

There are three ways to fight a fire. First, we can cool the burning material below its kindling temperature. Pouring water on the material is a good way to cool it. The water should be poured at the base of the fire, where the burning material is, and not at the flames.

Second, we can fight a fire by removing the oxygen. That is why it is wise to close doors if a building is on fire or if you are having a fire drill. This helps to keep fresh oxygen from getting to a fire. Some ***fire extinguishers*** spray a chemical or gas at the base of a fire to cut off the supply of air. Throwing shovels of loose dirt on a fire will both cool and smother the fire.

If your clothes ever catch fire, the best thing to do is to drop to the ground and roll. This both smothers and cools the fire. The worst thing you can do is run. Do you know why?

The third way to fight fires is to remove the fuel. "Where no wood is, there the fire goeth out" (Proverbs 26:20). Sometimes to stop a fire in a grainfield, a farmer will plow down some good grain ahead of the fire. Then when the fire gets to the plowed soil, it will go out because it has no more fuel. Men sometimes fight a forest fire by starting a backfire. This is a small, controlled fire that burns up the fuel ahead of the big fire so that it will go out when it reaches the burned area.

Fire is a blessing from God if we use it wisely. When He created the world, He made most materials with kindling temperatures high enough that they will not burn at ordinary temperatures. God also made the air with only about one-fifth oxygen so that fires do not easily get out of control. These things show the wisdom of God in caring for His creation.

For your safety in case of fire, remember the following rules.

- Leave a burning building immediately.
- Before opening an inside door, touch it to see if it is hot. The blast of hot air and sparks from a burning room can injure or kill.
- Crawl to avoid smoke. The best air is at the bottom of a room. There are poisonous gases at the top of a room filled with smoke.
- If your clothes catch fire, drop to the ground and roll. **DO NOT RUN.**

Exercises

1. What are the three things needed for a fire to burn?
2. Before a fuel will burn, it must be raised to its ——— ———.
3. What makes a match hot enough to start burning?
4. What is special about the chemicals in the head of a match?
5. Fire is a chemical reaction in which fuel combines with ——— to produce other materials.

Lesson 28 Answers

Exercises

1. fuel, oxygen, heat
2. kindling temperature
3. friction
4. They have a low kindling temperature.
5. oxygen

6. water and carbon dioxide
7. a. oxygen
   b. fuel
   c. soot
   d. kindling temperature
   e. ash
8. a. warm
   b. cook
   c. electricity
   d. gasoline
   e. light
9. under control
10. starting
11. during a dry spell (when the grass is very dry)
12. a. 1
    b. 1 *or* 3
    c. 2
    d. 3
    e. 1 *or* 3
    f. 2

Review

13. a. aluminum
    b. carbon
    c. copper
    d. oxygen
    e. hydrogen
    f. iron
14. a. water
    b. salt
    c. carbon dioxide
    d. sugar

6. What are the two main products of most burning?
7. Choose the correct word for each description.

a. Element in air that is needed for fire.	ash
b. Material such as wood that will burn.	fuel
c. Black unburned carbon.	kindling temperature
d. How hot a material must be to burn.	oxygen
e. Unburned white chemicals left by a fire.	soot

8. Write the missing words to tell how fire is useful.
   a. A house can be kept ——— with fire in a stove or furnace.
   b. Some people ——— food with fire on gas stoves.
   c. Much ——— is generated from the energy of burning coal.
   d. The explosion of ——— gives power to an automobile.
   e. Candles and lanterns give ——— from fire.
9. For fire to be useful, it must be (under control, out of control).
10. The best way to keep fire from destroying valuable crops and property is to keep the fire from ———.
11. When would a community have a "burn ban"?
12. Write the number of one way that each firefighting method helps to control or put out a fire.
    (1) cools the burning material
    (2) removes fuel from the fire
    (3) cuts off the oxygen source
    a. Pouring water at the base of a fire.
    b. Dropping to the ground and rolling.
    c. Lighting a backfire in front of a big fire.
    d. Closing the doors of a burning building.
    e. Using a fire extinguisher to spray a chemical on the fire.
    f. Plowing down dry grass in front of a fire.

Review

13. Name the elements represented by these chemical symbols.
    a. Al　　d. O
    b. C　　e. H
    c. Cu　　f. Fe
14. What compounds are represented by the chemical formulas below?
    a. H_2O　　c. CO_2
    b. NaCl　　d. $C_{12}H_{22}O_{11}$

Activities

1. This activity will show you how to put out a candle flame by cooling it. Wind some thin copper wire around a pencil to make a short spring on the end of the wire. Lower your copper spring on a candle flame. If you remove it quickly, the candle may burst back into flame. What does the copper spring do to the fire to make it go out?
2. This activity will show you how to put out a candle flame by cutting off the oxygen. Put a teaspoon of baking soda in the bottom of a quart jar. Add three tablespoons of vinegar. You know this is the way to make carbon dioxide. After the reaction has made plenty of carbon dioxide, gently pour the carbon dioxide from the jar onto the base of a candle flame. Be careful not to tip the jar so far that you pour the vinegar on the flame.
3. What happens when a fire uses all the oxygen in the air? You will need a candle about 4 inches long, a pan or dish with 2-inch sides, and a quart jar. Tip the lighted candle to let some melted wax form a small puddle in the center of the pan. Quickly set the bottom of the candle in the melted puddle, and hold it until the wax hardens.

   Put an inch of water into the pan. While the candle is burning, quickly cover the candle with the quart jar, with the mouth of the jar in the water. Can you explain what happens to the candle flame?

4. Here is another interesting activity. Hold a plastic bag filled with ice water in a candle flame. Does the plastic melt? Why or why not?
5. Examine the fire extinguishers in your home, shop, and school. Read the labels to find out what kind of fires they are designed to fight. Read how each fire extinguisher is to be used. Do not waste any of the material in the extinguisher, or it will not be there to fight a fire.

Activities

Note: A number of the activities require using matches. Students should do these activities only with adult supervision. Fire must be treated with respect, and young children should not have free access to matches.

1. Copper is an excellent conductor of heat. In fact, the copper coil conducts the heat away from the flame so fast that the fuel is cooled below its kindling temperature. You may want to try a few variations, though you should not let them obscure the main point of the demonstration. For instance, using a steel paper clip for the wire will not produce the same result. Steel is a much poorer conductor of heat. Also, if you preheat the copper coil by holding it above the flame, it may not conduct enough heat away to extinguish the flame when it is slipped down over the flame.
2. You can pour carbon dioxide out of the jar because this gas is heavier than air.
5. Fire extinguishers are marked with *A, B,* or *C* to indicate the type of fires for which they are intended: *A* for a burning solid, such as wood; *B* for a burning liquid, such as kerosene; and *C* for an electrical fire. Dry chemical extinguishers are marked *ABC* because they are effective against any kind of fire.

Lesson 29

Acids and Bases

Vocabulary

acid (as′·id), a sour chemical that turns blue litmus paper red.

base, a bitter, slippery chemical that turns red litmus paper blue.

litmus paper (lit′·məs), a paper used to test for acids and bases. Litmus paper turns red when touched to an acid, and blue when touched to a base.

neutral, neither acidic nor basic.

pH scale, a scale for measuring acids and bases, on which *1* indicates a strong acid, *7* indicates a neutral substance, and *14* indicates a strong base.

Acids

Many foods taste sour because they contain ***acids.*** Citrus fruits like oranges and lemons contain citric acid. Tomatoes and strawberries have acids in them. Even peaches are slightly sour from acid. The sourness of vinegar is from the acetic acid in it.

When milk turns sour, a bacteria produces an acid that gives it a sour flavor. The pleasant sour flavor of yogurt is from the lactic acid produced in this way. Of course, badly spoiled food becomes unpleasantly sour.

If you chew an aspirin, you will notice that it is very sour. That is because aspirin is acetylsalicylic (ə·sēt′·əl·sal′·i·sil′·ik) acid.

An acid produced in the walls of your stomach helps to digest your food. You have probably noticed a sour taste in your throat when you belched, or perhaps you know what heartburn is. Both of these are caused by the weak solution of hydrochloric acid in your stomach. If this acid is a strong solution, it is dangerous to use. It can even dissolve rocks and concrete.

Common items containing acid

Lesson 29

Lesson Concepts

1. Acids are the reason for the sour flavor in foods like oranges.
2. Bases are bitter and slippery.
3. Chemical tests are used to identify materials.
4. Blue litmus paper turns red in an acid.
5. Red litmus paper turns blue in a base.
6. Acids and bases react to neutralize each other.
7. Soil acidity is an important factor in healthy plant growth.

Teaching the Lesson

We use many acids and bases. Many common acids are found in the foods we eat. These acids are not very strong, of course, but they are still acids. We use bases in baking and as remedies for certain stomach ailments. Some of the stronger acids and bases are no longer as common in the household as they once were. But strong acids and bases are still used commonly by industry. If you handle any of the stronger acids or bases, be very careful with them. They can burn your skin or even blind you if they get into your eyes.

Be sure to get litmus paper and carry out the tests suggested in the lesson text. Oversee the testing of acids and bases. Be sure to exercise caution so that no one gets hurt, particularly when testing strong chemicals like those found in some cleaning agents. Litmus paper can be purchased from almost any scientific supply company for a reasonable price.

Acids can eat holes in your teeth. When you eat sweet food, small amounts stay between your teeth. Bacteria use the sugar for food and produce an acid. The acid acts on your teeth to cause cavities. Brushing your teeth after meals removes the food and bacteria and helps to prevent tooth decay.

Carbonated drinks have an acid in them. The acid is what gives them a sour, biting taste. But it also acts on the teeth to cause tooth decay. That is why it is not wise to drink a lot of soft drinks.

Some acids are very useful. Automobile batteries contain sulfuric acid. This is a very strong acid that can cause serious burns on the skin. Much sulfuric acid is used to make other chemicals such as fertilizer, paint, and explosives.

Important Acids	
hydrochloric acid	helps to digest food in the stomach
sulfuric acid	used in automobile batteries
acetic acid	gives vinegar its sour taste
citric acid	gives oranges and lemons a sour taste
lactic acid	found in yogurt and sour milk

Bases

A ***base*** is a chemical very different from an acid. It is not good for food, but if you have ever tasted bar soap, you know the bitter taste of bases. A base is also slippery. But most of the bases we use are so strong that you should not try to find out how they taste or feel.

Ammonia is a very common base used for cleaning. A little ammonia in water helps to get the windows clean. Another base that is a good cleaner is lye. Lye in drain cleaners helps to open clogged drains. But lye is too strong for ordinary cleaning.

If you put some lime in water, it will make a base. Limewater is not as strong as lye, but you would not want to drink it.

Acids and bases are opposite each other. If an acid is mixed with a base, the two will cancel each other and make a ***neutral*** mixture. For example, some people have too much acid in their stomach. They may take some milk of magnesia as a medicine. Milk of magnesia is a base. When this base goes into the stomach, it neutralizes some of the stomach acid.

Important Bases	
lye	used as a drain cleaner
ammonia	used as a household cleaner
lime	used to "sweeten" sour soil

Discussion Questions

1. How can we tell if a solution is acidic or basic?
   We could taste the solution, but that is neither safe nor reliable in most cases. Litmus paper provides a simple and reliable test.
2. Are all acids and bases of the same strength?
   No. Some acids and bases are stronger than others. Those that are strong may be diluted to weaken them. Chemical tests can be done to show just how strong an acid or a base is.

Common household bases

Farmers are familiar with another example of a base neutralizing an acid. Sometimes the soil has too much acid. Strawberries and blueberries grow well in slightly acid soil, but most plants do not. The farmer knows that if he puts lime on his soil, it will neutralize the acid and help crops to grow better.

Bee stings contain an acid that causes a burning feeling. Perhaps your mother makes a paste with baking soda to put on the bee sting. Baking soda solution is a base. This base neutralizes the acid in the bee sting and makes it feel better.

Testing Acids and Bases

The ***pH scale*** tells how acid or base a material is. It is a scale that begins with *1* and goes to *14.* The number *7* is in the middle. A pH of 7 is neutral. Ordinary water is neutral and has a pH of 7. The numbers below *7* are acid. The numbers above *7* are base. A strong acid would have a pH of 1. A weak acid may have a pH of 4 or 5. A special kind of test paper called hydrion paper turns different colors for the different pH numbers.

Another common test paper is ***litmus paper.*** An acid will make litmus turn red or pink. A base will make litmus turn blue. To take the tests, you need both red and blue

Find the pH of these items.

vinegar	tomato	ammonia	lime
baking soda	rhubarb	carrot	borax
orange	bar of soap	aspirin	soft drink
salt	sugar	yogurt	

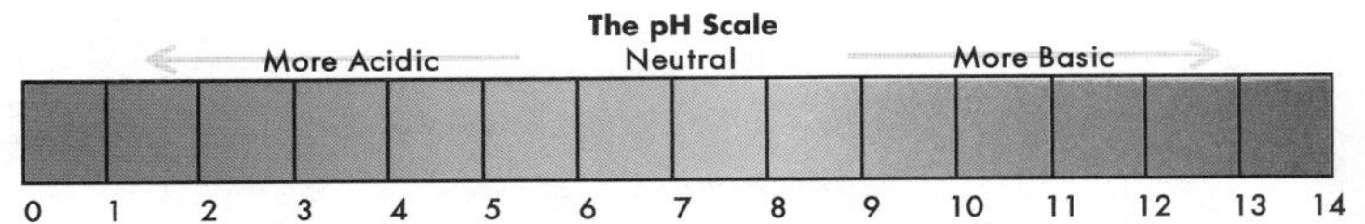

The color bar shows how some test paper reacts to different levels of pH.

litmus paper. If a chemical does not change the color of either red or blue litmus paper, it is neutral.

Now try your hand at some chemical testing, using hydrion paper, litmus paper, or both. Find the pH of each item listed near the bottom of page 172, or find out if they are acid, base, or neutral. A material must be wet to be tested with litmus paper. You will need to crush plant stalks to get the juice, or add distilled water to a powder to test for acid or base.

God has designed acids and bases to be useful to us. We often use acids in cooking. Our stomach acid helps to digest our food. Some bases are used for cleaning. Bases will neutralize acids. God made acids and bases to work together in a wonderful way.

Exercises

1. Copy and complete this table to compare acids and bases.

	Acid	Base
a. taste	———	———
b. litmus color	———	———
c. pH	———	———

2. Tell one way an acid is helpful in your body.
3. Tell one way an acid is harmful in your body.
4. If vinegar is mixed with ammonia, the result will be
   a. a stronger acid.
   b. a reaction that makes bubbles.
   c. a neutral solution.
   d. a stronger base.
5. Suppose you put red litmus paper in a solution and it stayed red, and you put blue litmus paper in the same solution and it stayed blue. Then you would know that the solution is (an acid, a base, neutral).
6. Tell whether each chemical is an *acid* or a *base*. Then write the number of its use.

a. lye	(1) neutralizing acid in soil
b. vinegar	(2) flavoring yogurt
c. lime	(3) cooking and baking
d. milk of magnesia	(4) cleaning windows
e. sour milk	(5) cleaning drains
f. ammonia	(6) neutralizing stomach acid

7. What is the same about pH values of 9 and 13?
8. What is different about pH values of 9 and 13?

Lesson 29 Answers

Exercises

1.

	acid	*base*
a.	sour	bitter
b.	red	blue
c.	below 7	above 7

2. digesting food
3. causing tooth decay
4. c
5. neutral
6. a. base, 5
   b. acid, 3
   c. base, 1
   d. base, 6
   e. acid, 2
   f. base, 4
7. Both are bases.
8. One is a weak base, and one is a strong base.

Review

9. about 100
10. atoms
11. a. H_2O
    b. NaCl
    c. CO_2
12. heat, oxygen, fuel
13. a. cooling the fuel
    b. removing the oxygen (also cooling the fuel)

Activities

3. Litmus paper coloring itself is obtained from various lichens. We should not be greatly surprised if other plant colorings can also be used as acid–base indicators. This is not a difficult activity, and the results make it worth trying.

   Well water is often acidic and will distort the results of pH testing. Therefore, distilled water should be used for these tests.

Review

9. About how many elements are found in the world?
10. A compound is made of two or more (molecules, atoms) of different elements.
11. Give the chemical formulas for the following compounds.
    a. water b. salt c. carbon dioxide
12. What three things are necessary for a fire to burn?
13. A fire may be extinguished in three ways: cooling the fuel, removing oxygen, and removing the fuel. Tell which way is used in each example below.
    a. Mary blows out a candle.
    b. John stamps on a bit of burning paper outside the incinerator.

Activities

1. Neutralize some acids and bases. Start with vinegar in a glass. Add a little ammonia. Test it with blue litmus paper. Is it still acid? Add a little more ammonia. Test it again with litmus paper. Continue to do this until you have a neutral solution. How will you know? If you have hydrion paper, test the pH. What should it be?

   Try neutralizing a soft drink with limewater. The result of this neutralizing will not hurt you. What does it taste like? Is it sour? Is it bitter?
2. Make your own test paper. Dissolve some turmeric in alcohol. Dip some strips of white paper in the turmeric solution. Let the strips of paper dry. Now you can use them to test for a base. Turmeric paper will turn reddish-brown in a base. Use it to test ammonia and limewater.
3. Another test material can be made from purple plants: red cabbage, purple flowers such as violets or irises, or the juice of elderberries, blueberries, or dark sweet cherries. Cut the flowers or cabbage leaves in small pieces. Put in one-half cup of hot water, and keep hot for half an hour. Pour off the colored liquid, and let it cool.

   Pour a little of the test water into a clear glass. Add a little vinegar, and observe the color change. Add some ammonia, and see how the color changes again. Can you explain the results? How can you tell if you have an acid? How can you tell if you have a base?
4. Phenolphthalein is an interesting test for bases. It can be purchased from a chemical supply house as a solution or as a powder. The powder can easily be made into a test solution by dissolving a little of it in alcohol. A few drops of phenolphthalein solution turn bright red in a base such as ammonia. If you neutralize the mixture with an acid, the color will disappear.

Lesson 30

Using Chemicals Safely

Vocabulary

antidote (an′·ti·dōt′), a substance taken to counteract a poison.

concentration, a chemical that is very strong because it is not diluted.

dilute, to make weaker, such as by adding water.

ore, rocky matter from which a metal can be taken.

plastic, a synthetic material that can be molded into many shapes.

pollution, the contaminating of air, soil, or water with harmful chemicals.

raw material, a source of chemicals as it is found in nature.

synthetic (sin·thet′·ik), put together by man; not found in nature.

toxic, poisonous; harmful to living things.

ventilation, a plentiful supply of fresh air.

Benefits of Chemicals

Chemicals are a wonderful gift from God. The one hundred elements can combine to form thousands of compounds. Each element or compound has properties that make it useful to us. Some of the materials we use can be found in nature. Oxygen (O_2), water (H_2O), salt ($NaCl$), and sugar ($C_{12}H_{22}O_{11}$) are a few of the many chemicals that we can get from natural sources.

Metals such as iron and copper are taken out of rocks that have compounds of these metals in them. Rocks from which metals can be taken are called ***ores.*** Bauxite is the name for an aluminum ore. Hematite is one kind of iron ore. All the metals you use were taken from ores.

A ***raw material*** is a source of chemicals just the way it comes from nature. Coal is a raw material. From coal, a special fuel called coke is made. Drugs, fertilizers, and roofing tar can be made from coal. Petroleum is a valuable raw material that is pumped out of the ground. Our gasoline comes from petroleum. Drugs, paint, and insecticides can be manufactured from petroleum.

People have studied God's laws of chemicals and have learned how to put chemicals together to form special materials. Such materials are called ***synthetic*** materials because man puts

Lesson 30

Lesson Concepts

1. Metals are taken out of ores that have compounds of the metals in them.
2. Some things we use are made from raw materials such as coal and petroleum.
3. Some materials are called synthetic materials because man puts them together. They are not found in nature.
4. Plastics are synthetic materials that can be molded into many shapes.
5. Pollution happens when chemicals make a part of the earth dangerous or unfit for living things.
6. Toxic chemicals are harmful if they are swallowed, if they get on the skin, or if the fumes are breathed.
7. Many concentrated chemicals must be diluted before they can be used safely.
8. Chemicals that give off toxic fumes must be used with proper ventilation.

Synthetic materials are useful for many common items.

them together. They are not found in nature. Some rubber is made from the sap of rubber trees. But much of the rubber in use today is synthetic. It is put together with chemicals taken from raw materials, such as coal and petroleum.

You may be wearing clothes made from synthetic fibers. Cotton, wool, silk, and flax are natural fibers that are used to make cloth. Nylon is used to make synthetic fibers from several raw materials, such as coal and petroleum. Rayon is a synthetic material made from wood or cotton. Polyester is a synthetic material made with a special alcohol that comes from petroleum.

A very important family of synthetic chemicals are the plastics. A ***plastic*** is a synthetic material that can be molded into many shapes. Today plastics are used to make bottles, raincoats, and even car bodies. Some plastic is soft, and some is hard. Some is clear, and some is dark. Some is made into foam for padding in seats. Some is rolled into thin sheets to make waterproof covers.

Many plastics are used to take the place of other materials. A plastic lunch box takes the place of a metal lunch box. Clear acrylic plastic can take the place of window glass. Many tabletops have a plastic layer that looks like wood. Plastic water pipes take the place of expensive copper pipes.

Signs on containers with poisonous chemicals help warn us of danger.

Teaching the Lesson

Our government has numerous regulations concerning the use and abuse of chemicals. Our goal should not be to get by with everything we can. Rather, it is the Christian's desire (1) to be a good steward of God's creation and avoid polluting it carelessly, and (2) to be obedient even to those restrictions for which we see no good reason. In addition, Christians are interested in safety whether or not the government requires it. Help your students to understand the importance of precautions designed to promote safety and prevent pollution.

In addition to the exercises and activities of this lesson, it would be wise for the pupils to spend extra time in preparing for the coming test. Review of the chemical symbols and formulas would be in order. Seek to rivet the concepts in their minds.

Discussion Questions

1. List a few common chemicals that can be dangerous.
   (Sample answers.) gasoline, alcohol, bleach, insecticides, strong cleaning fluids
2. What precautions should be taken with each of the items in number 1?
   (Individual answers.)
3. Discuss the importance of washing your hands after using a strong chemical.
   By washing your hands, you are less likely to get the chemical into your mouth or your eyes. Also, the chemical will not be absorbed into your skin.

People often call synthetic materials man-made because men have learned how to make them from raw materials. But even synthetic materials come from God. He placed the raw materials on the earth and established the laws of chemicals that make synthetic materials possible.

Problems With Chemicals

In using so many natural and synthetic chemicals, men have a problem with pollution. ***Pollution*** happens when chemicals make a part of the earth dangerous or unfit for living things. Sometimes in making chemicals, factories send poisonous smoke into the air or put poisonous wastes into rivers. The poisons in the air can give people lung diseases. Fish may die from the poisons in the water. Chemicals that are poisonous and harmful to living things are said to be ***toxic*** chemicals.

Some of the chemicals in our houses are toxic. Many cleaning chemicals would be harmful if swallowed. For example, ammonia is toxic and should never be swallowed. Lye is even more dangerous. The people who make toxic chemicals put warnings on the labels, such as "Caution" or "Danger." Often they have the warning "Keep out of reach of children." A small child may be seriously harmed or even killed if he puts a toxic chemical in his mouth.

Toxic chemicals can also do harm if they get on the skin or if the fumes are breathed. You should never play with toxic chemicals. Only older people who understand how to use the chemicals safely should work with them.

Labels of toxic chemicals also tell what antidote to use if the chemical is swallowed. An ***antidote*** is a substance that will help to work against the poison. Different toxic chemicals need different antidotes. Sometimes milk is given as an antidote.

Often the label says that much water should be used to work against the toxic chemical. Water will ***dilute*** many chemicals. Diluting a chemical with water makes it weaker so that it is less harmful. A chemical is strongest when it is a ***concentration.*** A chemical is most concentrated when it is not diluted with any water. Chemicals that farmers use on their crops would hurt the crops if they were sprayed in concentrated form. But the farmer dilutes the spray material with many gallons of water before spraying it on the field.

Some paints and glues give off toxic fumes. So the labels say there must be proper ventilation when they are used. ***Ventilation*** means that there is a plentiful supply of fresh air. The fumes from a gasoline engine contain a poisonous gas called carbon monoxide. For this reason, a gasoline engine should never be run inside a closed building. It should be run only where there is plenty of ventilation.

Many laws have been passed to protect the earth and the user of strong chemicals. As you grow older and use these strong chemicals, you should learn what these laws are and obey them for your own good and for the good of other living things.

Chemicals are helpful materials that are valuable for many things. But we must use them carefully and wisely. A chemical may seem like a quick and easy way to solve a problem, such as getting rid of weeds. But sometimes a chemical causes other problems that are just as bad or worse. Let us do our part in protecting and taking care of the world that God created.

How is this person using a chemical safely?

What problem with chemicals do you see?

Exercises

1. The iron that was used to make the metal parts of your desk was once in compounds of rock called ———.
2. Petroleum is called a ——— ——— because it is used to make many products such as gasoline, drugs, paint, and insecticides.
3. The rubber we use either came from the rubber ——— or is ——— rubber.
4. Give the names of two synthetic materials that are used to make fibers for cloth.
5. Which of the following is the best description of a synthetic material?
   a. A cheap imitation of a natural material.
   b. A material made by man and not by God.
   c. A material that man puts together from raw materials.
   d. A new product from the laboratories of modern science.
6. Today many products are made from ———, which is a synthetic material that can take the place of many different natural materials.

Lesson 30 Answers

Exercises

1. ore
2. raw material
3. tree, synthetic
4. (Sample answers.) nylon, rayon, polyester
5. c
6. plastic

7. Write the correct word for each description.
   a. A chemical that is very strong, with little water added.
   b. Plenty of fresh air.
   c. Poisonous; harmful to living things.
   d. To make weaker by adding water.
   e. A material that works against a poison.
8. Write three correct endings for the following sentence.
   A toxic chemical can harm you if you…
9. When using a strong chemical, why is it important to make the right dilution of the material before using it?
10. Why should a gasoline engine not be run inside a closed building?

Review

11. a. An ——— is the smallest possible particle of an element.
    b. An ——— is made of a single kind of atom.
    c. A ——— is made of two or more kinds of atoms chemically joined together.
12. Give the chemical formulas for the following compounds, and tell what elements they are composed of.
    a. water c. carbon dioxide
    b. salt d. sugar
13. If a substance changes in such a way that atoms are regrouped and new materials are formed, a (physical change, chemical change) takes place.

Activities

1. Plastics include a wide variety of synthetic materials used to make many different products. Do research to learn about the different kinds of plastics. Make a display of things made from plastic. Try to include different areas where plastics are useful, such as containers, covers, insulators, cases, paints, and tape.
2. Examine the labels of some dangerous chemicals. Check bottles of household cleaning chemicals, antifreeze, paint, and spray cans of insecticides. If you live on a farm, ask your father for the label booklets of some spray chemicals. Each student in the class could read the warning and antidote part of a different label. Notice how strongly the warnings are worded.
3. How can parents keep poisonous chemicals out of reach of children? How do the people who make the containers make it hard for children to get to the poisonous chemicals? Find several different examples.

7. a. concentration
   b. ventilation
   c. toxic
   d. dilute
   e. antidote
8. swallow it.
   get it on the skin.
   breathe its fumes.
9. Diluting a chemical makes it weaker so that it is less harmful.
10. The fumes from a gasoline engine contain a poisonous gas called carbon monoxide.

Review

11. a. atom
    b. element
    c. compound
12. a. H_2O—hydrogen, oxygen
    b. NaCl—sodium, chlorine
    c. CO_2—carbon, oxygen
    d. $C_{12}H_{22}O_{11}$—carbon, hydrogen, oxygen
13. chemical change

Activities

2. Screen any labels brought to school to make sure they are appropriate.

Lesson 31

Unit 5 Review

Review of Vocabulary

acid	dilute	pH scale
antidote	element	physical change
atom	fire extinguisher	plastic
base	fuel	pollution
chemical change	kindling temperature	property
chemical formula	litmus paper	raw material
chemical reaction	matter	soot
chemical symbol	molecule	synthetic
chemistry	neutral	toxic
compound	ore	ventilation
concentration		

Any material that you can see, touch, or feel is __1__. God made the natural world with only about one hundred different simple substances. Each of these is an __2__. An example is iron, which has the __3__ Fe. The study of how elements make up materials is called __4__. The smallest possible particle of an element is an __5__.

Sometimes several different kinds of atoms unite to form a new material, which is called a __6__. Each characteristic, or __7__, of the new material may be different from those of the original materials. One example is water, which has the __8__ H_2O. Two atoms of hydrogen and one atom of oxygen unite chemically to form one __9__ of water.

Materials often change. No new molecules are formed by changes like freezing, melting, and crushing. Such a change is a __10__. If atoms are regrouped to form new molecules and new materials, a __11__ has taken place. The process that causes a chemical change is called a __12__.

In order for a fire to burn, three things are needed: __13__, oxygen, and heat. When a fuel is raised to its __14__ in the presence of oxygen, it bursts into flame. Fire produces black __15__ when it does not have enough oxygen. If a fire gets out of control, a __16__ can help to put it out.

Lesson 31 Answers

Review of Vocabulary

1. matter
2. element
3. chemical symbol
4. chemistry
5. atom
6. compound
7. property
8. chemical formula
9. molecule
10. physical change
11. chemical change
12. chemical reaction
13. fuel
14. kindling temperature
15. soot
16. fire extinguisher

Lesson 31

This chapter is designed to give a basic understanding of the materials that God has created. There are myriads of different materials in the known universe. Yet as accurately as we can determine, everything is made of combinations of about one hundred elements. Thus, basic order can be rendered from the multitude of compounds in our world.

Read over the basic concepts listed for each lesson. Have your students mastered them? Be sure to give plenty of drill on the common chemical symbols and formulas. These will be beneficial now as well as in later years.

An _17_ is a chemical with a sour taste. It gives oranges and lemons their tangy flavor. A _18_ is a bitter, slippery chemical. Red and blue _19_ can be used to tell whether a substance is an acid or a base. When a substance is neither an acid nor a base, it is _20_. Such a substance has a value of 7 on the _21_.

A _22_ is a source of chemicals just the way it comes from nature. One example is the rocky matter from which a metal can be taken. Such matter is called an _23_. Many natural materials are put together to make _24_ (man-made) materials. One of these is _25_, which can be molded into many different shapes.

Some chemicals must be handled carefully because they are poisonous, or _26_. If they are swallowed, an _27_ will help to counteract the poison. If they give off strong fumes, good _28_ is needed. Some chemicals are harmful when used in strong _29_. Such chemicals are safer to use if we _30_ them with water. We must use chemicals carefully to avoid _31_, which makes part of the earth dangerous or unfit for living things.

17. acid
18. base
19. litmus paper
20. neutral
21. pH scale
22. raw material
23. ore
24. synthetic
25. plastic
26. toxic
27. antidote
28. ventilation
29. concentration
30. dilute
31. pollution

Multiple Choice

1. Which of these is the chemical symbol for oxygen?
   a. O c. Ox
   b. Y d. ON
2. Which of these is the chemical formula for water?
   a. CO_2 c. O_2H
   b. CH_4 d. H_2O
3. Which of the following statements about elements is **not** true?
   a. Some elements are metals.
   b. The smallest possible particle of an element is an atom.
   c. Some elements are compounds.
   d. About one hundred different elements have been discovered.
4. Atoms of which two elements are found in water?
   a. carbon and oxygen
   b. carbon and hydrogen
   c. hydrogen and oxygen
   d. sodium and oxygen
5. Which of these is the chemical formula for carbon dioxide?
   a. CO_2 c. H_2O
   b. CH_4 d. H_2O_2

Multiple Choice

1. a
2. d
3. c
4. c
5. a

6. b
7. d
8. b
9. d
10. b
11. c
12. a

6. How is a chemical change different from a physical change?
   a. The material has a different shape afterward.
   b. The atoms are regrouped, and new molecules are made.
   c. The material becomes cooler.
   d. The color of the material changes.
7. Which of the following is a physical change?
   a. a log burning in the fireplace
   b. iron rusting on a bicycle left outdoors
   c. food digesting in your stomach
   d. steel melting under a welding torch
8. Which of the following is a chemical reaction?
   a. laundry drying on a clothesline
   b. baking soda and vinegar making bubbles
   c. sugar dissolving in water
   d. limestone rock being crushed into dust
9. The three things needed for a fire to burn are
   a. fuel, hydrogen, and oxygen.
   b. heat, hydrogen, and fuel.
   c. fuel, nitrogen, and heat.
   d. fuel, oxygen, and heat.
10. We know that fire is a chemical reaction because
    a. fire is useful.
    b. the fuel combines with oxygen to form new materials.
    c. wood contains hydrogen and carbon.
    d. fire can get out of control.
11. All the following actions will put out a fire **except**
    a. cooling the fuel below its kindling temperature.
    b. removing the oxygen from the area.
    c. being very careful when burning trash.
    d. removing the fuel from the fire.
12. If your clothes catch fire, the best thing to do is
    a. drop to the ground and roll.
    b. beat the flames with your hands.
    c. run for the nearest water.
    d. call for help.

13. Some fruits are sour or tangy because
    a. they contain a base.
    b. they cause a chemical change in your mouth.
    c. they contain an acid.
    d. they are neutral.

13. c

14. If a chemical turns red litmus paper blue, it is
    a. an acid. c. a base.
    b. a compound. d. neutral.

14. c

15. If you mix a weak acid and a weak base, the mixture will be
    a. an acid. c. a base.
    b. a compound. d. neutral.

15. d

16. Chemicals that are toxic
    a. are poisonous. c. seldom cause pollution.
    b. explode easily. d. cannot be used safely.

16. a

17. People cause pollution
    a. every time they use chemicals.
    b. if they dilute a chemical too much.
    c. if they put poisonous wastes into a river.
    d. if they are good stewards of God's creation.

17. c

18. Which rule for using harmful chemicals is **not** stated correctly?
    a. Keep harmful chemicals away from children.
    b. Keep windows and doors closed when using chemicals with harmful fumes.
    c. Use a diluted form of a chemical whenever possible.
    d. Obey the laws for handling harmful chemicals.

18. b

Unit 6

Title Page Photo

God made our bodies to function in many wonderful ways. Our brain, nerves, and muscles all work together marvelously as we work and play. We thank God for this blessing by taking good care of our bodies.

Introduction to Unit 6

Our bodies are complex organisms that God has created to function in a wonderful way. The bones give support and protection to the body. The muscles move the body and operate its organs. The nervous system controls the activities of the body. The various systems of the body work together harmoniously and enable the body to move, grow, and heal in the way God intended.

This unit is an introduction to the wonders of the bones, muscles, and nervous system. Help your students to understand and appreciate the importance of good posture, vigorous exercise, and first aid in maintaining a healthy body. The bones, muscles, organs, and nervous system that God has given to us need proper care so they remain healthy. Stress the concept of being good stewards of the bodies God has given us.

Unit 6

Wonders of the Human Body

God has created our bodies in a wonderful way. Men have devised some remarkable inventions, but none of them work as marvelously as the human body that God designed.

In this unit we shall study three important parts of our bodies: the bones, the muscles, and the nervous system. In addition, we shall learn how to take good care of our bodies. Good posture and vigorous exercise help to keep our bodies healthy. First aid is necessary when we are hurt. Our bodies were created with an excellent ability to heal themselves; but if we care for our injuries properly, they will heal much faster and more completely.

We praise our Creator for the wonderful way that our bodies work. "I will praise thee; for I am fearfully and wonderfully made: marvellous are thy works; and that my soul knoweth right well" (Psalm 139:14).

Story for Unit 6

"Ouch, Mother; It Hurts!"

"Jerry, you may run and play for half an hour," said Mother one sunny afternoon when Jerry came home from school. "The exercise will be good for you. Then you must come in and do your homework."

"Oh, good," Jerry said. "I think I'll ride my bike. That will be lots of fun today." Then he asked, "May I have a glass of milk first?"

"Surely. And you may also have one of the cookies I baked today."

"Thank you!" said Jerry. He soon downed the milk and cookie and went to change his clothes.

"A whole half-hour to ride my bike. I'm glad we have a long driveway. That makes it seem more like a road."

Soon Jerry was happily riding his bicycle down the driveway. But the fun did not last long. As Jerry returned from his first trip to the road, he let the bicycle go really fast down the steep part of the lane. Suddenly the front wheel hit a big rock lying right along the edge. *Crash!* Jerry flew off the bicycle and landed on the hard, stony driveway.

Jerry was hurt. His hands were torn and bleeding. His pants were torn at the knee, and his leg was scratched too. But what hurt most was his left ankle. He winced as he sat up on the gravel driveway. Tears threatened to come. He was hurting terribly! "No, I'm not going to cry," he decided. "I'll just have to get my bike out of the lane and see if I can hobble back to the house."

Jerry pushed his bicycle to the edge of the driveway and let it drop into the grass. Foot by painful foot, he hobbled to the house. He went into the bathroom to wash his dirty, bleeding hands.

"Whatever happened to you?" Mother asked, coming to the door. "You're hurt."

"I–I had a bike wreck out on the lane," Jerry admitted sheepishly.

"Be sure you wash your hands well with soap and

Lesson 32

Bones for Support and Protection

Vocabulary

cartilage (kär′·təl·ij), a tough, rubbery material that pads the ends of bones at a joint.

femur (fē′·mər), the bone between the hip and the knee; the thighbone.

joint, a place where two bones are joined, which usually allows the bones to move in relation to each other.

ligament (lig′·ə·mənt), a tough band that holds bones together at a joint.

pelvis (pel′·vis), the bony structure at the bottom of the spine, to which the legs are attached.

rib, one of the twenty-four bones fastened to the spine and curving around the chest.

skull, the shell of bone that surrounds the brain, eyes, and ears.

sternum (stėr′·nəm), the flat vertical bone in the middle of the chest; the breastbone.

vertebra (vėr′·tə·brə), *plural* **vertebrae** (vėr′·tə·brē′), one of the bones that make up the spine.

"Thou hast clothed me with skin and flesh, and hast fenced me with bones and sinews" (Job 10:11). Job was talking to God. He recognized that God had made him and put his bones in place. Today we shall study the wonderful bones God has given to each of us.

Perhaps you have heard the expression "as dry as a bone." That might be true for a dead bone, but the bones of living creatures are far from dry. Bones are alive. They grow and heal as other parts of our bodies do. In fact, bones are constantly being dissolved and replaced with new bone material. This helps keep our bones healthy and strong.

Bones are filled with blood vessels. New blood cells are made inside the bones. Calcium and other minerals are stored in the bones as well. These minerals make a bone stiff and hard. They are either stored in the bone or released into the blood for the body to use.

warm water," Mother told him. "I'll get the Merthiolate and Band-Aids."

"Do we have to use Merthiolate? That stuff hurts!"

"I know it hurts," Mother replied, "but we must use it to kill the germs."

Jerry held his hands under the stream of warm running water. "Ouch, even water hurts. Mother, are you sure I have to use soap on these cuts? That will make them hurt even worse."

"It might hurt now, but it will keep infection from getting into your cuts," Mother answered. "Hurry and wash up. Here, let me help you."

Jerry bit his tongue to keep from crying as Mother cleaned his cut hands. "There, that should be good enough," she finally said. "Now we need to put on the Merthiolate."

"Mother, that's going to hurt!"

"I know it will hurt for a while, but it will be much better in the end. Hold still now, and be a brave boy so you don't make me spill the Merthiolate."

"There, that wasn't too bad, was it? You did very well. Now let's put on some Band-Aids. Oh, and what does your knee look like?"

"I think my knee will be all right. It isn't cut much. My ankle hurts worse. I can't even walk on it."

Mother carefully wrapped Jerry's ankle with a stretch bandage. "This will help keep it from swelling and hurting more," she told him as she fastened the end of the bandage in place.

"Now let's see your knee. That cut will need to be cleaned too."

"But it's just a little cut. It doesn't hurt much," Jerry said as he rolled up his pant leg.

"It still needs to be cleaned. Dirt in even a small cut can cause infection and make you sick. You don't want that to happen."

Gingerly Jerry finished rolling up his pant leg. But instead of one cut, there were two. Soon they were

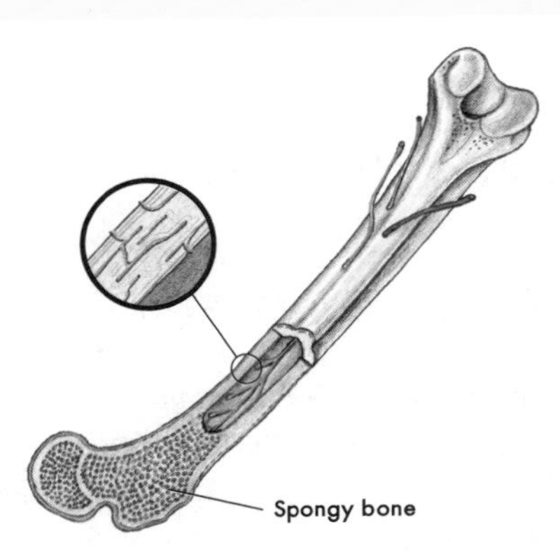

Bones are alive with blood vessels.

Bones Support the Body

Remember the exoskeleton that an arthropod has? It is the hard layer on the outside that gives an arthropod its shape. We are not like that. God made us with a skeleton of bones inside us. These bones support our bodies.

Our feet and leg bones support the weight of the rest of our bodies when we stand. Each leg has two bones below the knee and a ***femur*** (thighbone) above the knee. At the hip, the two femurs fasten to the ***pelvis.*** The pelvis is a broad, basin-shaped structure of bone that is fastened to the bottom of the spine.

The spine includes all the ***vertebrae*** in the backbone and neck. The spine of an infant has 33 vertebrae, but by adulthood some of the vertebrae near the pelvis have grown together. So an adult has fewer vertebrae than a child.

The vertebrae of the spine are stacked on top of each other to form a long, flexible column that supports the upper body. The ribs and shoulder bones are fastened to the spine, and the head rests on top of the spine. Our spines provide excellent support for our bodies as well as for the heavy loads that we carry sometimes. That such a flexible column can bear such heavy weights is a marvel of design.

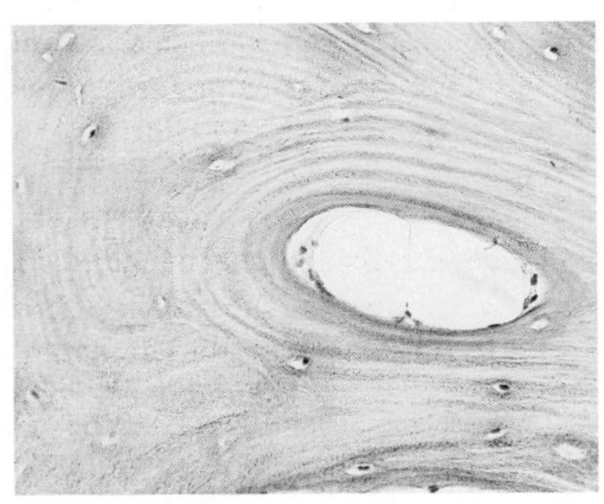
This magnified view of a bone shows tiny holes that serve as supply tunnels.

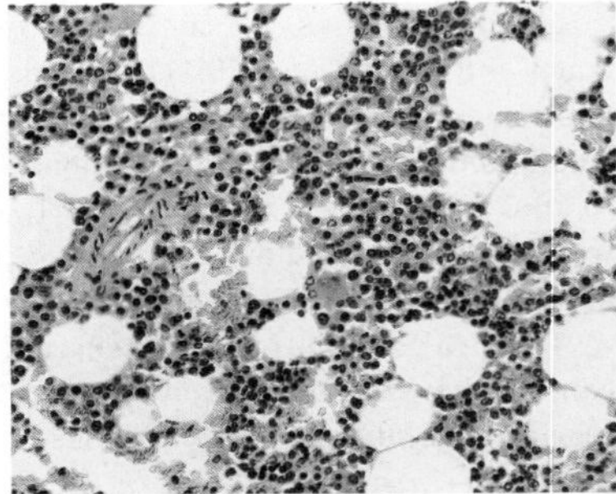
A magnified view of tissue inside a bone. What important cells are made inside bones?

washed, and Merthiolate was applied. Finally came the last Band-Aid, and Jerry was all fixed up.

"Now go and change into clean clothes. You got these all dirty. Then it will be time to sit down and do your homework."

"Know what, Mother?"

"What?"

"With all these Band-Aids on my hands, I won't be able to do any homework."

"You don't think so?" Mother did not sound convinced. "Go and change your clothes. Then we'll see."

Lesson 32

Lesson Concepts

1. Bones are made of living cells that grow and repair themselves.
2. The bones are designed to support and protect the body.
3. Joints allow the bones to move.
4. Different kinds of joints are designed to allow different kinds of movement.
5. The ends of the bones are specially designed to absorb shock. A layer of cartilage further protects the joints.
6. The bones of a joint are held together by strong bands called ligaments.

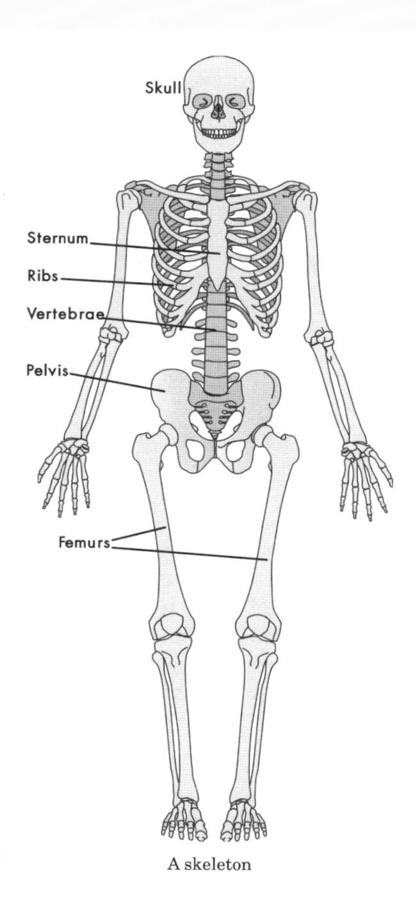

A skeleton

Our arms hang from our shoulders. Each arm has three main bones: one extending from the shoulder to the elbow, and two reaching from the elbow to the wrist. All these skeletal bones support our bodies and help to give them their shape. Without them, our bodies would be a flabby heap of flesh, and we could not sit or stand.

Bones Protect the Body

Bones provide not only support but also protection for our bodies. Many parts of our bodies would quickly be damaged or destroyed if they did not have bones to protect them.

God designed the ***skull*** to surround our brain, eyes, and ears. These body parts need special protection because they are very delicate. The rounded shape of the skull makes it very strong, just as the rounded ends of an egg make it surprisingly strong.

Long, slender bones called ***ribs*** are fastened to the spine in the back, and they curve around the sides of the chest. The 24 ribs make up the rib cage, which protects the lungs from injury. Because the rib cage is flexible, the chest can expand and contract as a person breathes.

The ***sternum*** is the flat, vertical bone in the front of the chest. It is often called the breastbone. The sternum lies directly over the heart and helps to protect it from harm.

The vertebrae in the spine form a bony tube, with the spinal cord running through. The spinal cord is attached to the brain and carries

Teaching the Lesson

Our bodies are not surrounded by exoskeletons like those of the crustaceans. We have bones inside that support our bodies and protect our organs. Today's lesson portrays the basic roles of bones in our bodies. Help the students to understand that these bones are real and alive. Further, they each have a complete set of them. Can they visualize that they are living skeletons with flesh and skin added? Try to make the lesson register with them, for they are studying one of God's greatest gifts.

Discussion Questions

1. In what ways are bones alive?
   They grow and heal like other parts of our bodies. They are filled with blood, and some blood is formed in the bones.
2. What would it be like to have no joints?
   We would be unable to bend our arms, legs, fingers, toes, or any other body part. We could not even open our mouths. We cannot imagine exactly what it would be like.
3. Why is it good that God gave us ball-and-socket joints in our shoulders?
   These joints allow our arms to swing freely and to move in almost any direction, which lets us work freely with our arms.

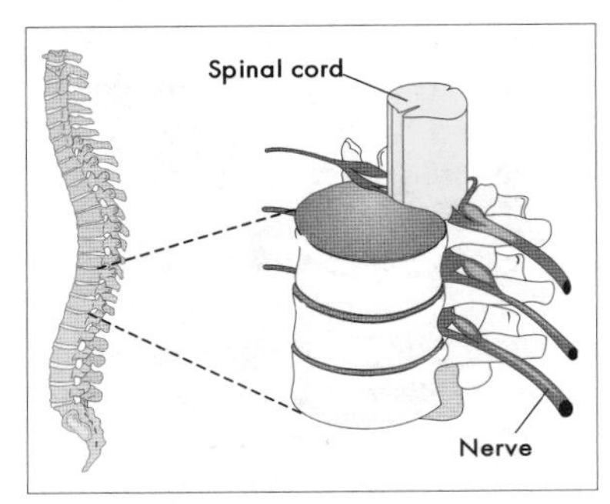

Vertebrae

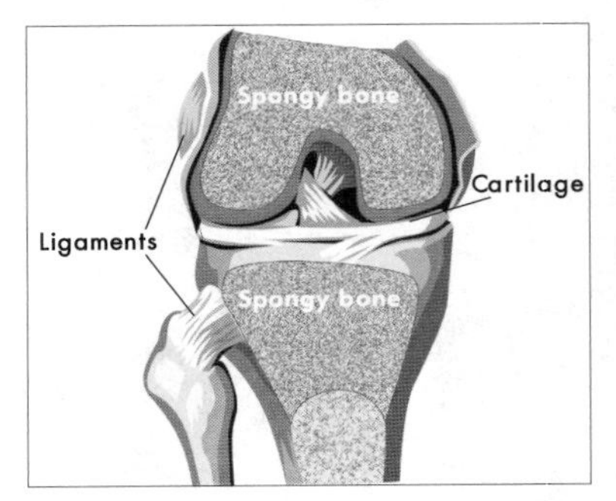

A knee joint

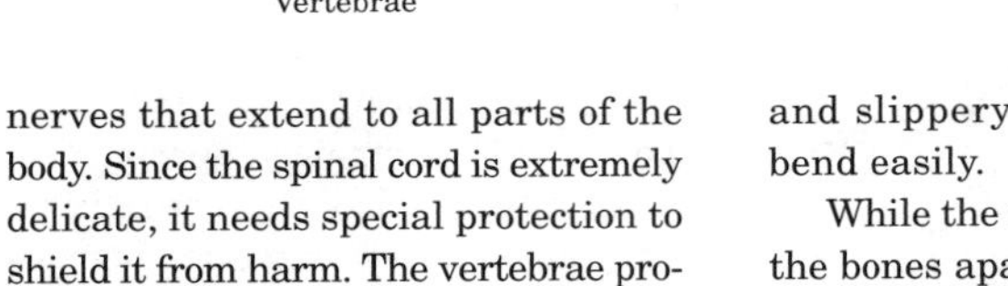

nerves that extend to all parts of the body. Since the spinal cord is extremely delicate, it needs special protection to shield it from harm. The vertebrae protect the spinal cord very well.

Joints Allow Bones to Move

Although your bones are stiff and hard, your body is not stiff. You have ***joints*** that allow your spine and arms and legs to move and bend. Joints are the places where bones are joined together.

At the joints, something is needed to keep the hard bones from grinding each other. God created our joints with a layer of ***cartilage*** to pad the bones and keep them from rubbing against each other. You have probably noticed this layer of cartilage many times at the dinner table. When you eat a chicken drumstick, you remove the "knuckle" of cartilage from the end of the bone. Cartilage is smooth and slippery so that the joints can bend easily.

While the cartilage in a joint keeps the bones apart, something is needed to hold the joint together. This is done by the ***ligaments.*** Some ligaments are strong bands that tie the bones together. Others are more like sheets of fiber. When a joint is struck with enough force to stretch or tear the ligaments, we call the injury a sprain.

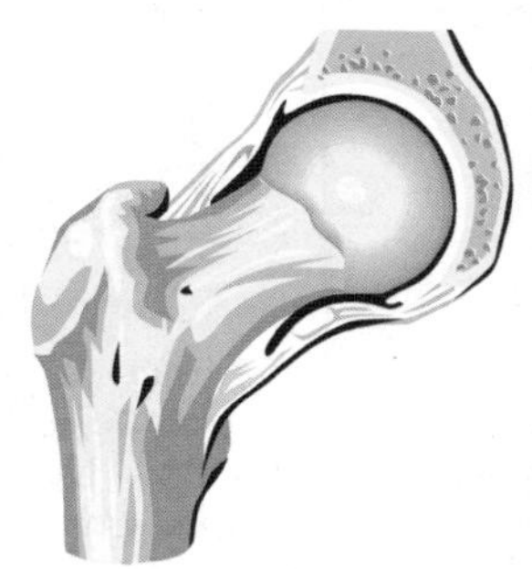

A ball-and-socket hip joint

Joints are designed to absorb bumps and jolts. As you have learned, there is cartilage between the bones to pad the joint. In addition, the ends of the bones often have a spongy section that absorbs the bumps and jolts. The knee and shoulder joints also have special fluid-filled sacs called bursas to provide extra protection and padding.

Different kinds of joints allow different kinds of movement. Hinge joints bend in only one direction (like a hinge). Knees are hinge joints. Pivotal joints allow a turning motion. The neck has pivotal joints that allow the head to turn from side to side. Ball-and-socket joints allow free movement in almost any direction. Hip and shoulder joints are ball-and-socket joints. They have a ball on the end of one bone and a rounded socket in the other bone. The ball fits into the socket, and the bone can swivel freely in many directions.

Kinds of Joints

Knee	Neck	Shoulder
Hinge joint	Pivot joint	Ball-and-socket joint

We must say with the psalmist, "I will praise thee; for I am fearfully and wonderfully made: marvellous are thy works; and that my soul knoweth right well" (Psalm 139:14).

Exercises

1. Choose the correct words from the list to match the descriptions.

cartilage	ligament	rib	sternum
joint	pelvis	skull	vertebra

a. A place where two bones join.
b. One of the bones in the spine.
c. The bony shell of the head.
d. A strong band that holds a joint together.
e. The bony structure that connects the legs to the spine.
f. Tough, rubbery material that pads the ends of bones at a joint.
g. The flat chest bone that protects the heart.
h. One of the slender, curving bones that protect the lungs.

Lesson 32 Answers

Exercises

1. a. joint
b. vertebra
c. skull
d. ligament
e. pelvis
f. cartilage
g. sternum
h. rib

2. a. skull
   b. sternum
   c. ribs
   d. vertebrae
   e. pelvis
   f. femur
3. a. ribs c. sternum
   b. skull d. vertebrae
4. cartilage
5. ligaments
6. a. knee
   b. neck joint
   c. hip, shoulder

Review

7. a false idea about why something happens; an idea based on fear or ignorance rather than on fact
8. egg, larva, pupa, adult
9. from material erupting from within the earth;
   from the crumpling of the earth's crust
10. the distance that light travels in one year
11. gravity, inertia, friction, molecular attraction

Activities

1. a. The adult human skeleton has 206 bones.
   b. A baby has more bones than an adult. However, some of a baby's bones are simply cartilage. They have not yet become hard and mineralized, so they do not even show up on x-rays. By maturity, some of the bones of the skull and lower spine have fused together, forming single bones. Because of the way the skeleton develops, scientists examining a skeleton can accurately estimate how old the person was when he died.
2. Broken bones usually heal so well that the break is undetectable. Bones are living, growing organs. The skeleton is constantly being renewed, whether or not a bone is broken. Your bones today do not contain exactly the same molecules that they had last year.

2. Write the name of the bone or bones marked by each letter on the diagram at the right. Choose from the following list.
   ribs
   pelvis
   skull
   sternum
   vertebrae
   femur

   a. ———
   b. ———
   c. ———
   d. ———
   e. ———
   f. ———
3. Name the bone or bones that protect each of these body parts.
   a. lungs
   b. brain, eyes, and ears
   c. heart
   d. spinal cord
4. In our joints, the ——— keeps the hard bones from grinding together.
5. What holds the bones together at a joint?
6. For each kind of joint, choose the joint that is an example of it. One kind will have two answers.
   a. hinge joint
   b. pivotal joint
   c. ball-and-socket joint

   hip
   knee
   neck joint
   shoulder

Review

7. Define the word *superstition*.
8. Name the four life stages of a butterfly.
9. In what two main ways has God formed mountains since the Creation?
10. What is a light-year?
11. List the four kinds of resistance to motion.

Activities

1. a. Find out how many bones are in the adult human skeleton.
   b. Does a baby have more bones than an adult, or fewer? Explain.
2. Have you ever broken a bone? Find out whether breaking a bone weakens it permanently. Also find out how a broken bone heals.

Lesson 33

Muscles That Move the Bones

Vocabulary

biceps (bī′·seps′), the large muscle at the front of the upper arm, which causes the arm to bend at the elbow.

contract (kən·trakt′), to become shorter; to draw together.

muscle, a mass of fibers that is able to contract and cause a body part to move.

skeletal muscle, (skel′·i·təl), a muscle attached to a bone of the skeleton.

tendon (ten′·dən), a tough cord in the body, which connects a muscle to a bone.

triceps (trī′·seps′), the large muscle at the back of the upper arm, which causes the arm to unbend.

Contraction of Muscles

Muscles are important to our bodies because they enable us to move. Muscles are made of many fibers arranged in bundles to work together. These fibers have the special ability to ***contract*** by using energy from the food we eat. As a muscle contracts, it becomes shorter and thicker. You can feel this thickening when you "make a muscle" by bending your arm and tensing the muscle of your upper arm.

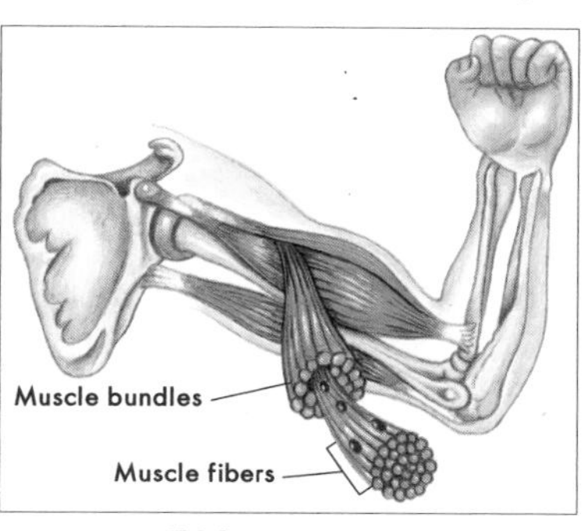

"Making a muscle"

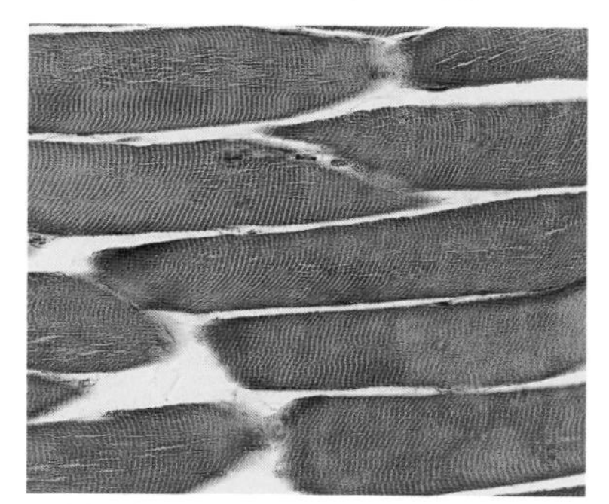
Muscle fibers enlarged by a microscope

Lesson 33

Lesson Concepts

1. Muscles contain special cells that use food energy to contract.
2. Muscles are connected to bones by tendons.
3. The muscles moving each part of an arm or a leg are in the part next closer to the body.
4. Muscles are arranged in pairs so that joints can bend and unbend.
5. The arrangement of the muscles is highly complex to permit a great variety of motion.

Teaching the Lesson

Muscles enable us to move. Bones have joints that allow us to move, but muscles produce the motion. Even though muscles seem common and ordinary, yet they are miracles of function. They are the link between food energy and action. Without muscles, joints would be completely useless.

This should be a practical lesson. Do you often see horses working in a field or pulling carts or wagons? Point them out to your students. Watching the muscles rippling beneath the hide of a horse is one of the best ways to see muscles in action.

When a muscle applies only a slight force, only some of its cells are contracting. But if a muscle applies great force, most of the cells are contracting and pulling together.

There are several kinds of muscles. The ***skeletal*** muscles are fastened to the bones of the skeleton. These muscles move body parts like arms and legs when we are active. Other muscles enable the heart, stomach, intestines, and other organs inside us to do their work. In this lesson we shall study the skeletal muscles.

Skeletal Muscles

Skeletal muscles move the bones of the skeleton. Tough cords called ***tendons*** attach a muscle to two bones with a joint between them. When the muscle contracts, it causes one bone to move.

The muscles that move a body part are usually located on the part next closer to the trunk of the body. For instance, the muscles that bend the fingers and wrist are located in the part of the arm just above the wrist. The muscles that bend the elbow to lift the forearm are in the part of the arm above the elbow. And the muscles that make the arm bend at the shoulder are on the chest, shoulder, and back.

Skeletal muscles are arranged in pairs. One of the muscles bends a joint, and the other one unbends the joint. A good example is the two muscles that you use to bend your arm at the elbow. These muscles are located in your upper arm, just above the elbow. The ***biceps*** is the large muscle on the front of your upper arm. It is used to bend your arm at the elbow. The ***triceps*** is the large muscle on the back of your upper arm. It is used to unbend your arm. The biceps and the triceps both move the same bone. You need both muscles to operate your arm properly.

Your body has many different muscles for many different motions. For example, you have at least thirty-five muscles that move your hand. Because of all these muscles, your hands are very nimble and useful.

Your eyes are surrounded by muscles. When you move your eyes, the muscles on one side relax just as the muscles on the other side tighten. This allows your eye to move very smoothly and steadily. With your eye muscles, you can make your eyes roll in many different directions. Also amazing is the way your two eyes move together so perfectly! You do not even need to concentrate to make them move together. You just turn your eyes to look at something, and they automatically move together.

Your face also has many muscles. Did you know that you use the muscles in your face when you smile or frown? Face muscles help to show how you feel. People use their face muscles to knit their brows in a frown or to

Discussion Questions

1. What would it be like if you had no muscles?
   You would be like a person who is completely paralyzed—unable to work, play, or do anything else. The only way you (or any part of you) could move would be if an outside force caused you to move.
2. Why do we need muscles that work in pairs?
   Muscles can only pull (contract); they cannot push. So we need paired muscles to bend and unbend our joints.

squint their eyes when they laugh. Muscles pull dimples into some people's cheeks when they smile. The muscles in your face help you to look like you.

Your arms, legs, back, and many other parts of your body are well supplied with muscles. They help you work and play. God gave you all these muscles because your bones need muscles to make them move. They cannot move by themselves. Once again we must say with the psalmist, "I will praise thee; for I am fearfully and wonderfully made: marvellous are thy works; and that my soul knoweth right well" (Psalm 139:14).

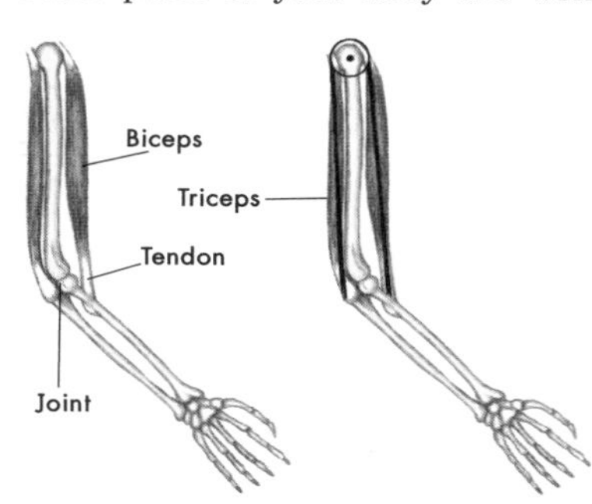

Muscles work like pulleys by pulling one side and relaxing the other.

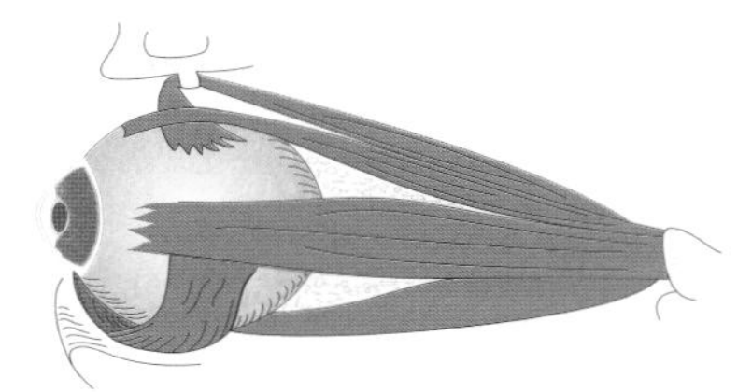
Eye muscles can turn your eyeballs in many different directions.

Exercises

1. What special ability do muscle fibers have?
2. Muscles that move bones in the skeleton are called ——— muscles.
3. What is the purpose of tendons?
4. Where are the muscles that move the wrist and fingers?
5. For what reason did God arrange muscles in pairs?
6. Name the muscle that raises your forearm.
7. Which muscle makes your forearm move down?
8. How do face muscles affect your appearance?

Review

9. Name the bones described below.
   a. the bone that protects the heart
   b. the bony shell that protects the brain
   c. the bones that protect the lungs
   d. the bones that protect the spinal cord

Lesson 33 Answers

Exercises

1. the ability to contract by using energy from food
2. skeletal
3. Tendons connect muscles to the bones they move.
4. They are located in the forearm, the body part next closer to the body.
5. so that one muscle bends a joint and the other unbends it
6. biceps
7. triceps
8. Face muscles help you smile or frown. They pull dimples into some people's faces. They help you to look like you.

Review

9. a. sternum c. ribs
   b. skull d. vertebrae

10. Cartilage pads a joint and keeps the bones apart.
    Ligaments hold the bones of the joint together.
11. hinge joints, pivotal joints, ball-and-socket joints

Activities

1. There are more than 600 muscles in the whole human body.

2–4. Help your students to observe muscles. Let them observe workhorses in action if there are any nearby. Point out the muscles that stand out to you. Roll up your sleeve, and show them the finger and wrist muscles in your forearm. Since the muscles in an adult are better developed, the students will probably be better able to observe them in your arm than in their own. Encourage them to go home and ask Father to wiggle his fingers. A sinewy, work-toughened hand and forearm would be a good study for today's lesson.

5. As this activity illustrates, many times the paired, opposing muscles differ in strength. This may be due both to design and exercise. There is no need to open our jaws with force, but crushing food requires great force. Since the closing requires much more effort, those muscles are exercised and strengthened much more than the opposing muscles.

   The biceps and triceps are much more evenly matched. In fact, the triceps are probably stronger than the biceps (contrary to what your students may think). For an objective test, you might compare pushups and chin-ups—though this will not be exactly fair, since the toes support some weight when doing pushups.

10. Explain how cartilage and ligaments differ in what they do.
11. Name three different kinds of joints.

Activities

1. There are at least 35 muscles that control the motion of the hand. Find out how many muscles are in the whole human body.
2. Observe your lower arm while you wiggle your fingers. Can you see muscles moving in your lower arm? Wiggle your fingers one at a time. Can you see how different muscles operate different fingers? Feel the back of your hand while wiggling your fingers. Do you feel the tendons that connect the bones of your fingers to muscles in your forearm?
3. You can observe the operation of your biceps and triceps as well. Lift a rock by bending your arm at the elbow. Can you see the muscle in your upper arm working as you lift? Now hold the rock beside your ear. Push the rock upward from that position, unbending your elbow. Can you feel the triceps tightening as you lift?
4. Observe a working horse. Notice the muscles that ripple under its skin as the horse pulls. The word *muscle* comes from the Latin word for "little mouse." Can you guess why muscles have that name?
5. Because skeletal muscles are arranged in pairs, we can exert force in two different directions. We can squeeze with our hands, and we can also open our hands. Of course, the paired muscles may not be equal in strength. Our hands can squeeze with great force, but they do not open with great force. Allow a classmate to squeeze your fist. Can you open it? Now hold his fist closed while he tries to open it. There is quite a difference, isn't there?

   It is not hard to stand on your toes, for your calf muscles are strong enough to support your weight easily. The opposing muscles raise the front part of your foot. Are they just as strong? Have a classmate about your size stand on your toes. Can you easily lift him?

   One set of muscles closes a dog's mouth. A paired set opens its mouth. Which set is stronger? An alligator can close its jaws with terrific force, yet a man can easily hold an alligator's jaws shut with his hands.

   Can you explain why in each case one set of muscles is stronger than the opposing set? Is this always true? Consider the biceps and the triceps, the paired muscles that move the lower arm. Is there a big difference in their strength?

Lesson 34

Muscles in Organs

Vocabulary

artery (är′·tə·rē), a tube in the body that carries blood away from the heart.

diaphragm (dī′·ə·fram′), the dome-shaped sheet of muscle beneath the lungs, which makes breathing possible.

digestive tract, the passage in the body where food is digested, beginning with the mouth and including the esophagus, stomach, and small and large intestines.

esophagus (i·sof′·ə·gəs), the tube leading from the back of the mouth to the stomach.

heart, the muscular organ in the chest that pumps blood throughout the body.

peristalsis (per′·i·stôl′·sis), the wavelike action of muscles in the esophagus, stomach, and intestines, which pushes food along in the digestive tract.

pyloric valve (pī·lôr′·ik), the valve at the lower end of the stomach through which food passes from the stomach into the small intestine.

The Heart

The ***heart*** is a fist-sized, muscular organ that pumps blood by contracting and relaxing time after time. The heart has four chambers—two receiving chambers and two pumping chambers. Blood fills the two receiving chambers of the heart when it relaxes. Then the receiving chambers contract and push the blood into the pumping chambers. A moment later, the powerful muscles of the pumping chambers contract and squeeze the blood into the ***arteries,*** which carry it to all parts of the body. This double contraction makes up one heartbeat. After each heartbeat, the heart muscles relax and the heart once more fills with blood.

Why does the blood not flow backward through the heart? The heart has four valves that allow the blood to flow in only one direction. These valves keep the blood from flowing back into the heart when the heart muscles relax.

Lesson 34

Lesson Concepts

1. The heart is a muscular organ that pumps blood by alternate contraction and relaxation.
2. The size of the arteries is regulated by the muscles in their walls.
3. The muscles of the digestive tract move food by the wave motion of peristalsis.
4. The stomach muscles churn the food during digestion.
5. Various muscular valves, such as the pyloric valve, control the flow of fluids.
6. The diaphragm is a sheet of muscle that enlarges the chest cavity to take in air.

Teaching the Lesson

The heart, stomach, and diaphragm are muscular organs that serve vital functions in our bodies. With the exception of the diaphragm, these muscles cannot be controlled consciously. God has created the autonomic nervous system to keep these organs functioning. They function well to meet the needs of the body without our conscious control.

While these organs cannot be touched with our hands, their presence is easy to detect. We have all taken our pulse, felt hunger, and observed the rise and fall of the chest in breathing. These are some of the evidences of the heart, stomach, and diaphragm that you can point out to your students.

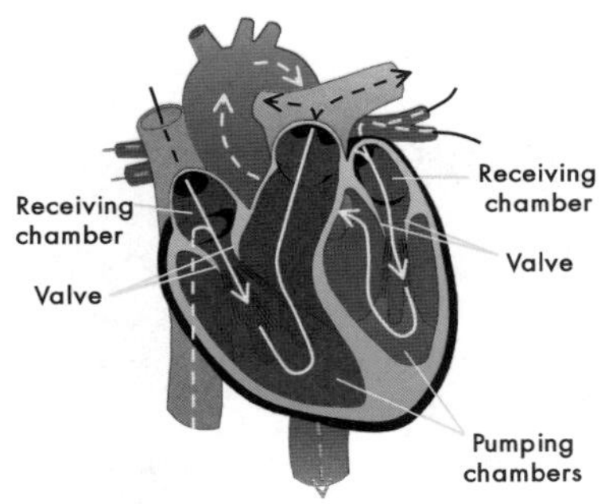

How the heart pumps blood

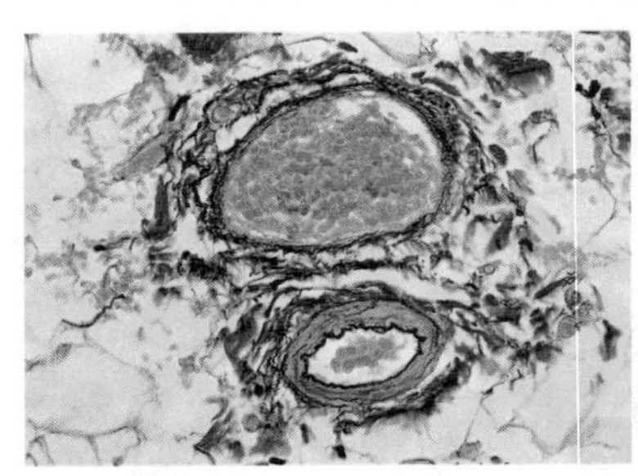
In this cross section of blood vessels, the top one is swollen with blood. The walls of the bottom one are constricted.

The heart is a very steady pump. It pumps automatically at a rate of 60 to 120 beats per minute. It does this day and night through your whole life. The only time it rests is during the brief pauses between beats. God designed the heart as a wonderful pump to keep our blood flowing through our bodies. In one day the heart pumps enough blood to fill twenty-five to thirty-five 55-gallon drums. It takes a powerful muscle to do that!

The arteries also help to pump blood. When the heart pumps blood into them, they swell to receive the new supply. As the heart relaxes to fill its chambers with more blood, the muscles in the walls of the arteries contract, helping to send the blood on its way. You can feel the swelling and contracting of the artery walls at your wrist when you take your pulse.

The heart pumps enough blood in one day to fill about thirty 55-gallon drums.

Discussion Questions

1. Why is the heart so important?
   (Sample answers.)
   The heart pumps blood through the lungs and throughout the whole body.
   In this way it sends food and oxygen to every part of the body.
   If your heart were to stop beating, you would die.
2. What is peristalsis? Can you describe it?
   Be sure the students understand what it is. It is fundamental to understanding the digestive tract. Use the illustration of squeezing a toothpaste tube to explain peristalsis. (See Activity 3 in the pupil's book.)

The muscles in the artery walls also control the size of the arteries. This controls the amount of blood that can flow through the artery.

The Digestive Tract

The ***digestive tract*** is the passage in the body where food is digested. The digestive tract begins with the mouth, and it includes the ***esophagus,*** stomach, small intestine, and large intestine.

Most of the digestive tract is made of muscular tubes that operate in a special way. When we swallow, the muscles of the esophagus contract with wavelike motions that push the chewed food down into the stomach. This wavelike motion of the muscles is called ***peristalsis.*** Because of peristalsis, the food you swallow moves into your stomach even if you are lying down with your head lower than your stomach.

The stomach and intestines also use peristalsis to move food through the digestive tract. Unlike swallowing, the peristalsis of the stomach and intestines is automatic. We cannot control it.

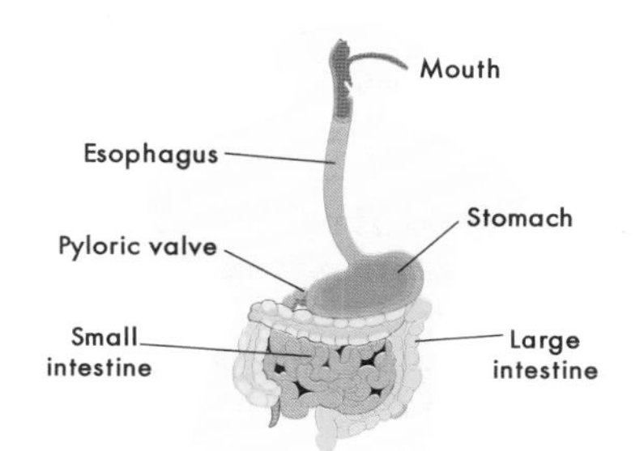

The digestive tract

As food is swallowed and enters the stomach, the stomach adds digestive fluids to it. Then the stomach muscles churn the food thoroughly so that it is well mixed with digestive fluids. At the lower end of the stomach is the ***pyloric valve,*** which keeps the food in the stomach until it is well mixed. Then the valve opens automatically, and the

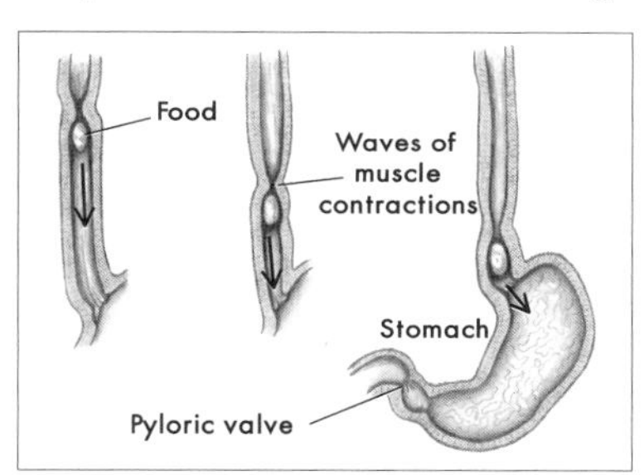

Peristalsis pushes food along, much as you would squeeze toothpaste out of its tube.

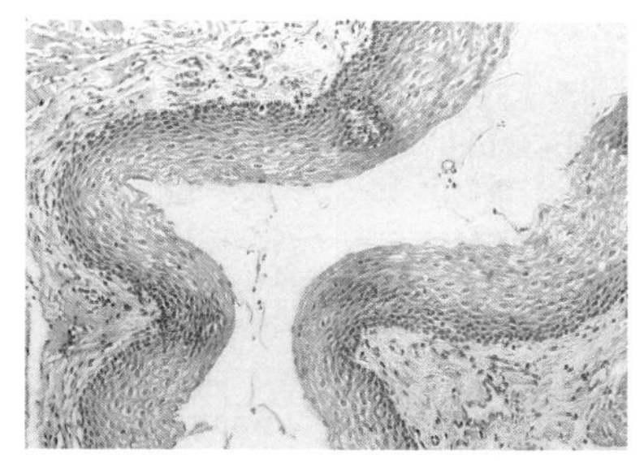
A cross section of the esophagus. Muscle tissue surrounds the wavelike lining.

muscles in the stomach squeeze the food into the small intestine where digestion continues.

The Diaphragm

The ***diaphragm*** is a dome-shaped sheet of muscle under the lungs and heart in the chest. This special muscle extends from the backbone to the sternum and lower ribs. When the diaphragm contracts, its dome shape flattens out, thus increasing the size of the chest. Other muscles move the ribs outward at the same time, which increases the size of the chest even more. This causes the lungs to stretch and draw in air. When the diaphragm relaxes, the chest becomes smaller and the stale air is pushed out of the lungs.

The motion of breathing is partly automatic. We breathe without even thinking about it. When we fall asleep, our breathing continues. Yet we can hold our breath or breathe faster if we want to.

God has created us with muscular organs that pump our blood, digest our food, and help us breathe. These special muscles are so important that without them we would soon die. Truly God has made us in a wonderful way.

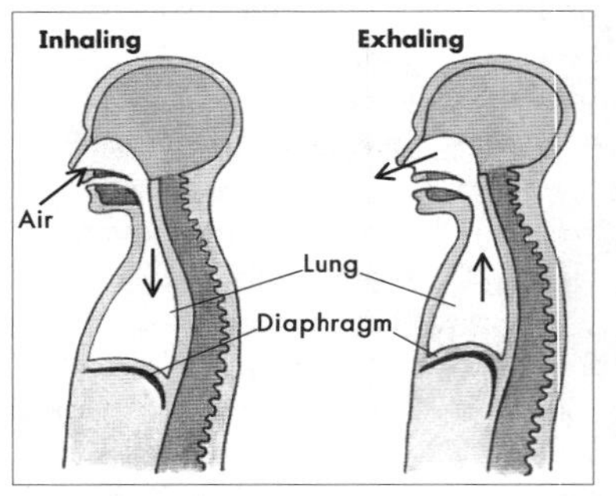

The diaphragm contracts for inhaling, but it relaxes for exhaling.

Exercises

1. a. The heart fills with blood when it is (contracting, relaxing).
   b. The heart pumps blood through the body when it is (contracting, relaxing).
2. What keeps the blood from flowing back into the heart again after each heartbeat?
3. When does the heart rest?
4. How do the arteries help to pump blood?
5. The ——— ——— is the passage in the body where food is digested.
6. How does peristalsis help us swallow?
7. How does peristalsis in the stomach and intestines differ from the peristalsis in the esophagus?
8. What is the main purpose of the muscles in the stomach?
9. Name the valve that keeps food in the stomach till it is properly mixed with digestive fluids.

Lesson 34 Answers

Exercises

1. a. relaxing b. contracting
2. valves
3. in the brief pauses between beats
4. They swell when the heart pumps blood into them. When the heart relaxes to fill with blood again, muscles in the arteries contract. This squeezes the blood on to the rest of the body.
5. digestive tract
6. Peristalsis is a wavelike motion of the muscles in the esophagus. This wavelike motion squeezes the food along the esophagus till it reaches the stomach.
7. Peristalsis in the stomach and intestines is automatic; we cannot control it.
   Peristalsis in the esophagus (swallowing) is not automatic; we must choose to swallow.
8. to churn and mix the food with digestive fluids
9. the pyloric valve

10. When the diaphragm flattens out,
    a. air is pushed out of the lungs.
    b. the chest becomes smaller.
    c. air is drawn into the lungs.
    d. blood is pumped to the lungs.
11. What happens when the diaphragm relaxes?

Review

12. Tell where each of these bones is found.
    a. skull c. sternum
    b. ribs d. vertebrae
13. Which bones protect the following body parts?
    a. brain, eyes, and ears c. heart
    b. lungs d. spinal cord
14. Since bones are stiff, what allows our bodies to move?
15. What is the purpose of tendons?
16. Why are many skeletal muscles arranged in pairs?

Activities

1. If the heart beats an average of 70 times per minute, figure out how often it would beat in one year. You need to find how many beats that is per hour and then how many beats per day. Multiply that number times 365 to find the number of beats in one year. How many heartbeats would that be in a lifetime of 70 years?
   Your heart pumps about 5 quarts of blood every minute. How many gallons is that in one day? In one year? In 70 years?
2. Sometimes just before mealtime, your stomach growls. The sounds come from beginning contractions in your stomach, which will soon begin digesting your food. Listen carefully sometime when this happens. What does it sound like? What does it feel like? Can you feel a squeezing motion in your stomach?
3. You can demonstrate peristalsis. Cut the bottom off an empty toothpaste tube. Put a lump of very soft Play-Doh in the end of the tube, and pinch the end shut. Now squeeze the Play-Doh forward. Continue squeezing until the Play-Doh pushes out through the neck of the tube. This is how the esophagus works. It is lined with muscles that squeeze the food down into the stomach.

10. c
11. The chest becomes smaller, and the stale air is pushed out of the lungs.

Review

12. a. head
    b. chest, sides, and back
    c. chest
    d. backbone
13. a. skull
    b. ribs
    c. sternum
    d. vertebrae
14. joints between bones
15. Tendons connect muscles to the bones they move.
16. One muscle bends a joint, and the other muscle unbends the joint.

Activities

1. Here are the calculations for this activity. This illustrates what a tireless pump the heart is—a marvel of God's creation.
   70 × 60 minutes × 24 hours × 365 days = 36,792,000 beats in 1 year
   70 × 36,792,000 = 2,575,440,000 beats in 70 years
   5 quarts = 1¼ gallons, × 60 minutes × 24 hours = 1,800 gallons per day
   365 × 1,800 gallons = 657,000 gallons per year, × 70 = 45,990,000 gallons in 70 years (With leap years, the amount exceeds 46 million gallons in 70 years.)

Extra Activity

You may wish to have one student lie on his back so the rest can observe the rise and fall of his chest as well as the sinking and swelling of his abdomen as the diaphragm pushes down into the abdominal region. Boys tend to breathe more with the stomach (diaphragm) than girls do, so it might be better to demonstrate with a boy.

Be sure to keep all activities modest and decent.

Lesson 35

The Nervous System

Vocabulary

brain, the nerve center in the head that controls the actions of the body.

coordinate (kō·ôr′·dən·āt′), to cause to work in a smooth, controlled manner.

involuntary (in·vol′·ən·tar′·ē), not controlled by conscious thought.

nerve, a stringlike bundle of fibers that carry messages between the body and the brain.

spinal cord, the thick cord of nerves attached to the brain and passing through the vertebrae.

voluntary (vol′·ən·tar′·ē), controlled by conscious thought.

The Brain

God has created you with a mind that can think and make choices. The ***brain*** is the place where this thinking takes place. Although we do not fully understand how the brain works, yet we know God created it in a wonderful way because it works very well.

Besides allowing you to think, your brain also controls your body. When you walk and talk and use your hands, your brain is telling your body muscles what to do. Your brain also controls breathing, heartbeat, and digestion.

Your brain is connected to all parts of your body by ***nerves.*** One kind of nerve carries messages from your body to your brain. Whether you feel a hot stove or smell peppermint tea or see a sunset, nerves carry those messages to your brain. Nerves carry an astounding number of messages all the time. Each eye has 100 million light-sensitive cells connected to more than 500,000 nerve fibers, all going to the brain. Nerve fibers from all five sense organs—such as your ears, tongue, and skin—continually carry

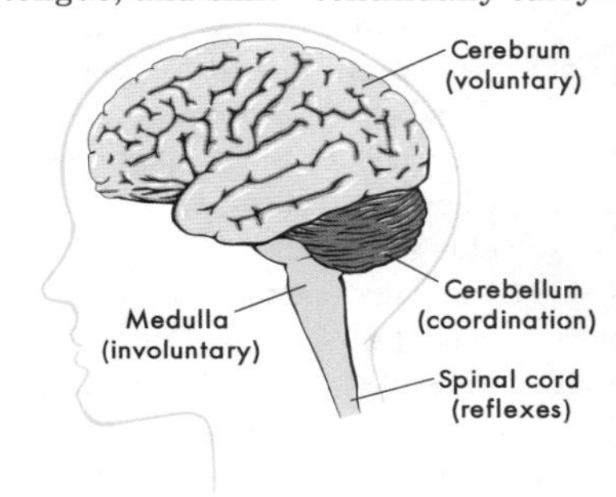

The brain

Lesson 35

Lesson Concepts

1. The brain is the main nerve center that allows you to think and make choices.
2. Messages are sent to and from the brain through nerves.
3. The skeletal muscles are voluntary, controlled by conscious thought.
4. The cerebellum coordinates the voluntary muscles to make smooth and even motion.
5. The action of organ muscles is involuntary, controlled by the brain.

Teaching the Lesson

The nervous system is a wonderful creation. Each of us possesses one of these delicate wonders. Its intricate design is unparalleled by human achievement. Many things about its chemical circuitry are not understood even by those at the cutting edge of medical research. Some questions about the mind, the brain, and human personality will never be answered by scientists. But some of the basics of the system have been discovered.

As before, do not let the lesson become too abstract. Make sure your students understand that they are studying a part of their own bodies.

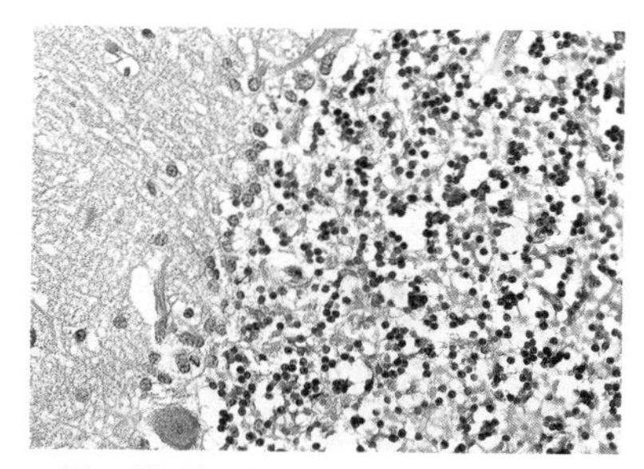

Magnified brain tissue. The brain has many nerve networks and many special nerve cells.

reports from your body to your brain.

Another kind of nerve carries messages from your brain to your muscles. These messages control the movements of your muscles. Different parts of your brain control the different parts of your body, but every part of your body is controlled by your brain.

Your ***spinal cord*** is a thick bundle of nerves that is attached to the base of your brain. It runs down through your neck and backbone and carries nerves that branch out all over your body. If the spinal cord is cut or badly damaged, the brain is no longer properly connected to the body. Usually the body becomes paralyzed and can move very little or not at all. That is why a broken neck or back is a very serious injury.

Voluntary Muscles

As you have seen, the brain controls all parts of the body. But different parts are controlled in different ways. Some parts are controlled by conscious thought, while others are controlled automatically. The skeletal muscles are controlled voluntarily. You move your arms or legs when you choose to move them. That is why those are called ***voluntary*** muscles.

The cerebellum (ser′ə·bel′əm) is a special part of the brain that ***coordinates*** voluntary motions, making them smooth and even. This smooth and accurate muscle control must be learned. A baby cannot control the motions of his arms, hands, and legs right away. It usually takes from ten to fourteen months before a baby can control his muscles well enough to walk. But after we learn good control of our muscles, walking becomes so easy that it is almost automatic. We do not need to do much thinking in order to walk, write, or pick up something. We just decide to do these things, and the cerebellum makes our motions smooth and coordinated.

Involuntary Muscles

We do not have voluntary control of some muscles in our bodies. Our heart, lungs, stomach, and other organs are automatically controlled by the brain. These body parts operate by ***involuntary*** muscles. We do not need to remember to breathe or to keep our hearts beating, because the brain controls these things without our thinking about it. Control of voluntary and

Discussion Questions

1. Just how important is the nervous system?

   Since it controls the whole body, the nervous system is very important. Without it, the body is unable to function. It is necessary for life.

2. What is meant by voluntary and involuntary muscles?

   When you volunteer to do something, you do it because you want to. That is like voluntary muscles. You control the muscles in your hands and arms by deciding to move them.

   By contrast, involuntary muscles operate without any thought on our part. One example is our hearts, whose work is completely automatic. We do not even need to think about it.

Voluntary control

Involuntary control

involuntary motions is done by different parts of the brain. But the brain controls the whole body because God designed it that way.

"I will praise thee; for I am fearfully and wonderfully made: marvellous are thy works; and that my soul knoweth right well" (Psalm 139:14).

Lesson 35 Answers

Exercises

1. enables us to think and make choices;
   controls the movements of the body
2. nerves
3. b
4. voluntary
5. the cerebellum
6. These muscles are automatically controlled by the brain. We do not make them work by conscious thought.

Review

7. a. vertebrae
   b. femur
   c. skull
   d. sternum

Exercises

1. What are the two main things that the brain does for us?
2. Bundles of fibers called ——— carry messages between the body and the brain.
3. The spinal cord is important because
   a. it is attached to the brain.
   b. it connects the brain with many parts of the body.
   c. it passes through the vertebrae.
   d. it is a voluntary muscle.
4. The muscles that move the legs and arms are (voluntary, involuntary) muscles.
5. Which part of the brain was designed to coordinate our movements?
6. The heart, stomach, and lungs operate by involuntary muscles. Why do we call them that?

Review

7. Give the names for the bone or bones described below.
   a. sections of the backbone
   b. leg bone above the knee
   c. shell of bone in the head
   d. breastbone

8. For each kind of joint, choose the joint that is an example of it. One kind will have two answers.
   a. hinge joint
   b. pivotal joint
   c. ball-and-socket joint

   hip — neck joint — knee — shoulder
9. Write *biceps* or *triceps* for each description.
   a. Bends the elbow and raises the forearm.
   b. Located on the back of the upper arm.
   c. Located on the front of the upper arm.
   d. Unbends the elbow and lowers the forearm.
10. How do the arteries help in pumping blood?
11. What is peristalsis?

Activities

1. a. Find out if the size of a person's brain has anything to do with how intelligent he is.
   b. The brain is divided down the center into two halves, the right hemisphere and the left hemisphere. One hemisphere is usually dominant, that is, it has more control than the other half. In a right-handed person, which hemisphere is dominant?
2. One interesting thing that God designed into our nervous systems is reflexes. If you accidentally brush a hot light bulb, you jerk away before you have time to think what to do. If a chip of wood comes flying toward your face, you blink your eyes immediately, without thinking. These immediate actions are called reflexes. Reflexes save you from many dangers. How do they work?

   Reflex actions are fast because they do not wait for the brain to decide what to do. Normally, a sensor sends a message to the brain, and the brain returns a message to the muscles telling them what to do. But in a reflex action, when the message reaches the spinal cord, a response message is immediately sent to the muscles. This way the muscles can respond fast enough to avoid the danger. Your hand does not get burned. Your eye is safe. We can thank God for reflexes.

   An interesting activity is to test your patella reflex. Sit on a chair, and cross your legs so that one leg is dangling. Take a ruler, and tap the area right below your kneecap. When you tap the correct spot, your leg automatically jerks forward. This is a reflex action.

8. a. knee
   b. neck joint
   c. hip, shoulder
9. a. biceps
   b. triceps
   c. biceps
   d. triceps
10. As the heart pumps blood into the arteries, they swell to receive it. As the heart relaxes before the next beat, the arteries contract and help to push the blood through the body.
11. the wavelike action of muscles in the digestive tract, which pushes food through the system

Activities

1. a. The size of a person's brain has nothing to do with how intelligent he is.
   b. The left hemisphere is dominant in a right-handed person.
2. Reflexes are briefly explored in this activity. While reflexes are not directly studied in the lesson, they do provide an additional glimpse into the wonderful working of the nervous system.

Lesson 36

Posture and Exercise

Vocabulary

fatigue (fə·tēg′), weariness from exercise; tiredness.

muscle tone, the slight tension or contraction of a resting muscle, which makes it firm.

God created us to stand upright. He designed our bodies so that they work best when they are in an upright position. Good posture is simply holding our bodies in the position that God intended.

Good posture is healthful. First of all, it keeps our organs from being cramped together. Our heart, lungs, and other organs need sufficient room to work properly. Good posture also keeps our bodies balanced so that there is less strain on our muscles. It makes us feel better and look better too. It is the position that God designed for our bodies.

Good Posture

Good posture needs to become a habit. Study the following diagrams and descriptions of good posture, and practice until what is healthful becomes a habit.

Good standing posture. Your shoulders should be square, your back straight, and your head upright and balanced. Raise your chest, and let your arms hang naturally. Hold your abdomen flat, and position your legs to throw your weight forward onto the front part of your feet. Stand evenly on both feet, and keep your legs straight.

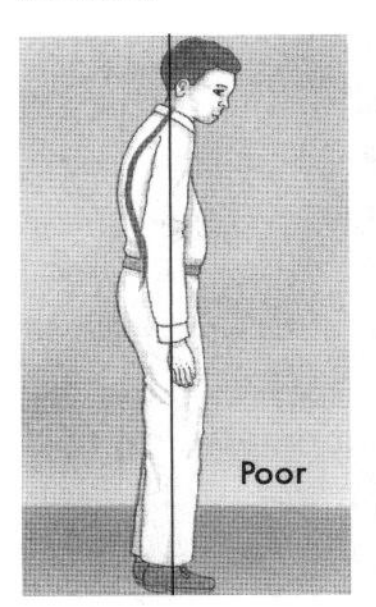

Standing posture

Sitting posture

Lesson 36

Lesson Concepts

1. God created man to stand upright, a position that is most healthful for the body.
2. Good posture gives the internal organs room to function.
3. Good posture is a result of developing good habits of sitting, standing, and lifting.
4. Feelings like guilt and worry are hard on the body and tend to cause poor posture.
5. Regular vigorous exercise helps to produce good muscle tone, a condition of slight muscle tension.
6. Proper rest is needed to avoid excessive fatigue.

Teaching the Lesson

Good posture is a habit. Help the students to gain an appreciation for good posture. If they cultivate this habit while they are young, it will be a blessing to them throughout life. Good posture is one stamp of a respectable person. Since we want all our students to be respectable people, let us teach them good posture.

This lesson is a very practical one. Be sure to demonstrate good posture. The students will learn better if they have a good example.

Do not teach good posture as an end in itself. Poor attitudes and poor posture often go hand in hand. Correcting the posture will not correct the attitude, but it makes the attitude easier to deal with. This lesson should help teach the students how to use their bodies to the glory of God.

Walking posture

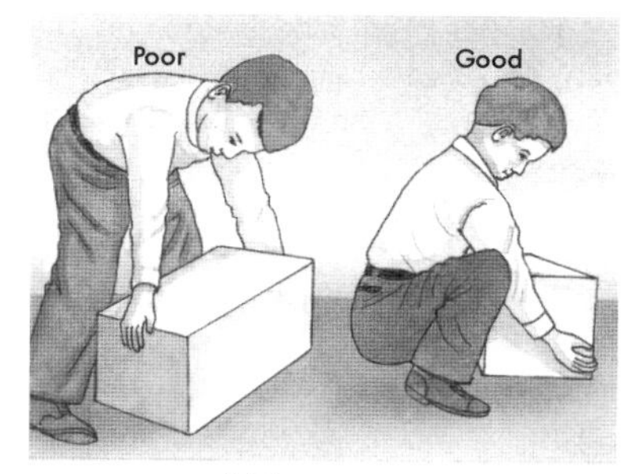

Lifting posture

Good sitting posture. You should sit upright on a firm seat. Sit back into the seat, and sit up straight. If you are reading, hold your book so that you can read without hunching your back or dropping your head too far forward. Check yourself—are you sitting properly right now?

Good walking posture. Hold your shoulders square when you walk. Swing each arm briskly in time with the leg on the opposite side. Keep your chin lifted so that your head is upright. Do not shuffle along, but lift your feet as you walk.

Good lifting posture. When you lift something heavy, bend your knees and lower yourself with your legs. Keep your back straight and upright. Hold the object close to yourself, and then lift with your legs.

Poor Posture

Poor posture often results from carelessness or laziness. But posture may also show how we feel. A cheerful person tends to square his shoulders and walk briskly. A person who is worried or guilty may sit in a slumped position. A rebellious person may express his attitude by slouching in his seat or shuffling as he walks.

Lack of sleep, insufficient food, and illness also tend to bring poor posture. The body may be too weak to hold itself in an upright position. Lack of exercise may weaken the muscles so that poor posture results.

Vigorous Exercise

In order to be healthy, the body needs plenty of exercise. God created us to work and designed our bodies so that they are most healthy when they are active. Exercise keeps our bones and muscles healthy and strong.

Vigorous exercise produces good ***muscle tone.*** Muscle tone is the slight tension of a resting muscle. If the biceps and triceps of your upper

Discussion Questions

1. What are the benefits of good posture?
   Good posture helps keep the body healthy.
   It is the most comfortable position once a person is accustomed to it.
   It gives internal organs the room they need to function well.
   It can prevent many backaches, headaches, and even stomachaches.
2. Discuss the importance of plentiful exercise and sufficient sleep.
   Vigorous exercise produces good muscle tone. It helps us feel our best, and it promotes better sleep at night.
   Sufficient sleep allows our tired bodies to restore themselves. Without it, we are more likely to become sick. We also cannot react as quickly and are more likely to have accidents.

arm have good tone, they contract slightly even when they are relaxed. Good muscle tone is a sign of healthy muscles.

Vigorous exercise also makes us feel good. We are usually happier when we work hard than when we have little to do. When we get sufficient exercise, we rest better at night. "The sleep of a labouring man is sweet" (Ecclesiastes 5:12).

Like most good things, however, exercise can be overdone. Too much exercise causes ***fatigue.*** Our muscles become exhausted, we cannot react as quickly, and we are more likely to have accidents. Now rest becomes the important thing. If we neglect to rest, fatigue will rob our bodies of good health. It is especially important that young children get enough sleep each night.

Our bodies need good posture, vigorous exercise, and plenty of rest to stay healthy. God wants us to take good care of our bodies.

Exercises

1. God created our bodies to be the most healthy when they are in an ——— position.
2. In what two ways is good posture healthful?
3. Write whether the person in each picture has *good* or *poor* posture.

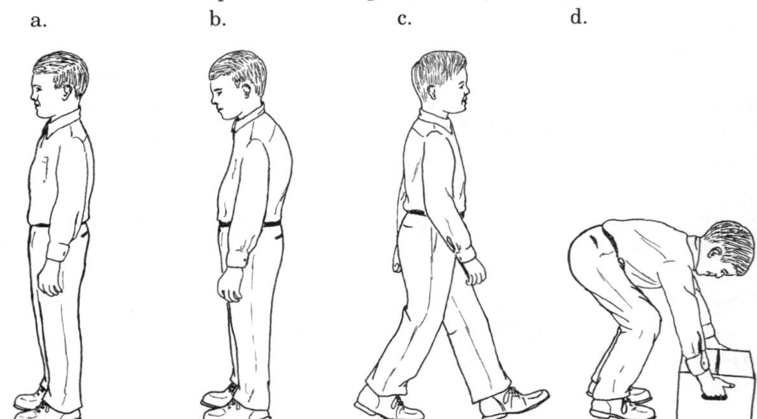

4. For each picture in number 3 that shows poor posture, explain what is wrong with the posture.

Lesson 36 Answers

Exercises

1. upright
2. It keeps our organs from being cramped together.
   It keeps our bodies balanced so that there is less strain on our muscles.
3. a. good
   b. poor
   c. good
   d. poor
4. b. The head is not upright. The shoulders are not squared. The back is not straight. The chest is not raised. The abdomen is not held flat.
   d. The knees are not bent sufficiently to lift the body and the object with the knees. The back is not straight and upright. The object is not held close to the body.

5. When you want to lift something heavy, you should
   a. keep your legs straight and bend over to pick it up.
   b. bend your knees, squat to grasp it, and then lift with your legs.
   c. bend at the waist but keep your back straight as you lift.
   d. bend your knees as you lift it with your back muscles.
6. For good posture, in what position should the following body parts normally be held?
   a. your head
   b. your back
   c. your shoulders
7. List four causes of poor posture.
8. Healthy, well-exercised muscles have a natural tension that is called ——— ———.
9. Why is it important to get sufficient rest?

Review

10. True or false? Bones are living organs that are filled with blood vessels and actually produce new blood cells.
11. True or false? Bones are hard because of the calcium that is stored in them.
12. Moving parts on machinery must be greased regularly, or they will wear out. What prevents joints from wearing out in seventy years of use?
13. What purpose do the ligaments serve?
14. What is the name of the large muscle that enlarges the chest cavity, drawing air into the lungs?
15. Suppose skeletal muscles were not arranged in pairs. What disadvantage would this design have?
16. a. Where do you find the muscles that move your fingers?
    b. Where do you find the muscles that move your lower arm?
    c. Where do you find the muscles that move your upper arm?
    d. In general, the muscles that move each body part are located where?

Activities

1. It is an amazing fact that most of us spend one-fourth to one-third of our lifetime unconscious of what is happening around us. This unconscious state of sleeping is important to our health.
   a. If a man spends 8 hours every day sleeping, how many years would he sleep in a 72-year lifetime?
   b. How can you tell if you are getting enough sleep?

5. b
6. a. upright
   b. straight
   c. back *or* square
7. (Any four.) carelessness; laziness; worry; guilt; rebellion; illness; or insufficient rest, food, or exercise
8. muscle tone
9. If we neglect to rest, fatigue will rob our bodies of good health.

Review

10. True
11. True
12. A layer of cartilage pads the joints. The ends of some bones are spongy, allowing them to absorb shock. Some joints have fluid-filled bursas that reduce friction.
13. Ligaments hold the bones together at the joints.
14. diaphragm
15. We would be greatly hindered in our ability to do things with our hands and feet.
16. a. in the lower arm
    b. in the upper arm
    c. on the shoulders, chest, and back
    d. on the body part next closer to the trunk of the body

Activities

1. a. Eight hours is one-third of a 24-hour day. One-third of 72 years is 24 years.
   b. Here is a sure way to tell if we are getting enough sleep: If we easily fall asleep during the day when we sit quietly (reading a book, studying lessons in school, or listening to a sermon), we are not getting enough sleep.

   According to one source, a baby needs 2 hours of sleep for every hour he is awake. An adult needs 30 minutes of sleep for every hour of wakefulness. Children's needs are somewhere between these two figures. The average adult needs 7 or 8 hours of sleep per day. Some do well on 5 or 6 hours; some may need 9 or 10. Fifth grade children should probably sleep 9 hours a day. Of course, all these figures are averages.

   How much sleep do *you* need? Here is a way to tell. For one week, go to bed at the same time, and sleep until you wake up naturally, without any alarm clock or other disturbance. In a week or so, you should be waking up at about the same time, fully rested.

2–4. Help the students develop the habit of good posture. Drill them on posture in class. Demonstrate the proper way to walk, sit, lift, and stand. Challenge them to always follow the rules of good posture.

2. Practice good posture in class. Can you stand properly? Do you know how to walk properly? Can you sit and lift correctly? Help each other remember to use good posture.
3. Practice good posture elsewhere too. Sit straight, and try to follow the rules of good posture. Be careful how you stand, walk, and lift. Tell your friends to remind you if they see you with poor posture. Good posture is an excellent habit to develop while you are young.
4. Use the following test of standing posture. Have a person stand as he normally does, with his shoulder next to a weighted string hanging from above his head to the floor. If you view him from the side, the line should pass through the center of his ear, the center of his shoulder, and in front of the middle of his knee. His lower back should not be curved too much. If the test reveals poor posture, have the person make adjustments so that his posture is correct.

Lesson 37

First Aid for Injuries

Vocabulary

antiseptic (an′·ti·sep′·tik), a medicine used to kill germs in an open wound.

bandage, a strip of cloth or other material used as a protection for an injury.

first aid, simple care given to an injured person before a doctor or nurse can help him.

fracture (frak′·chər), a crack or break in a bone.

splint, a piece of stiff material fastened to a broken arm or leg to keep it from bending.

Lesson 37

Lesson Concepts

1. Being safety-conscious will prevent many accidents and resulting injuries.
2. Knowledge of first-aid techniques can mean the difference between life and death.
3. Small cuts and scratches should be washed, treated with an antiseptic, and covered with a sterile bandage.
4. Heavy bleeding should be stopped by applying pressure directly to the wound.
5. Fractures should be immobilized with a splint to prevent the ends of the bone from doing more damage to the surrounding muscles.
6. In any case where the back or neck might be broken, no attempt should be made to move the victim lest damage be done to the spinal cord. (The only exception is the rare case in which a victim is in immediate danger of his life—such as when a vehicle catches fire after an accident. In such a case, one must simply do the best he can.)
7. In a case of serious injury, the average person should not go beyond simple first-aid techniques, but should obtain the services of a doctor or an ambulance immediately.

Preventing Accidents

It is important to be safety-conscious. Avoiding injuries is much better than taking care of them after they happen. Being safety-conscious means being careful at home, at school, at work, at play, and wherever you are.

Even common objects around the house may be quite dangerous. For instance, knives are made to cut, and they can be very dangerous if they are not used safely. Never run while holding a knife or other pointed object. When cutting a hard object such as a stick, cut away from your body so that you will be safe even if the knife slips. When handing a knife to another person, give it with the handle first.

Also use other tools with care. Drills, saws, axes, ladders, and many other tools can cause serious injuries or even death if they are used carelessly.

Play activities can be dangerous too if we become careless. Many children have been seriously injured while using bicycles, sleds, baseball bats, and ice skates. Such things can give us much enjoyable exercise, but they must always be used with caution. Using common sense and courtesy will help to avoid many accidents while playing.

Even though people are very careful, they do sometimes get hurt. So we must be prepared to give ***first aid*** when an accident happens. First aid is the simple care given to an injured person before a doctor or nurse can help him. It is not always possible for an ambulance to reach the person soon enough. First aid is very important—it can save lives.

First Aid for Open Wounds

An open wound is an injury in which the skin is broken. Both small scratches and deep cuts are wounds.

Small cuts and scratches. Most small cuts and scratches are not very serious, but even they must be cleaned and cared for properly. First wash them thoroughly with soap and water. All grit and grime should be cleaned out so that dangerous infection does not set in. Next, apply an ***antiseptic*** like iodine or Merthiolate to kill any germs that might be left in the wound. Finally, cover the wound with a clean ***bandage*** to keep out dirt and to soak up any blood that might ooze out. If the bandage gets wet or dirty, replace it with a fresh one.

Deep cuts. Deep cuts require different treatment. Because the person may bleed to death, the most important thing is to stop the bleeding promptly. To do this, place the palm of your hand directly on the cut and press firmly. If you have a piece of clean gauze or cloth available, place it over the cut first to help soak up the blood. Keep pressure on the wound until the bleeding stops or you can get medical help.

Teaching the Lesson

Stress being safety-conscious. That concept is not only popular in today's society, but it is also Scriptural. We are stewards of the bodies God has given us. Preventing injuries is a major part of caring for our bodies.

For a more complete background reference for teaching the lesson, visit your local Red Cross unit. Ask them for the latest material on first-aid techniques. They will likely have some pamphlets and posters that you can use effectively in today's lesson.

Playground accidents provide a means of reinforcing your teaching on first aid. As you care for scuffed hands and apply Band-Aids, identify the steps of first aid as you go. Teach first aid as a practical, useful tool in helping others.

Discussion Questions

Prepare a few scenarios, and ask your students what the proper response would be in each case. Here are two samples.

a. Henry falls on the gravel driveway and cuts his hands. It is a little muddy, and his hands are well plastered. What should you do?
   Wash his hands thoroughly (even if it hurts!). Apply an antiseptic, such as Merthiolate. Cover the cuts with a clean bandage.

b. Sarah is swinging high when she falls off and lands on her back. She has terrible pain in her neck and back. What should you do?
   Do not try to move her, since her back may be broken. Call an ambulance.

Point out that if an adult is available, he should handle emergency situations—especially one like the second case above.

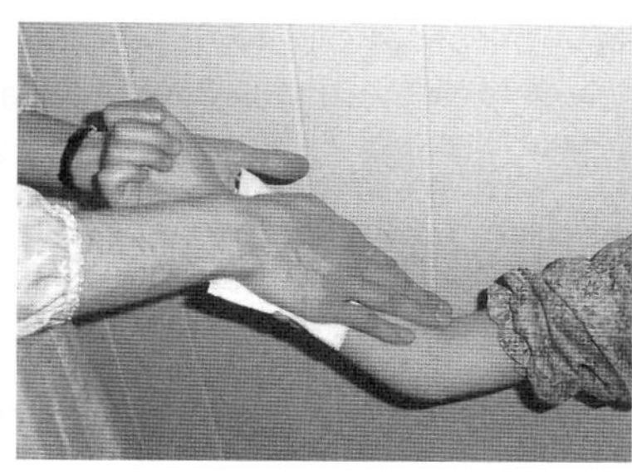

Apply pressure to a wound to stop bleeding.

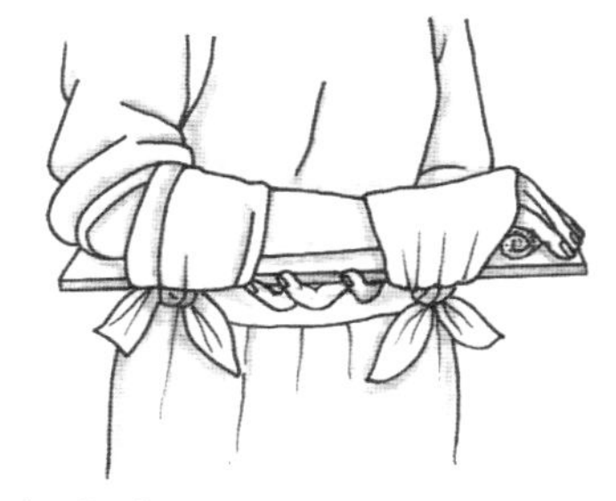

A splint keeps a broken limb in a fixed position. Include padding to improve comfort.

Do not try to wash out a deep cut or even apply an antiseptic to it. That could cause it to bleed more heavily again. Simply try to get the blood flow stopped until you can get help from a doctor or nurse.

If someone has a deep cut on his arm or leg, have him lie on his back and raise the cut limb above the level of his heart. Raising the injured limb will help to slow the blood flow to the cut. Lying on his back also helps to prevent him from fainting and falling because of losing too much blood. If he faints while he is standing, he could be hurt even worse.

First Aid for Fractures

Many people suffer broken bones in automobile accidents, bad falls, and other accidents. Some bone ***fractures*** are hairline cracks, and some are complete breaks. Fractures are usually very painful and must be treated carefully. The important thing to remember is that the ends of the broken bone must not be allowed to move. If they move, they may do further damage to the muscles around them.

Before a person with a fracture is moved, a ***splint*** should be fastened to the broken limb to prevent the ends of the bones from moving. The splint may be a piece of wood or some other stiff material. It must be long enough to keep the broken limb stiff and straight.

A broken bone should not be splinted if it has pushed through the skin. The person should not even be moved more than necessary. Make the person as comfortable as possible, and call an ambulance. Such a fracture should be handled only by specially trained people.

If you think a person's back or neck might be broken, do not move him at all. Moving him could damage his spinal cord, which might paralyze or even kill the person. Specially trained workers know the best way to move someone with a broken back. Always call for an ambulance rather than trying to move the person yourself.

With serious wounds and fractures, never do more than give first aid—the simple care needed right away for someone who is injured. If he is hurt badly, he needs specially trained workers and doctors to give him proper care. Call an ambulance; and when it arrives, stand back and let the trained workers take over. Be ready to answer any questions they might ask.

Exercises

1. Which is better—providing excellent first aid for someone who is injured, or being safety-conscious so that the injury is avoided?
2. What two things help to make play activities safe?
3. Why is first aid so important?
4. What are the three steps in caring for a small cut or scratch?
5. How should you stop the bleeding from a deep cut?
6. Why should a deep cut not be washed out or have an antiseptic applied to it?
7. What is the purpose of a splint?
   a. to keep a deep cut from bleeding so heavily
   b. to keep a person with a deep cut from fainting
   c. to keep a fractured bone from damaging the surrounding muscles
   d. to protect a fractured bone from dirt and infection
8. If you think a person's neck or back is broken, you should
   a. carefully put a splint on the injury so that you do not damage the spinal cord.
   b. not try to move him; just quickly call an ambulance.
   c. be sure to keep his head in line with his body as you move him.
   d. ask someone to help you move him.
9. When someone suffers a severe injury,
   a. call for medical help and stand by until it comes.
   b. give proper first aid and call for medical help.
   c. give proper first aid to avoid calling an ambulance unnecessarily.
   d. call an ambulance and tell the workers what to do for the injured person.

Review

10. What two main things do bones do for us?
11. What does the body use muscles for?

Lesson 37 Answers

Exercises

1. being safety-conscious so that the injury is avoided
2. common sense and courtesy
3. Proper first aid can save someone's life.
4. (1) Wash the wound thoroughly with soap and water.
   (2) Apply an antiseptic to kill any germs that might be left in it.
   (3) Cover the wound with a clean bandage.
5. Place the palm of your hand directly on the cut, and press firmly.
6. That could cause the cut to bleed more heavily.
7. c
8. b
9. b

Review

10. They support the body and protect the body.
11. to move the bones

12. Write *biceps* or *triceps* for each description.
    a. Located on the back of the upper arm.
    b. Located on the front of the upper arm.
    c. Bends the elbow and raises the forearm.
    d. Unbends the elbow and lowers the forearm.
13. What is the muscular organ that pumps blood throughout your body?
14. Explain what peristalsis is.
15. What is the purpose of the diaphragm?
16. The main nerve center in the body is the ———.

Activities

1. Direct pressure may not always control bleeding from a deep cut. In such a case, it is helpful to press on the cut as well as the supplying blood vessels. The diagram shows four pressure points, where you can press to stop bleeding in the arms and legs. See if you can find the pressure points on your arms.

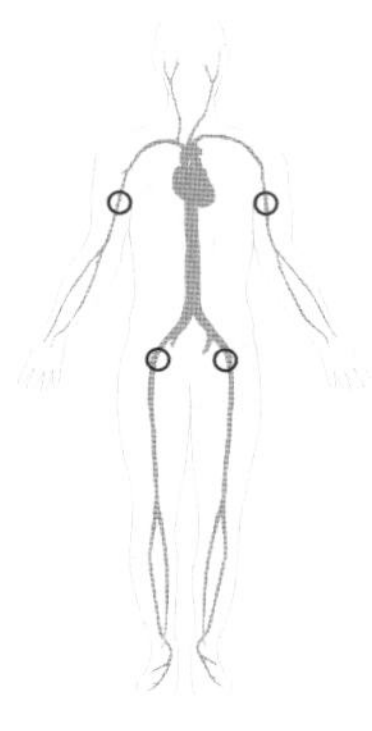

2. First aid for choking is one of the most important methods to know. If someone is eating and suddenly cannot breathe or talk, ask him if he is choking. (He might nod or give the choking sign by putting his hand to his throat.) If he is choking, check for food or other material stuck in his throat. You may be able to remove it with your fingers.

   If the person still cannot breathe, use the abdominal thrust. Stand behind the person with your arms around his waist. Place your fist against the person's stomach just below the ribs. Hold your fist with your other hand, and give three hard thrusts upward and inward. Your hands and arms should be just at the bottom of the ribs and slip up under them as you pull back. If the maneuver is successful, it will dislodge whatever is in his throat, and he will be able to breathe again.

   Practice the abdominal thrust as a class activity.

12. a. triceps
    b. biceps
    c. biceps
    d. triceps
13. heart
14. the wavelike action of muscles that push food along in the digestive tract
15. The diaphragm contracts to draw air into the lungs.
16. brain

Activities

Be sure to make this lesson practical with some hands-on activities. A book such as the Red Cross First-aid Manual should be read by every teacher (as well as parents). It provides many valuable diagrams on first-aid techniques, pressure points, and other things.

1. To stop bleeding by using the pressure points in the arms, pinch the artery against the bone where you feel the pulse. For the legs, have the victim lie flat on his back and use the heel of your hand to pinch the artery against the pelvis. When pressure points are used, the blood flow should be cut off only long enough to stop the bleeding.

   For modesty's sake, avoid practicing the pressure points for the legs.
2. Any first-aid book will explain the abdominal thrust and should give directions for administering it if the patient is lying down. The method works because there is always some air in the lungs, and this air is forced out to dislodge the foreign object.

Lesson 38

Unit 6 Review

Review of Vocabulary

antiseptic	digestive tract	ligaments	skull
arteries	esophagus	muscle	spinal cord
bandage	fatigue	muscle tone	splint
biceps	femur	nerves	sternum
brain	first aid	pelvis	tendons
cartilage	fracture	peristalsis	triceps
contract	heart	pyloric valve	vertebrae
coordinate	involuntary	ribs	voluntary
diaphragm	joints	skeletal muscles	

God created bones to support our bodies. The bones in our arms and legs make them stiff and strong. The __1__ supports the main part of the body and connects the legs to the backbone. This bony structure is supported by the __2__, or thigh-bone, in the upper part of each leg.

We have __3__ where our bones meet. They allow our bones to move so that we can work and play. Where two bones join, there is a layer of __4__ between the bones to pad them. Tough bands called __5__ hold the joints together.

Bones also protect various parts of our bodies. The __6__ protects the brain, eyes, and ears. __7__ surround the chest and help to protect the lungs. The __8__ is the flat, vertical bone in the center of the chest that protects the heart. The spinal cord is protected by a column of __9__ in the back and neck.

A __10__ is a special mass of fibers that causes a certain body part to move. It uses energy from food to __11__, thereby moving a bone or causing an organ to do its work.

The __12__ are muscles connected to bones by tough cords called __13__. These muscles are arranged in pairs, as illustrated by the two muscles in your upper arm. The __14__ bends your elbow and raises your forearm. The __15__ unbends your elbow and lowers your forearm.

Some organs also have muscles. The __16__ is a muscular organ that pumps blood to the rest of the body. Tubes called __17__ help to move the blood through the body. The __18__ is a sheet of muscle that makes breathing possible.

Lesson 38 Answers

Review of Vocabulary

1. pelvis
2. femur
3. joints
4. cartilage
5. ligaments
6. skull
7. Ribs
8. sternum
9. vertebrae
10. muscle
11. contract
12. skeletal muscles
13. tendons
14. biceps
15. triceps
16. heart
17. arteries
18. diaphragm

Lesson 38

This unit is a basic study of the skeletal, muscular, and nervous systems, several major organs, and proper care for the body. We study the body so that we can better care for it. An awareness of how the body functions will aid us in being good stewards of God's wonderful gift to us.

Clinch the unit by helping the students review this material. Try to rivet main concepts and key vocabulary terms. See that they have a working understanding of good posture and first aid.

The __19__ is made of a number of organs, including the stomach and the intestines. One part is the __20__, a muscular tube that moves food from the mouth to the stomach by a wavelike motion called __21__. Muscles in the stomach contract to mix the food with digestive juices. When the food is well mixed, it passes through the __22__ at the lower end of the stomach and goes into the small intestine.

The nervous system controls the body. The __23__ is the main center of control. Messages to and from this center are carried by stringlike bundles of fibers called __24__. A thick bundle of these fibers is attached to the brain and extends through the neck and backbone. This bundle is called the __25__.

Your skeletal muscles are called __26__ muscles because you control them consciously. The cerebellum of the brain helps to __27__ the movements of these muscles, or make them smooth and even. You also have __28__ muscles, such as those in the heart and stomach, which you do not control consciously. The brain controls these organs automatically.

We need good posture and vigorous exercise to stay healthy. Good posture keeps our organs from being cramped together. Vigorous exercise produces strong muscles. Healthy muscles have good __29__, a condition in which the muscles contract slightly even if they are relaxed. Too much exercise causes __30__ in our muscles. We need to rest so that they can work properly again.

__31__ is the care given to an injured person before a doctor or nurse can help him. Small cuts and scratches should be washed thoroughly with soap and water. Next, you should apply an __32__ to kill any remaining germs. Finally, you should cover the wound with a clean __33__ to keep out dirt.

A __34__, or broken bone, is very painful and must be treated carefully. If an arm or a leg is broken, a __35__ should be fastened to the limb to keep the broken ends of the bone from doing more damage to the surrounding muscles.

19. digestive tract
20. esophagus
21. peristalsis
22. pyloric valve
23. brain
24. nerves
25. spinal cord
26. voluntary
27. coordinate
28. involuntary
29. muscle tone
30. fatigue
31. First aid
32. antiseptic
33. bandage
34. fracture
35. splint

Multiple Choice

1. Which of the following things do bones provide for the body?
   a. movement and protection
   b. movement and support
   c. support and protection
   d. support and exercise
2. The ligaments in our joints
   a. connect the bones to the muscles.
   b. pad the ends of the bones.
   c. lubricate the muscles.
   d. hold the bones of the joints together.
3. The cartilage in our joints
   a. makes blood for the body.
   b. pads the ends of the bones.
   c. holds the bones of the joints together.
   d. connects the bones to the muscles.

Multiple Choice

1. c
2. d
3. b

4. a
5. d
6. b
7. b
8. a
9. c
10. d
11. a
12. c

4. Muscles contain special cells that
   a. use energy from food to contract.
   b. send messages to the brain.
   c. carry messages to the body.
   d. store minerals and make blood.
5. Muscles are arranged in pairs
   a. because they are stronger that way.
   b. so that there is a muscle on each side of the body.
   c. to make our movements smoother.
   d. because one muscle bends a joint and the other unbends it.
6. The muscles that move each part of a limb are located
   a. in the next part away from the body.
   b. in the next part closer to the body.
   c. in that part itself.
   d. in the main part of the body.
7. Muscles are connected to the bones by
   a. ligaments.
   b. tendons.
   c. cartilage.
   d. nerves.
8. Blood is pumped through the body by
   a. the heart and arteries.
   b. the heart and nerves.
   c. the heart and esophagus.
   d. the heart and lungs.
9. Peristalsis can best be described as
   a. the contractions and relaxations of the heart.
   b. the motion that carries messages to the brain.
   c. the wavelike motion of muscles in the digestive tract.
   d. the digesting of food in the stomach.
10. The diaphragm
    a. controls how much food is let out of the stomach.
    b. helps the heart pump blood.
    c. churns and mixes food in the stomach.
    d. enlarges the chest cavity to draw air into the lungs.
11. Messages are sent to and from the brain through
    a. nerves.
    b. tendons.
    c. muscles.
    d. ligaments.
12. The brain has voluntary control over
    a. the heart.
    b. the stomach.
    c. the skeletal muscles.
    d. the pyloric valve.

13. Which of the following things goes with poor standing posture?
    a. Shoulders are back.
    b. Abdomen is held flat.
    c. Head is held upright and balanced.
    d. Most of your weight is on one foot.

13. d

14. When you are lifting a heavy object, you should
    a. bend your knees, hold the object close to yourself, and lift with your legs.
    b. bend at your waist, grasp the object, and stand up straight.
    c. keep your back straight as you bend over to pick up the object.
    d. lift it with your back and arms.

14. a

15. Things that cause poor posture include
    a. carelessness, worry, and a healthy body.
    b. lack of sleep, lack of food, and cheerfulness.
    c. guilt, thankfulness, and tension.
    d. lack of sleep, carelessness, and illness.

15. d

16. If we get plenty of vigorous exercise,
    a. we will not get sick.
    b. we will not need as much rest.
    c. our bones and muscles will grow stronger.
    d. our bodies will wear out more quickly.

16. c

17. Which of the following things is most important?
    a. preventing accidents
    b. buying a good first-aid kit
    c. memorizing the doctor's telephone number
    d. taking a course in first aid

17. a

18. The proper way to care for small cuts and scratches is to
    a. wash them, bandage them, and then apply an antiseptic.
    b. apply pressure to stop the bleeding and then bandage them.
    c. wash them and then bandage them.
    d. wash them, apply an antiseptic, and then bandage them.

18. d

19. In which case would it be safe to splint the broken bone?
    a. John broke his arm when he lost control of his bicycle. A piece of broken bone pushed out through the skin.
    b. Marlin broke his leg when he fell through a hay hole. He also has pain in one arm.
    c. Susan fell down the stairs and broke her leg. She also has severe pain in her back.
    d. Elsie slipped and fell on the ice on the sidewalk. She broke her arm, and her neck hurts terribly.

19. b

Unit 7

Title Page Photo

One of the wonders of sound is the ability that God has given us to sing His praises. God wants us to glorify Him with our voices.

Introduction to Unit 7

Sound. We rely on it. We set a timer. *"Ding!"* and it reminds us. We leave our keys in the car ignition as we climb out. *"Buzz!"*—another reminder. We want to communicate thoughts to others. No problem; we speak. We are happy, we want to praise God, so we sing. We go out for a nature hike. Sounds are all around us.

What makes sound? Where does it come from? What is the nature of sound? How do we hear? This unit addresses these questions. Let's learn about sound.

Unit 7

The Wonder of Sound

"Make a joyful noise unto the LORD, all ye lands. Serve the LORD with gladness: come before his presence with singing" (Psalm 100:1, 2). The Bible commands us to make sounds. It even tells us what kind of sounds to make. We are to praise the Lord in song. God wants us to use our voices to worship Him.

God has created us with voices that can make many different sounds. We use these sounds to sing and talk. Other people know what we are saying because God has given us ears so that we can hear the sounds around us.

What is sound? How is it produced? How can sounds move from one place to another? You will find answers to these questions as you study this unit.

Story for Unit 7

The Big Silence That We Could Feel

We were repairing an old, tumbledown schoolhouse at the edge of a large tract of empty land. While trees and brambles clogged the area right around the schoolhouse, the surrounding country was a mass of grassy hills. The area had been strip-mined for coal some years past, and almost nothing was left to show that people had once lived there. This empty land, however, held numerous treasures. Many kinds of wildlife scampered and scurried through the tall grass.

But the most impressive treasure of that wonderful place was its great vacancy—its powerful silence, empty of the noises of men and modern machinery. There were only a few trees to rustle and sigh in the wind. Acres and acres of flowing grass covered the hills.

We sat on top of a hill to eat our lunch one day. The silence of the grassland was so mighty that it seemed to squeeze in upon us. We decided to yell and listen to our echoes. We yelled at the top of our voices and waited. Nothing. We yelled again. Nothing. It seemed like the wide-open space completely absorbed our voices. We whistled. We shouted. We clapped boards together. More silence. Amazing! Here was a vast emptiness that we could almost feel. What we needed was thunder. Surely God's noises would echo and fill these hills with sound!

One day a storm did come. Torrents of rain fell. The wind blew so hard that we feared the rickety old schoolhouse would come crashing down upon us. Then, sure enough, a startling stab of lightning struck close by. A powerful *BOOM!* rattled the building. Echoes? There were a few, but only faint ones. It was the strangest thunder I had ever heard. Streak after streak, boom after boom—they were all the same. Just one tremendous *CRASH!* and it was all over. The familiar, drawn-out rumble was gone, absorbed by vast stretches of flowing grass.

The storm passed. The sky cleared. Out came the tiny sounds of crickets and birds and rustling grass. More silence. Silence with all those little noises? Yes, silence. God's wonderful hills held a silent bigness that we could feel.

I took some photographs of those impressive hills to show my friends. "What's so special about those grassy hills?" they wondered. If they had been there to hear the silence, they would not have asked.

Lesson 39

What Causes Sound

Vocabulary

larynx (lar′·ingks), the voice box, containing the vocal cords that vibrate to produce the voice.

resonator (rez′·ə·nā′·tər), a hollow device that strengthens a sound and gives it a certain quality.

vibration (vī·brā′·shən), a fast back-and-forth motion.

"They shall lift up their voice, they shall sing for the majesty of the LORD, they shall cry aloud from the sea" (Isaiah 24:14). God wants us to make sounds that are joyful and that praise Him. He has created us so that we can make and hear sounds. Sound is a wonderful gift from God.

What Is Sound?

Energy is required to produce sound. We use the effort of our muscles when we speak. A tape player needs electricity to produce sound from its speaker. When a book falls, some of its energy of motion changes to sound as it lands on the floor.

Sound itself is energy because it has force and can do work. Sound can move things. Perhaps you have felt the bench at church vibrate while everybody was singing. Sound from the low bass voices caused the bench to vibrate. Or perhaps you remember a strong thunderstorm. When thunder rumbles, it is often strong enough to make the windows rattle. "The voice of thy thunder was in the heaven: the lightnings lightened the world: the earth trembled and shook" (Psalm 77:18). Sound is energy because it accomplishes work by moving things.

Sound is ***vibration.*** Vibration is a fast back-and-forth motion. When air vibrates, we hear it as sound. All sound, whether it comes through air, water, or any other material, is a vibration.

Sound is vibration.

Lesson 39

Lesson Concepts

1. Energy is required to produce sound, and sound itself is a form of energy.
2. Vibration is a fast back-and-forth motion.
3. Sound is produced by vibrating objects.
4. God designed the larynx to produce the voice for singing and communication.
5. Resonators help to increase sound and give it certain qualities.
6. Sound can be decreased by absorbing it or by reducing vibrations at their source.

Teaching the Lesson

The key concept of today's lesson is that sound is vibration. Prepare to demonstrate this concept in various ways. Let the pupils see or feel for themselves that a tuning fork, a twanging rubber band, or a bell vibrates to produce sound. Demonstrate how to feel the vibrations from the larynx while talking. Show them how to hum with closed lips and feel their lips vibrate as the sound resonates in their mouths. We are surrounded with sounds. Help your students appreciate the marvels of sound.

How Is Sound Produced?

Sound is produced when something vibrates. When a buzzer vibrates, it causes the air around it to vibrate. These vibrations travel through the air, and we hear them as sounds. Vibrating objects cause sound.

Did you ever touch the speaker of a tape player while it was playing? If so, you probably felt a vibration. Electricity made the speaker vibrate so it produced sound. Tuning forks, bells, loose guards on machinery, and many other things make sound because they vibrate.

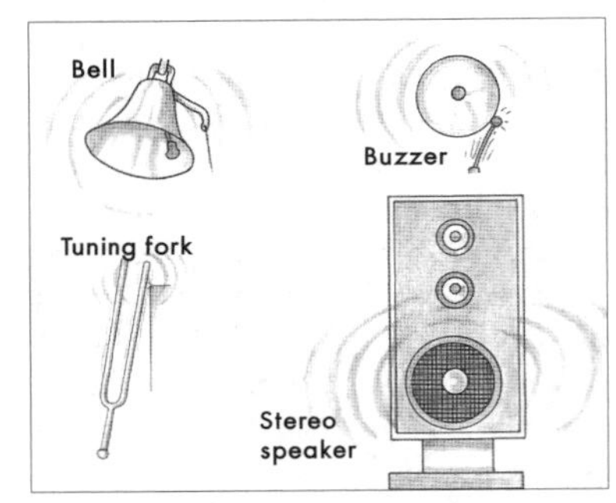

Vibration produces sound.

Your voice is produced by vibrations from your ***larynx,*** sometimes called the voice box. Your larynx is in your throat, at the upper end of your windpipe. All your breath passes through your larynx, which contains two vocal cords. When you talk, you draw the vocal cords tight and force air from your lungs between them. The air causes the vocal cords to vibrate and make sound. If you put your hand on your throat while talking, you can feel the vibrations from your larynx.

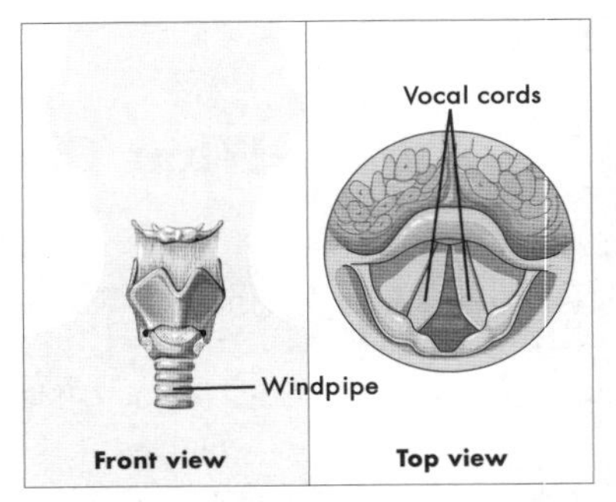

The larynx

Your voice is not produced by the vocal cords alone. Your nose and mouth are ***resonators*** that give richness and quality to your voice. Grandfather clocks have specially made cases that act as resonators to make the sound of the chimes louder and more musical. The tube of a whistle is also a kind of

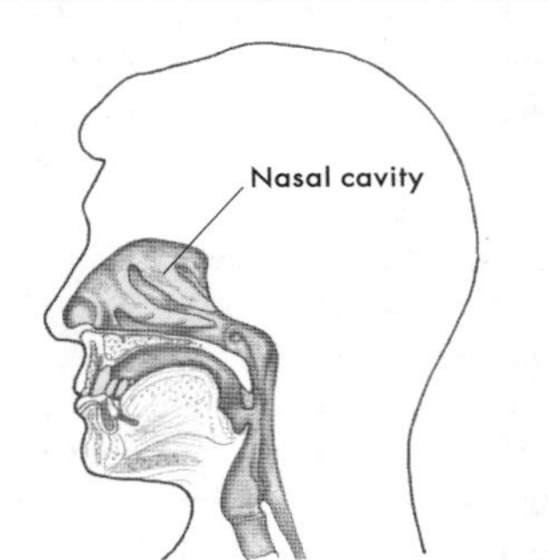

The nasal cavity serves as a resonator.

Discussion Questions

1. What is sound?
   Stimulate the students' thinking. They all know what sound is, but can they define it? An old riddle asks, "If a tree falls in the forest with no ears to hear it, does it produce sound?" The answer depends on one's definition of sound. Scientifically speaking, sound applies to vibrations that could be heard if an ear were nearby, and also to vibrations above and below the normal range of human hearing.
2. What makes us able to talk? Where does the voice come from?
   Discuss the workings of the larynx (voice box). Have the pupils feel their throats for vibration when they talk.
3. Why did God create us with voices?
   so that we would be able to praise Him and communicate with each other

Early churches did not have microphones and loudspeakers to amplify the preacher's voice. Instead, a sounding board was sometimes built above the pulpit to make his voice louder by resonating.

resonator. It gives the whistle a loud, clear sound. A tuning fork held against a wooden door will cause the door to resonate. An entire class can easily hear the tuning fork vibrate when a resonating door makes the sound louder.

How Can Sounds Be Reduced?

Often we are more interested in reducing sound than in increasing it. This may be accomplished in two ways. First, we may use soft materials to absorb sound. Curtains, carpets, and padded furniture make our houses quieter. (Have you ever noticed the strange hollow sounds when you walk into a room that is completely empty?) Insulation in the walls of a house helps to keep out wind noises and other sounds from outside. All these soft materials help to reduce sound by absorbing it.

Second, we may reduce vibrations to decrease sound. If a loose guard on a piece of machinery is vibrating, it makes a noise. By fastening the guard well and stopping the vibration, we can reduce the noise. The engines in most cars are held in place with rubber motor mounts. The rubber is soft and absorbs the vibrations from the engine. This reduces engine noise

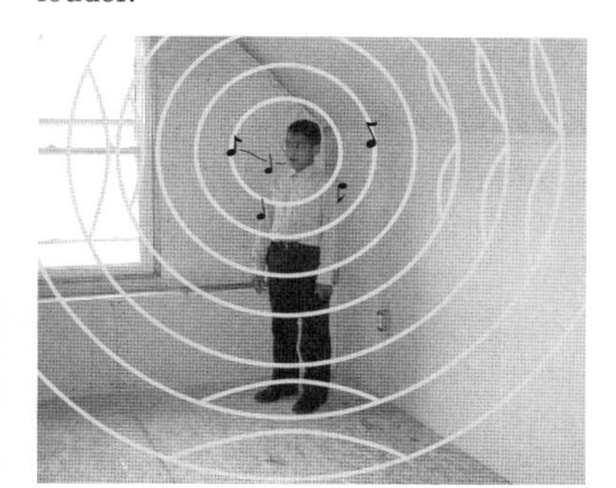

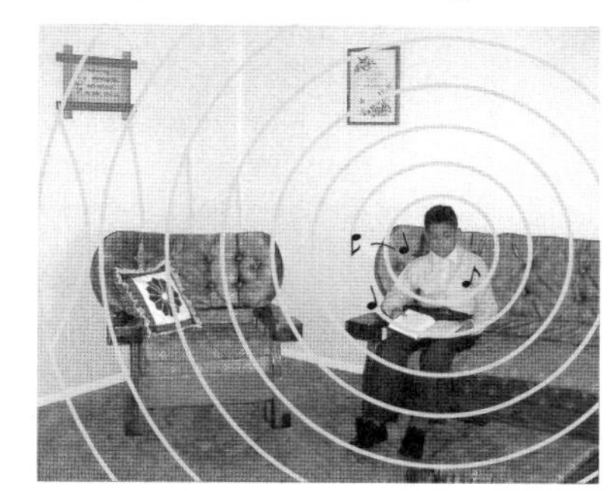

In which room are more sound waves being absorbed?

Lesson 39 Answers

Exercises

1. (Sample answers.)
   tape player—needs electricity;
   falling book—energy of motion;
   speech—uses muscle energy
2. energy
3. (Sample answers.)
   thunder rattling windows;
   bass voices vibrating church bench;
   noise from an explosion shaking a building
4. vibration
5. The larynx makes sound when we tighten the vocal cords and force air to pass between them. This makes the vocal cords vibrate and produce the sound of our voices.
6. resonators
7. by using soft materials to absorb the sound;
   by tightening a vibrating object or using rubber mounts

Review

8. support and protection
9. brain
10. a. 2 main body parts and 4 pairs of legs
    b. 3 main body parts and 3 pairs of legs
11. from material erupting from within the earth;
    from the crumpling of the earth's crust
12. gravity, inertia, friction, molecular attraction
13. One muscle bends a joint, and the other unbends it.

Activities

As always, try these demonstrations yourself before class to see how they work.

1–2. These activities help the students to explore sound as vibration.

inside the car. Reducing vibration is one of the best ways to control unwanted sound.

We are surrounded by sounds. If all sounds were similar, the world would be unpleasant and confusing. Resonators strengthen some sounds, and soft materials reduce other sounds.

God has created us with sound-producing voices and sound-detecting ears to use for Him. "O come, let us sing unto the LORD: let us make a joyful noise to the rock of our salvation. Let us come before his presence with thanksgiving, and make a joyful noise unto him with psalms" (Psalm 95:1, 2).

Exercises

1. Give two examples to show that energy is required to produce sound. For each example, tell what form of energy produces the sound.
2. We know that sound is ——— because it has force and can do work.
3. Give an example of sound moving something. (See if you can think of one that is not mentioned in the lesson.)
4. All sound is produced by a fast back-and-forth motion called ———.
5. Write several sentences to tell how your larynx makes sound.
6. The nose and mouth serve as ——— to make our voices stronger and give them certain qualities.
7. List two ways that we can reduce sound.

Review

8. What two main purposes do bones have in the body?
9. The main nerve center that controls the body is the ———.
10. How many main body parts and pairs of legs do the following creatures have?
    a. arachnids
    b. insects
11. In what two main ways has God formed mountains since the Creation?
12. List the four kinds of resistance that hinder motion.
13. Why are the skeletal muscles arranged in pairs?

Activities

1. Sound comes from vibration. Bring things to class that vibrate to make sound. Tuning forks, bells, rubber bands, and many other objects vibrate enough that you can easily see or feel the vibrations. Do not forget to feel your larynx. Put your finger against your throat while you talk. Can you feel the vibration?

2. See and feel vibration. Strike a tuning fork gently so that it produces a tone. (A kitchen fork will also work.) Touch the tip of one prong lightly. Do you feel the vibration? Now touch the vibrating prongs to a calm water surface. Do you see proof of vibration?
3. A resonator strengthens a sound by vibrating with it. Stand in front of the classroom, and strike your tuning fork. Can those in the rear of the room hear the tone? Now strike it again, and hold the stem of the fork against the chalkboard. The sound should be much louder. Why? Vibration travels down the stem of the fork and causes the chalkboard to vibrate as well.
4. This activity will let you see sound doing work. Get a cardboard tube (or PVC pipe) 2 to 4 inches in diameter and about 2 feet long. Lightly stretch a piece of rubber from a balloon over the one end and fasten it with a string or rubber band. Use a sheet of paper and tape to make a paper cone, and fit it over the other end of the tube. The small opening of the cone should be about ¼ inch in diameter.

   Put the paper cone as near as possible to a candle flame. Make a low-pitched sound close to the rubber, such as by striking the bottom of a wastebasket. What happens to the candle flame? Can you explain why?

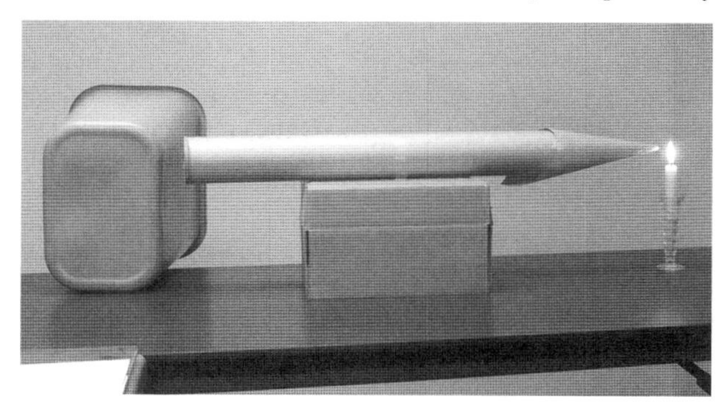

5. Can sound do work such as drilling a hole or breaking a rock? See what you can learn from an encyclopedia.

3. Experiment with different ma-terials (such as a door or a desktop) to find the best resonators. If the tuning fork is touched to a sound-absorbing material, the sound will not be increased.
4. The candle flame should flicker.

   The sound waves are focused by the tube and cone, and this makes the effect of the sound waves more noticeable.
5. Sound has been used to break up rocks in drilling deep oil wells. Sound is used to clean small parts. It has been used to drill holes in metal and in teeth. Sound can pasteurize milk and detect submarines. Most of these special uses of sound require vibrations either higher or lower than audible sound.

Lesson 40

How Sound Travels

Vocabulary

echo (ek′·ō), a repeated sound caused by reflection from a surface some distance away.

wave, a pulse of energy in the form of vibrations moving through a material.

What Are Sound Waves?

Have you ever played with a Slinky? In the picture below, suppose one boy compresses a few coils of the Slinky at one end. If he suddenly releases them, the compression travels rapidly across the spring to the second boy and then returns to the starting point. This moving compression is a ***wave.*** Notice that the spring coils themselves did not move from one boy to the other; it is only the wave that traveled that distance. The wave is a pulse of energy moving through the spring. The energy moves back and forth between the boys, but the spring itself does not.

What will happen if one boy releases some compressed coils?

Sound moves through air like the waves in a Slinky. This is how it works. When a rubber band vibrates, it strikes the air molecules around it and causes them to vibrate as well. These vibrating air molecules strike other air molecules which then strike still others. Waves of vibrations pass through the air molecules as the sound moves out, away from the vibrating rubber band. In this way the sound energy is carried from the vibrating rubber band to your ear by vibrating waves in the air.

Be sure to notice that sound waves do not move the air molecules very far. The air molecules move rapidly back and forth, but they do not travel from the source of sound to your ear. Instead, a wave of energy moves from the source of sound to your ear.

Sound waves do not travel only through air. They can also pass

Lesson 40

Lesson Concepts

1. Sound travels in a wavelike motion through the molecules of a material.
2. Since molecules are necessary for sound to travel, sound will not travel through a vacuum.
3. Sound travels 750 miles per hour (1,200 km/h) through air, faster through a liquid, and even faster through a solid.
4. Sound waves are reflected from surfaces like cliffs or walls.
5. If the reflection of sound is from a distant object, it results in an echo.

Teaching the Lesson

Sound travels in waves. This basic concept helps us to understand many of the workings of sound. Why does sound travel faster through some materials than others? Why does sound bounce? What causes an echo? The study of sound waves helps to answer these questions.

By all means, use a Slinky to demonstrate the traveling of sound waves. The longitudinal waves of the Slinky coils illustrate the mechanics of sound waves much more accurately than the up-and-down waves seen in water or ropes.

Since sound consists of waves passing through molecules, it must have a material to travel through. It cannot pass through a vacuum. If there are no molecules to vibrate, sound cannot be transmitted. (Light also travels in waves, but those waves do have the ability to pass through a vacuum. Thus, light is a greater mystery than sound.)

Reflections of sound waves are possible when the

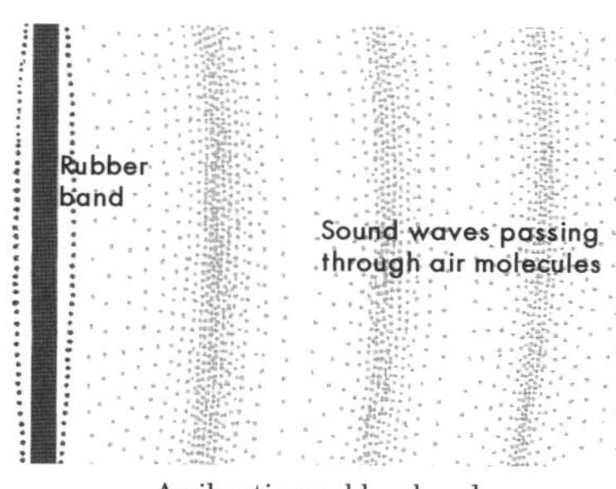

A vibrating rubber band

through water, steel, rocks, and any other material. In fact, sound waves travel better through water and steel than through air. Ask a friend to tap the water pipes in one room while you listen in the next room. Which way can you hear the tapping more easily, by air or through the pipes?

Since sound waves move by means of vibrating molecules, they must have a material to move through. Sound cannot travel through a vacuum, where there are no molecules.

How Fast Does Sound Travel?

Do sound waves move as fast as waves in water? Do they move as fast as waves in a Slinky? Sound waves travel much faster than either of these other waves. If they moved as slowly as waves in water, it would take a long time for sounds to reach you.

The speed of sound waves varies depending on what material they are passing through. Sound waves in air move about 750 miles (1,200 km) per hour. That is faster than most modern jet airplanes can fly. Sound waves in water move about 3,300 miles (5,300 km) per hour. That is the speed of a bullet from a high-powered rifle—over 4 times as fast as sound waves traveling through air. Sound waves move even faster through solids. Through steel, sound waves travel about 11,250 miles (18,100 km) per hour. That is

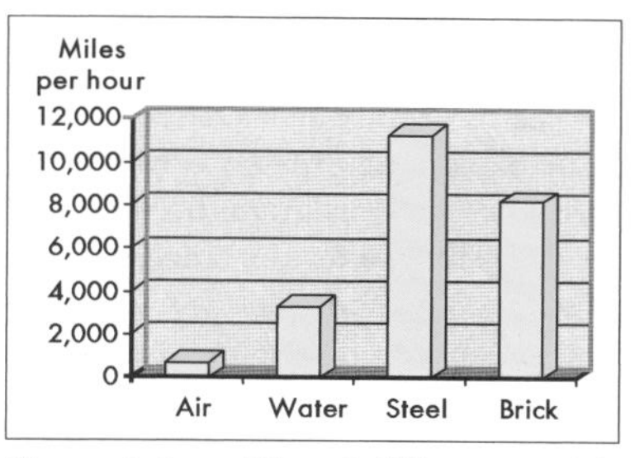

The speed of sound through different materials

over 3 times as fast as in water and 15 times as fast as in air. Sound waves travel very swiftly.

What Causes an Echo?

Have you ever noticed what happens when a wave of water hits a large rock? The wave bounces off the rock and then moves away from the rock. Sound waves work in the same way. When they pass through air and strike a wall or ceiling, they are reflected.

densities of two adjoining materials differ greatly. Such is the case when sound bounces off a distant cliff to produce an echo. The cliff is much harder than the adjoining air. Therefore the sound wave is reflected and returns through the air, back toward its source.

Perhaps you live in an oil-drilling community. Introduce the children to seismographic surveying—the practice of using sound echoes for locating oil and gas deposits. In some communities, the children will be familiar with the dynamite charges and seismographic trucks used in the process.

The peculiar acoustics of the Statuary Hall in the Capitol Building in Washington, D.C., may raise some questions. Why can you hear a whisper spoken on the opposite side of the hall while it is inaudible in the middle of the hall at half the distance? The dome-shaped ceiling in the hall focuses the sound waves at the opposite side of the hall.

Discussion Questions

1. What is a wave? What does it do?
   Remind students of the facts about waves in Lesson 14 ("Mighty Oceans"). Review the difference between waves and currents, noting that waves only move water up and down; they do not move it from one place to another. A similar principle applies to sound waves.
2. What is an echo?
   Discuss briefly what an echo is and how it works. We hear an echo when a sound is reflected from a surface some distance away.

Water waves hitting a rock will bounce off.

The Statuary Hall is a room inside the Capitol of Washington, D.C. In this hall, words spoken in a whisper on one side can be heard at the opposite side. But in the middle of the hall, the whisper cannot be heard. The shape of the room causes sounds to be reflected to some places but not to others.

The reflecting of sound waves often makes the sound seem louder. When your class sings inside a room, the voices are reflected from the walls and ceiling. This makes the singing sound much louder than when you sing outdoors. Some buildings are designed to reflect sound in a special way so that people can hear better. Big auditoriums often have high, rounded ceilings to help reflect the speaker's voice to the audience.

When sound is reflected from objects nearer than 30 feet (9 m), we hear the reflected sound at nearly the same time as the original sound. But when a sound is reflected from a more distant object such as a building or a cliff, the reflected sound has to travel farther to get back to us. Because of

Echoes are reflected sound waves.

this, it reaches our ears a short while after the original sound. That delayed sound is an ***echo.*** To produce an echo, an object must be at least 30 feet (9 m) away from the source of the sound.

Sometimes a hill is positioned in such a way that you can say a short sentence before the echo comes back. Distant rock cliffs may echo well enough to repeat a long sentence to you quite clearly. Echoes make thunder beautiful. They take the sharp *Crack!* of lightning and turn it into a delightful string of rumbles. Sound is truly an interesting wonder that God has made.

Exercises

1. Which statement about sound waves is correct?
   a. Sound waves move energy in one direction and molecules in another direction.
   b. Sound waves move molecules only a short distance back and forth.
   c. Sound waves carry molecules from the source of the sound to the listener's ear.
   d. Sound waves can pass only through air.
2. When a cricket sings, how are the vibrations carried from the cricket to your ear?
3. Sound cannot pass through a vacuum. Why not?
4. Sound waves in air move
   a. about as fast as a car.
   b. nearly as fast as a jet airplane.
   c. faster than a jet airplane.
   d. three times as fast as a high-powered rifle bullet.
5. Sound moves at different speeds through different materials. Write the phrases *sound in steel, sound in air,* and *sound in water* in order on your paper, from slowest to fastest. After each phrase, write the speed in miles or kilometers per hour.
6. When sound waves strike a wall or ceiling,
   a. they stop.
   b. they are reflected.
   c. they move faster.
   d. they sound more pleasant.
7. a. When sound reflects from a surface 30 or more feet away, it produces an ———.
   b. Why does this happen?
8. Why is no echo produced if the distance is less than 30 feet?

Lesson 40 Answers

Exercises

1. b
2. The singing cricket vibrates the air around it. These vibrations spread out in waves through the air farther away until the waves reach your ears.
3. Sound waves are carried by vibrating molecules. Since there are no molecules in a vacuum, sound cannot pass through.
4. c
5. sound in air—750 mph (1,200 km/h);
   sound in water—3,300 mph (5,300 km/h);
   sound in steel—11,250 mph (18,100 km/h)
6. b
7. a. echo
   b. Since the reflected sound travels farther than the original sound, we hear it a little while after we heard the original sound.
8. The original sound and the reflected sound are so close together that we hear only one sound.

Review

9. vibrates
10. a. A resonator increases the loudness of a sound by vibrating with it.
    b. (Any two.) mouth, nose, clock case, whistle tube
11. a. Carpet the hallway.
    b. Tighten the bolts.

Activities

Study the student activities, and choose one or two as a demonstration for the class. Prepare for them beforehand to reduce the amount of time needed for the demonstration. Where distances need to be measured, step them off beforehand and mark them with stakes or ribbons.

Review

9. Sound is produced when something ———.
10. a. What does a resonator do?
    b. Give two examples of resonators.
11. How could you reduce the sound in the following cases?
    a. The children's shoes are very noisy in the school hallway.
    b. The tractor hood rattles when the engine is running.

Activities

1. You can find the distance to a cliff or large building by timing the echo. Shout "Hi!" and then time how long it takes for the echo to return. Sound travels 1,100 feet (335 m) each second, so multiply the number of seconds by 1,100 (by 335 if you are using meters). Then, since the sound has to go out and back, divide that answer by 2. The second answer will be the approximate distance in feet or meters.
2. Have someone stand 550 feet (168 m) away and strike a board with a hammer. You should hear the sound one-half second later. If you live near a railroad track, have someone move 1,100 feet (335 m) down the track and strike the rail with a hammer. The sound should arrive in one second through the air, but almost instantly through the steel rail. Why?
3. Another way to see how well sound travels through a solid is by using a yardstick. Hold a yardstick at arm's length, and have someone scratch it lightly. Can you hear it? Now put one end of the yardstick up against your ear. Have someone scratch the far end of the yardstick lightly. Can you hear it better? Why?
4. Sound is transmitted molecule by molecule. The molecules move very little, but the wave of sound energy is transferred through them. How is this possible? This activity should help you to understand.

   Get six checkers. Put five of them in a straight row, one against another. Now snap the sixth checker sharply into one end of the row. Notice how the checker at the other end shoots away from the rest. The four middle checkers moved very little, yet a wave of energy passed through them. The energy was transferred from one checker to another even though the checkers moved very little.

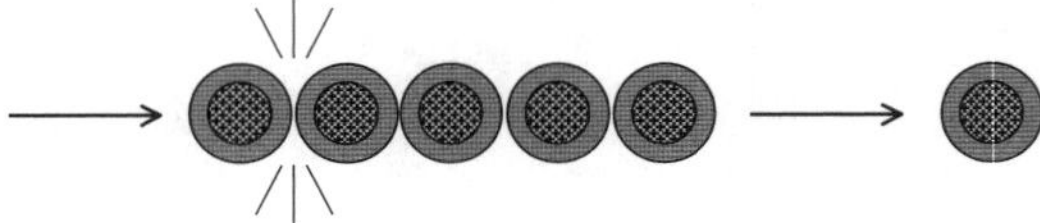

Lesson 41

Pitch and Volume of Sound

Vocabulary

frequency (frē′·kwən·sē), the number of vibrations per second in a sound wave, which determines the pitch of the sound.

pitch, the highness or lowness of a sound, determined by the frequency of sound waves.

volume, the loudness or softness of a sound, determined by the strength of sound waves.

Sounds are not all the same. We can tell who is talking by the sound of a person's voice. God has created sounds to be different so that we can pray, sing, and talk with others. Some sounds are peaceful, and others warn us of danger. Some sounds are beautiful, and others are harsh and unpleasant. God has created sound to be useful for us.

Sounds Differ in Pitch

By ***pitch,*** we mean how high or low a sound is. A chirping cricket makes a sound with a high pitch. Thunder rumbles at a low pitch. Men's voices have a different pitch from that of women's voices. Women sing high-pitched soprano, and men sing low-pitched bass. If we hear someone speaking, the voice pitch can tell us whether the person is calm or excited.

What causes differences in pitch? Remember that sound is caused by vibration. But not all vibrations are the same. Some have a high ***frequency,*** and some have a low frequency. The faster something vibrates, the higher its frequency and the higher the pitch of the sound.

The frequency of a pitch is usually measured in vibrations per second. Humans can hear a range from about 20 vibrations per second to about 20,000 vibrations per second. Of

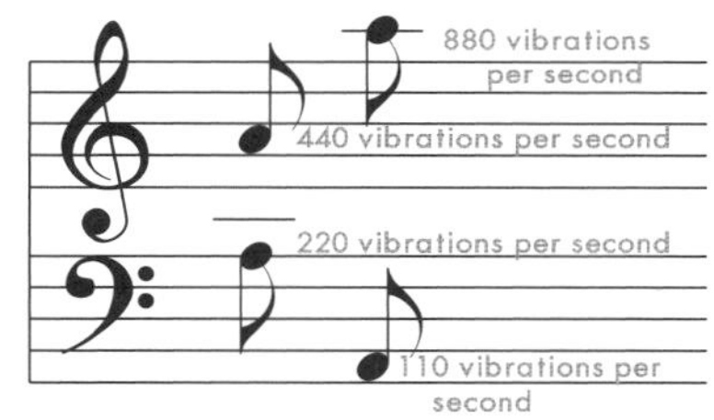

High notes have a high frequency, and low notes have a low frequency.

Lesson 41

Lesson Concepts

1. The frequency of vibration determines the pitch of the sound produced.
2. The shorter, tighter, and thinner a string is, the higher the pitch.
3. The volume of a sound is determined by the amount of energy in the sound.
4. Differences in sounds make it possible for us to sing and communicate with spoken words.
5. Various combinations of pitch and volume help us to recognize voices and to tell what a vibrating object is made of.

Teaching the Lesson

Begin by introducing the lesson in a flat monotone. Continue to use the monotone until the pupils' curiosity is stirred. Then suddenly begin modulating your voice and make it ring with interest. Ask them what was wrong with your voice at first. Help them to see the connection between your voice "problem" and today's lesson. Show them the importance of differences in pitch and volume.

course, you cannot sing at a pitch nearly as high as 20,000 vibrations per second. A frequency of 880 vibrations per second would be a very high note for you to sing. And a pitch with a frequency of 110 vibrations per second is low indeed for someone your age.

When we sing, we do not use the same pitch the whole way through a song. That would be chanting, not singing! Instead, we vary the pitch of our voices to follow the notes of the music. But how do we change the pitch of our voices? How can we change the frequency at which our vocal cords vibrate? There are three laws of vibrating strings that will help you understand how this works.

1. The tighter a vibrating string is, the higher the pitch will be. The vocal cords in your larynx work like that. When you sing a higher note, muscles in your larynx tighten your vocal cords. The tighter cords vibrate faster and produce a higher pitch. When you sing a lower note, you loosen your vocal cords and they produce a lower pitch. Frequent singing will strengthen the muscles of your larynx, and this will help you to sing higher and lower notes as well as to sing the notes more accurately.

2. The longer a vibrating string is, the lower the pitch will be. Differing lengths is another reason for voices of varying pitch. If the vocal cords are short, as children's are, they will vibrate with a higher frequency and produce a higher pitch. If the vocal cords are longer, they will vibrate at a lower frequency and make low-pitched sounds.

Women's vocal cords are shorter than men's vocal cords. That is why women's voices are higher pitched than men's voices. When a boy's voice changes to a man's voice, his larynx grows rapidly for a time. Because of the rapid change, a boy sometimes cannot control his vocal cords very well, and his voice may be unsteady or

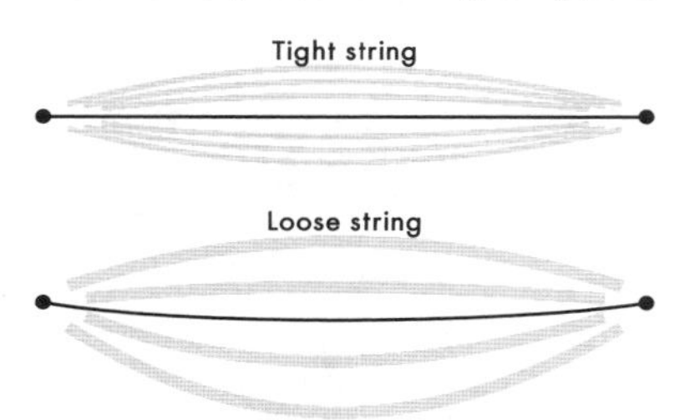

A tight string produces a higher pitch than a loose string.

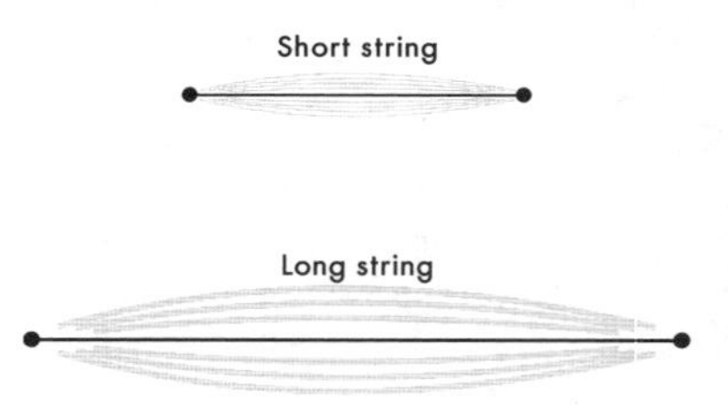

A short string produces a higher pitch than a long string.

Discussion Questions

1. Why are differences in pitch so important?
   Varying pitch makes speech interesting and meaningful.
   It also makes singing possible.
2. Can you tell who is talking if you cannot see the speaker? How can you do this amazing thing?
   You might ask for two volunteers to demonstrate this for the rest of the class. Have the two stand at the front of the room. Tell all the others to close their eyes, and then have one of the volunteers count to five. Ask the rest of the class who it was that talked.
   We can recognize different voices because people are different in the way their vocal cords are made, as well as in the resonators that affect their voices. These differences cause discernible variations of pitch and volume in people's voices.

may have sudden changes in pitch. Later, when the larynx has reached its full adult size, he learns how to use his new voice box. Then his voice becomes firm and steady once again.

3. The thicker a vibrating string is, the lower the pitch will be. Sometimes when we have a cold, our vocal cords become thick and inflamed. Then they cannot vibrate as rapidly, and our voice becomes lower. Sometimes the vocal cords become so thick that they produce no sound at all!

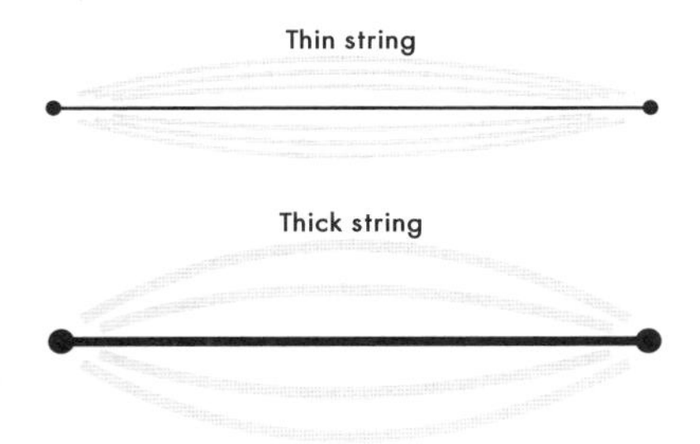

A thin string produces a higher pitch than a thick string.

Sounds Differ in Volume

The ***volume*** of a sound is how loud or soft a sound is. One example is the difference between talking and shouting.

Sounds with strong volume have more energy than sounds with weak volume, and it takes more energy to produce them. We use more effort to

It takes more energy to produce loud sounds than soft sounds.

shout than to speak in a normal voice. Because a loud sound has more energy, it is able to travel farther. This is why you shout when you try to talk to someone far away. You give your voice strong volume so that the sound has enough energy to reach the other person.

The volume of a sound wave does not affect its speed, or pitch. Loud sounds and soft sounds all travel at the same speed. Both loud sounds and soft sounds can have high or low pitch.

Sound Differences Are Useful

Differences in pitch and volume are helpful in our speaking and singing. They make our speech interesting and meaningful. If our voices always had the same pitch and volume, our speech would sound dull and we could not sing at all. God created our voices to be useful and beautiful.

The pitch and volume of sounds can vary in subtle ways. For example, two girls may try to say the same sentence with the same pitch and volume, yet you can easily tell which one is speaking. Why? It is because of differences in their vocal cords as well as in the resonators that affect their voices. These differences cause tiny variations in pitch and volume, which we recognize as different-sounding voices. Your voice is uniquely yours; you sound just like yourself because of the way you are made. Differences in people's voices are another benefit of sound as God made it.

Different kinds of sound also help you identify different materials. Imagine a glass marble rolling down a wooden stairway. You can hear the marble rolling from one wooden step to the next. If it falls off the bottom step and lands on concrete, can you hear the difference? Of course you can! When a marble falls on concrete, it produces a different combination of pitch and volume than it does when it falls on wood. We often use sound in that way to tell what something is made of.

Sounds	
A window breaking	noise
A rippling brook	musical
A motor humming	musical
A clap of thunder	noise
A bird singing	musical
A door slamming	noise

A bouncing ball sounds different on different surfaces.

God created our voices to be useful and beautiful.

Different combinations of pitch and volume can produce beautiful music. Our voices blend to glorify God when we sing. The psalmist often praised God, and we must praise Him too. "It is a good thing to give thanks unto the LORD, and to sing praises unto thy name, O most High" (Psalm 92:1). "Make a joyful noise unto the LORD, all the earth: make a loud noise, and rejoice, and sing praise" (Psalm 98:4). "Praise ye the LORD: for it is good to sing praises unto our God; for it is pleasant; and praise is comely" (Psalm 147:1).

Exercises

1. The highness or lowness of a sound is its ———.
2. The ——— of a sound is the number of vibrations per second in that sound.
3. The faster the vibration,
   a. the louder the sound will be.
   b. the softer the sound will be.
   c. the higher the frequency and pitch will be.
   d. the lower the frequency and pitch will be.
4. When you tighten your vocal cords, will the pitch of your voice be higher or lower? Why?
5. Why are women's voices higher in pitch than men's voices?
6. The loudness or softness of a sound is its ———.
7. If a sound wave has much energy, its volume is (loud, soft) and the sound will travel a (long, short) distance.
8. How can we keep our voices from having a flat, dull sound?
9. How are we able to recognize who is speaking even when we cannot see the speaker?
10. If our voices could not vary, we would not be able to sing. Why not?

Review

11. All sound is produced by ———.
12. How does your larynx produce sound?
13. Why is sound unable to pass through a vacuum?
14. Explain what causes an echo.
15. Name two resonators in your head which change the sound produced by your vocal cords.

Lesson 41 Answers

Exercises

1. pitch
2. frequency
3. c
4. higher;
   The tighter a vibrating string is, the higher the pitch will be.
5. Women's vocal cords are shorter than men's. The shorter a vibrating string is, the higher the pitch will be.
6. volume
7. loud, long
8. We can use our God-given ability to change the pitch and volume of our voices in an appealing way.
9. A speaker's voice has a certain combination of pitch and volume, which helps us to recognize that voice.
10. Our voices must vary to produce the correct pitch for each note in a song. Without this variation, singing would be a monotonous chant.

Review

11. vibration
12. The vocal cords in the larynx vibrate as a stream of air passes between them.
13. Sound waves are carried by vibrating molecules. Since there are no molecules in a vacuum, sound cannot pass through.
14. An echo is caused by the reflecting of sound from a surface some distance away.
15. mouth, air passages in the nose

Activities

1. If possible, bring several tuning forks of different keys to class. Perhaps one or more of the pupils can bring a tuning fork from home. Tuning forks and pitch pipes will serve well to demonstrate the points in today's lesson. Be sure the students do not blow too forcefully into a pitch pipe, for that can cause the sound to become distorted.

Activities

1. See if you can find the frequency number on the side of a tuning fork. The number you see will depend on the pitch that the fork produces. A fork labeled *A* will produce a frequency of 440. A fork labeled *C* has a frequency of 523.3. Which do you suppose is a higher pitch? You can also look on the case of a pitch pipe, which is usually labeled with the frequency of A.
2. Try singing the frequencies labeled on the staff shown in the lesson. Are you able to sing a pitch with a frequency of 110 vibrations per second? a frequency of 220? 440? 880?

   Notice that each time the pitch rises one octave, the frequency doubles. High *do* on the scale has exactly twice as many vibrations per second as low *do*.
3. Would you like to hear the difference between pitch and volume? Blow into one hole of a pitch pipe, and then blow into a hole nearby. The difference you hear is a difference in pitch. Now blow gently into a hole, and notice the soft sound. Then blow harder into the same hole. The difference you hear this time is a difference in volume.
4. Get eight identical tall, thin drinking glasses. They must be made of glass. Fill the first almost to the top with water. Use a pencil to tap it lightly. Now tap an empty glass. Which has the higher pitch? Can you tell why?

   Call the lower pitch *do* on the musical scale. Can you build the entire scale by adding different amounts of water to the remaining glasses? (You may not be able to build the entire scale. Perhaps you could get from *do* to *sol* or *la*.)
5. Even when blindfolded, you can recognize your classmates by the sounds of their voices. But do you know what your own voice sounds like? You do hear yourself talk, but the sound you hear is different from what your classmates hear. What you hear is not only the sounds that come through the air to your ears, but also the vibrations that travel through your head. Try recording a classroom conversation or a reading class. You may find that you recognize all the voices except one!

Extra Activity

Do the demonstration described in Discussion Questions, number 2.

Lesson 42

Ears for Hearing Sound

Vocabulary

anvil, the second of the three tiny bones in the middle ear, so-called because it is shaped like an anvil.

auditory nerve (ô′·di·tôr′·ē), the nerve that carries signals from the ear to the brain.

cochlea (kok′·lē·ə), the spiral tube of the inner ear that looks like a snail shell, where vibrations are changed into nerve signals for the brain.

ear canal, the tube that leads from the outer ear to the eardrum.

eardrum, the thin membrane at the end of the ear canal, which vibrates when sound strikes it and which separates the outer ear from the middle ear.

hammer, the first of the three tiny bones in the middle ear, so-called because it is shaped like a hammer.

inner ear, the part of the ear that lies deepest inside the head, where the cochlea is.

middle ear, the part of the ear lying between the outer ear and the inner ear, which contains the hammer, anvil, and stirrup.

outer ear, the fleshy outside part of the ear, along with the ear canal.

stirrup, the third of the three tiny bones in the middle ear, so-called because it is shaped like a stirrup.

Hearing is a wonderful gift from God. You would be greatly handicapped without it. You use your ears to listen to others. You use your ears to enjoy birds, waterfalls, and many other things that make sounds. With your ears, you can hear the instructions of parents and teachers. Most importantly, your ears make it possible to hear God's Word. "I will hear what God the LORD will speak: for he will speak peace unto his people, and to his saints" (Psalm 85:8).

God designed your ears in a wonderful way. You can hear faint sounds as well as loud sounds. You can hear both high-pitched and low-pitched sounds. This helps to make your ears useful for many things.

God designed your ears so that they would be safe from harm. The delicate inner parts of your ears are deep inside your skull. They are surrounded by the

Lesson 42

Lesson Concepts

1. Hearing is a wonderful gift from God that we should use wisely.
2. The delicate parts of the ear are surrounded by the bone of the skull for protection.
3. Sound waves are transferred from the eardrum through the three small bones of the middle ear to the cochlea.
4. The ear changes sound waves into nerve impulses that are sent to the brain.
5. Having two ears gives us the ability to tell the direction of a sound.

Teaching the Lesson

Our ears are wonderful gifts from God. He has wisely given us the sense of hearing so that we can communicate with one another. Today's lesson describes how our ears work. The ears are delicate and complicated organs; but if you spend some time with today's activities and discussion questions, the pupils should gain a better understanding of their ears.

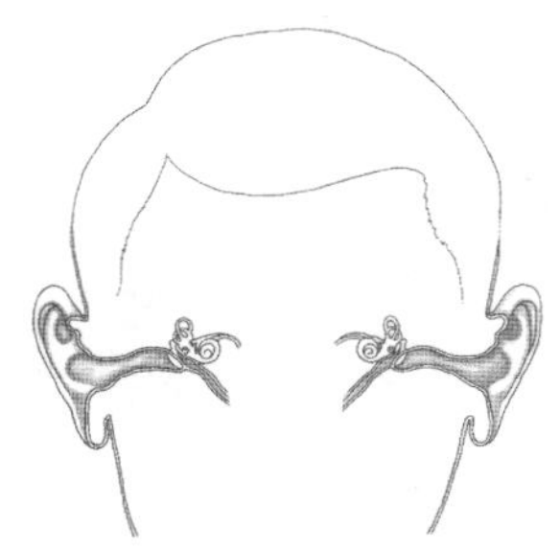
The inner ear is deep inside the skull.

hardest bone in your body. Your ears are well protected.

You studied sound waves in the last several lessons. Your ears detect sound waves and send nerve signals about them to your brain. Sound enters your outer ear, is amplified in your middle ear, and is changed to nerve signals in your inner ear.

The Outer Ear

The ***outer ear*** has two parts. It begins with the part of the ear that we can see on the outside. It also includes the ***ear canal,*** the tunnel leading in to the ***eardrum.*** The eardrum is a membrane ¼ inch across, which covers the end of the ear canal.

The outer part of your ear directs sound waves into your ear canal. When the sound reaches the eardrum, it makes your eardrum vibrate.

The Middle Ear

The ***middle ear*** contains three tiny bones that carry sound from the outer ear to the inner ear. Because of the lever action of these bones, they make the vibrations 22 times stronger as the sound moves through them.

The first little bone is the ***hammer,*** which is connected to the eardrum. When the eardrum vibrates, the hammer sends the vibration on to the ***anvil,*** the next little bone in line. The increased vibration passes from the anvil to the ***stirrup,*** the third little bone. The stirrup connects to the oval window of the inner ear.

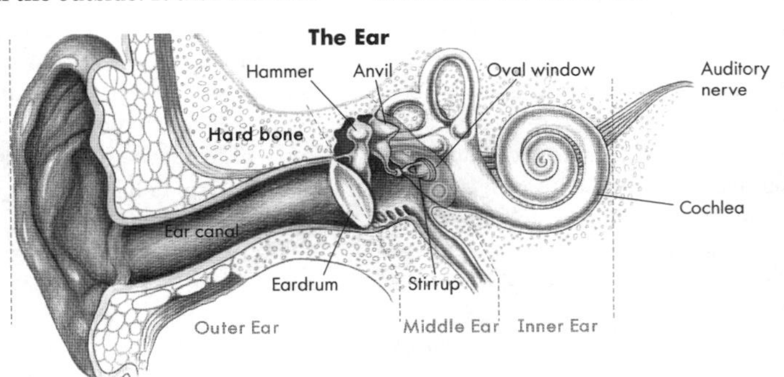

Discussion Questions

1. Where are your ears?

   Your pupils may wonder why you ask this question. Quite likely, most of them could point to their ears since they were a year old. But do they know where the organ of hearing actually is? Is it the fleshy part of the ear outside the head? No. The true organ of hearing is deep within the lower part of the skull. You might illustrate with a diagram on the board.

   After studying this lesson, the pupils should understand ears better than did a teacher of the early 1900s—who tried to wash a bean out of a pupil's ear by pouring water into the other ear.

2. What happens in your ears?

   Sound waves are changed to nerve signals, which are sent to the brain.

The Inner Ear

In the ***inner ear,*** we find the oval window, an opening covered by the base of the stirrup. When the stirrup vibrates, it produces sound waves that enter the ***cochlea.*** This is a tiny snail-shaped organ filled with fluid and containing over 15,000 tiny nerve hairs. The nerve hairs pick up the vibrations in the fluid, change them to nerve signals, and send the signals to your brain through the ***auditory nerve.*** Your brain recognizes these signals as sounds.

The inner ear also contains an organ that helps you keep your balance. The organ of balance tells you whether your head is upright or leaning to one side, and in this way it helps you to keep from falling. If you have an inner ear infection, you may become dizzy because your organ of balance is not working properly.

The Value of Two Ears

God made you with two ears so that you can hear well. Because you have an ear on each side of your head, you can hear sounds coming from any direction around you.

Having two ears also helps you tell what direction a sound is coming from. Because one ear is closer to the source of sound than the other, the sound gets to that ear a little sooner than to the other ear. The sound is also slightly louder in the ear that is closer to the source. Of course, since your ears are not far apart, there is only a tiny difference in the time and volume of the sound received by the two ears. But your brain can detect the difference, and this is how you can tell where the sound is coming from.

Caring for Your Ears

Imagine living in a silent world with no music, no bird songs, and no voices. Imagine being unable to join freely in a family discussion. If we value the gift of hearing, we need to take good care of our ears. Here are some rules for doing that.

1. Never strike a person on the ear. Air forced into the ear by a blow may burst the eardrum.
2. Never pull forcefully on the ears. This may damage the tiny, delicate parts inside the ear. Thomas Edison said he began to lose his hearing after a conductor grabbed him by the ears to help him onto a moving train.
3. Do not put small objects into your ears. It is all right to clean your ears with your little finger inside a washcloth, but not with anything smaller. Cleaning out the ears with toothpicks, hairpins, Q-tips, or other small objects may damage the ear canal or eardrum.
4. Wear earmuffs or earplugs if you need to work in a very loud place. It is true that loud noise does not cause you to lose your hearing immediately. But if a person works around loud machinery day after

Lesson 42 Answers

Exercises

1. a. (Sample answers.)
   hearing sounds of nature;
   hearing the instructions of parents and teachers;
   hearing God's Word
   b. (Sample answers.)
   listening to filthy talk or ungodly music;
   listening to false teachers;
   failing to listen carefully when instructions are given
2. He has placed the delicate parts of our ears deep within the skull, where the hardest bone is.
3. a. middle ear
   b. inner ear
   c. outer ear
4. a. ear canal e. stirrup
   b. eardrum f. cochlea
   c. hammer g. auditory nerve
   d. anvil
5. a. ear canal e. auditory nerve
   b. hammer f. eardrum
   c. anvil g. stirrup
   d. cochlea h. oval window
6. Two ears enable us to hear sounds coming from any direction.
   Two ears help us to tell what direction a sound is coming from.
7. (Any three.)
   a. Never strike a person on the ear.
   b. Never pull forcefully on the ears.
   c. Do not put small objects into your ears.
   d. Wear ear protection if you need to work in a very loud place.

day without protecting his ears, he does gradually become hard of hearing. How soon do you want to wear a hearing aid?

"He that hath ears to hear, let him hear" (Matthew 11:15). God gave us ears, and He expects us to be wise in using and caring for them.

Exercises

1. a. List two ways we can use our ears wisely to glorify God.
   b. What is one way that we can use our ears foolishly?
2. How has God provided for the safety of our ears?
3. Read each sentence, and write whether it tells what happens in the *outer ear, middle ear,* or *inner ear.*
   a. Sound is increased in volume.
   b. Vibrations are changed to nerve signals for the brain.
   c. Sound is gathered and directed inward.
4. Write the vocabulary words that fit these descriptions.
   a. The tube leading in to the eardrum.
   b. The membrane between the outer ear and the middle ear.
   c. The tiny bone connected to the eardrum.
   d. The second tiny bone in the middle ear.
   e. The tiny bone that vibrates in the oval window of the inner ear.
   f. The tiny organ in the inner ear that changes vibrations to nerve signals.
   g. Something that carries nerve signals from the ear to the brain.
5. Write the name for the part of the ear indicated by each letter.

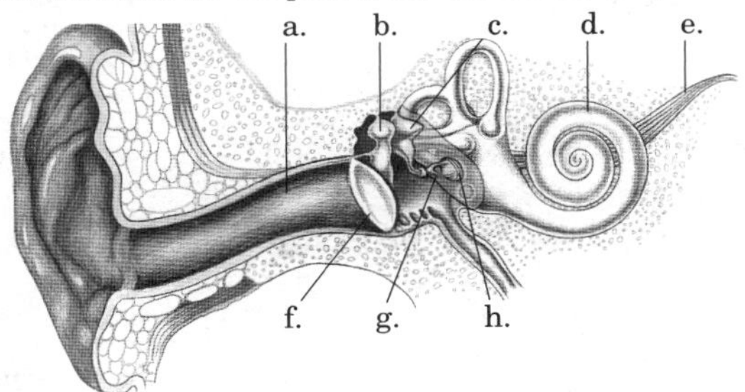

6. We can hear better with two ears than if we had only one. Give two reasons why this is true.
7. The ear is a valuable gift from God. Give three rules for taking good care of our ears.

Review

8. Sound is (vibration only, energy only, both vibration and energy).
9. The ——— of a sound is how high or low the sound is.
10. The vocal ——— found in the ——— vibrate to produce sound.
11. The number of vibrations per second is called the ——— of sound.
12. Does sound travel faster through air or through water?

Activities

1. Your eardrum vibrates when sound waves strike it. You can demonstrate how this works. Cut out the top and the bottom of a small tin can. Stretch a piece of aluminum foil over one open end, and hold it in place with a rubber band. Now talk into the other end. Can you hear the foil vibrate? Put your finger lightly against the foil, and talk again. Can you feel the vibration?

   Now go one step further. Gently push a toothpick through the foil so that it fits snugly into the foil. Touch the end of the toothpick as you talk into the can. Can you feel a stronger vibration? The toothpick is like the hammer that is fastened to your eardrum. It helps make the sound vibration stronger.
2. Blindfold someone, and have him plug his ears. Have him sit in a chair in the middle of the room. The rest of the class should stand around him, forming a loose circle about ten feet away. Everyone should move around so that he does not know where anyone is standing. Then tap him on the shoulder as a signal that he should unplug one ear. Have one person say "Here." The blindfolded person should point to the person who talked. Does he know where the sound came from? Try the test again, this time letting him unplug both ears. Can he do better than before?
3. Outer ears catch the sound and channel it into the ear canal. Could you hear better if your ears were several times larger? Cup your hands behind your ears to enlarge the receiving surface. Can you hear better?
4. Sometimes we feel pressure inside our ears. We say they go shut.
   a. When does this happen?
   b. Yawning or swallowing helps to solve the problem. See if you can find out why.

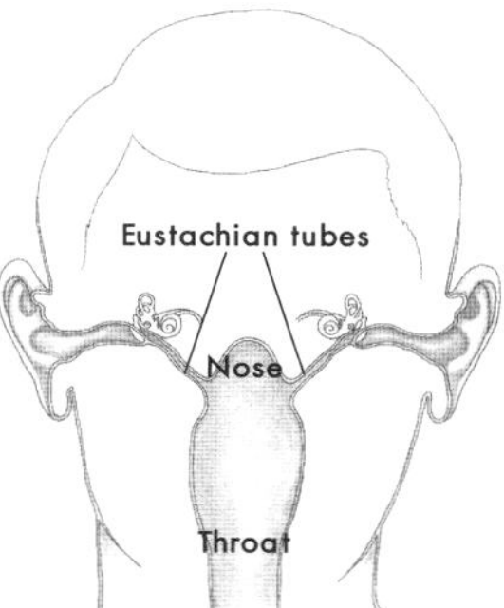

Review

8. both vibration and energy
9. pitch
10. cords, larynx
11. frequency
12. through water

Activities

1–3. Plan for and experiment with these demonstrations before class. Be sure you know how to perform them and that you have all the needed supplies.

4. a. This usually happens when changes in atmospheric pressure occur rapidly, such as when we are driving up or down a mountain. As the outside pressure falls or rises, the eardrum bulges outward or inward. This produces the uncomfortable feeling.
   b. The eustachian tube leading from the middle ear to the back of the mouth provides a means of equalizing the pressure. But this tube is not very large, so it may need to be opened by the flexing that yawning and swallowing provide.

Lesson 43 Answers

Review of Vocabulary

A. 1. vibration
2. larynx
3. resonators
4. waves
5. echo
6. frequency
7. pitch
8. volume
9. outer ear
10. ear canal
11. eardrum
12. middle ear
13. hammer
14. anvil
15. stirrup
16. inner ear
17. cochlea
18. auditory nerve

Lesson 43

Sound is a wonder. Have your students grasped that? Do they understand the main concepts of this unit? Mentally review the main points listed for each lesson. Have they been covered? Scan the test briefly. Does the material look familiar? Help the students review and clinch the unit before taking the test. If you have been reviewing regularly, you will need to spend less time reviewing now before the test.

Lesson 43

Unit 7 Review

Review of Vocabulary

anvil	echo	middle ear	stirrup
auditory nerve	frequency	outer ear	vibration
cochlea	hammer	pitch	volume
ear canal	inner ear	resonators	waves
eardrum	larynx		

Part A

Sound is the __1__ of molecules in a material. Sound can travel through the air as well as through solids and liquids, but it cannot travel through a vacuum. When you speak, the vocal cords in your __2__ vibrate to make sound. Then your nose, throat, and mouth serve as __3__ to strengthen and change the sound.

When sound travels through a material, it travels as __4__. Sound moves very swiftly from one place to another. When it reflects from a distant surface, we can hear an __5__.

Not all sounds are the same. Some sounds are high, and some are low. A high-pitched sound has a very high __6__, which is measured in vibrations per second. We say that it has a high __7__. Some sounds are loud, and some are soft. We say that a loud sound has a strong __8__. Loud sounds have more energy than soft sounds.

Our ears are a wonderful creation of God. The fleshy part of the ear that you can see from the outside is called the __9__. This part of the ear also includes the __10__, a tube that leads to the __11__. The second main part of the ear is the __12__. This part contains three tiny bones: first the __13__, then the __14__, and then the __15__. These three tiny bones carry sound vibrations to the oval window in the third main part of the ear, which is called the __16__. The third main part of the ear contains the snail-shaped __17__, where sound vibrations are changed to nerve signals and sent to the brain by the __18__.

Part B

Name the part of the ear indicated by each number.

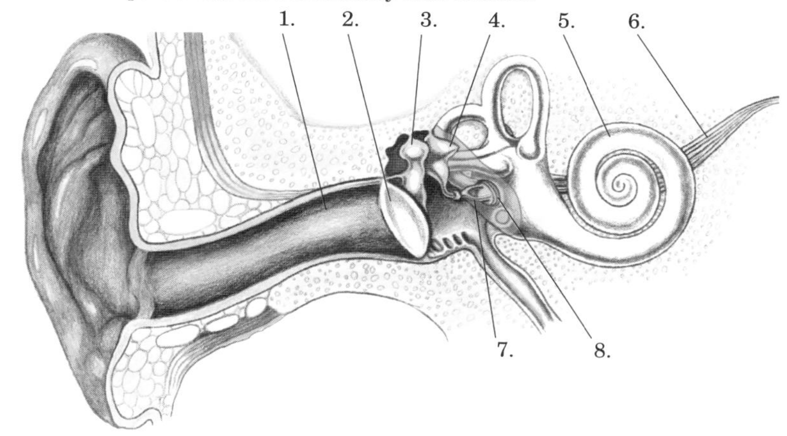

B. 1. ear canal
2. eardrum
3. hammer
4. anvil
5. cochlea
6. auditory nerve
7. stirrup
8. oval window

Multiple Choice

1. Which of the following would **not** produce sound?
   a. a vibrating rubber band
   b. a bouncing marble
   c. a stone lying on a gravel road
   d. a towel flapping on a clothesline
2. In number 1, three of the things named
   a. are not vibrating at all.
   b. cause the air around them to vibrate.
   c. are vibrating because the air around them is vibrating.
   d. produce sound without vibrating.
3. When we speak,
   a. sound waves make our vocal cords vibrate.
   b. our breath makes our vocal cords vibrate to produce sound.
   c. our lungs make our breath vibrate to produce sound.
   d. our vocal cords vibrate and our larynx is a resonator.

Multiple Choice

1. c
2. b
3. b

4. c
5. d
6. b
7. c
8. a
9. b
10. c
11. a

4. All of the following are good ways to reduce sound **except**
   a. reducing the vibration at its source.
   b. absorbing sound with soft materials.
   c. plugging your ears.
   d. resting the vibrating object on soft supports.
5. Sound travels only through
   a. air.
   b. a vacuum.
   c. a solid.
   d. a material.
6. Sound waves move most rapidly through
   a. air.
   b. steel.
   c. water.
   d. a vacuum.
7. Sound travels through a material in much the same way that
   a. wind travels across the land.
   b. water travels through a pipe.
   c. compression waves travel through a Slinky.
   d. a current travels through the ocean.
8. An echo is produced when
   a. sound is reflected from a distant object.
   b. sound travels faster through one material than another.
   c. one sound moves faster than another sound.
   d. sound is reflected from a nearby object.
9. Sound traveling through air moves
   a. a little faster than a car.
   b. a little faster than most jet airplanes.
   c. faster than the bullet of a high-powered rifle.
   d. almost as fast as light.
10. If something vibrates more swiftly,
   a. the sound will have a lower pitch.
   b. the sound will have a lower volume.
   c. the sound will have a higher pitch.
   d. the sound will have a higher volume.
11. If a string is long and loose, it will produce
   a. a low-pitched sound.
   b. a high-pitched sound.
   c. a soft sound.
   d. a loud sound.

12. Which one of these activities would be impossible if our voices could not vary in pitch and volume?
    a. praying
    b. singing
    c. talking
    d. playing

12. b

13. Sound waves are changed to nerve signals in the
    a. oval window.
    b. eardrum.
    c. cochlea.
    d. ear canal.

13. c

14. The order of the three tiny bones of the middle ear from outside to inside is
    a. hammer, stirrup, anvil.
    b. hammer, anvil, stirrup.
    c. anvil, stirrup, hammer.
    d. stirrup, anvil, hammer.

14. b

15. We can tell what direction a sound is coming from because
    a. our ears are on the side of our heads.
    b. the delicate parts of our ears are deep within our skulls.
    c. we can turn our heads to listen to the sound from a different angle.
    d. we have two ears.

15. d

Unit 8

Title Page Photo

A damp, shady area is full of living wonders. Moss and ferns grow here, and billions of bacteria are busily turning this dead log back into soil. The following lessons will tell you the wonderful ways these organisms live and grow.

Introduction to Unit 8

Biologists at one time considered algae, bacteria, fungi, ferns, and mosses as all belonging to the plant kingdom. Today many consider only ferns and mosses as true plants. This pupil's text refers to ferns and mosses as plants, but it describes algae and fungi as plantlike.

You should not find it hard to stimulate the students' interest in this unit. All except the most arid sections of the country exhibit specimens of these wonderful living things. A partly stagnant watering trough, an old fence post, a moldy piece of bread, a shady spot by the creek—these are places that teem with life. Algae, bacteria, and mold are too small to be studied individually without magnification. But mosses, ferns, mushrooms, and colonies of mold and algae are large enough for us to see with the unaided eye.

Many of these special living organisms are small or tiny in a physical sense, but they are ecological and economical giants. Their impact on their surroundings is great, in a positive way as well as a negative way. Many times, they are the crucial link between other species and survival. They are not to be ignored.

Unit 8

Wonders of Ferns, Mosses, and Plantlike Organisms

Often when we speak of living things, we are thinking of trees and grass and cows and dogs and other large creatures. We forget sometimes that there are billions of small living things all around us and on us—and, yes, even inside us. Bacteria are tiny organisms (forms of life) that you can find almost anywhere—if you can see them. You need a microscope to see them because they are very tiny. Algae will grow almost anywhere in the world, wherever there is moisture and sunlight. Fungi also grow almost anywhere. Mosses and ferns are delicate living things that grow in nearly every wooded section of the country.

These wonderful living things are important even if we often forget about them. Algae is responsible for providing ocean and freshwater creatures with food. Bacteria help keep the earth clean. They turn dead animals and plants back into soil. Some algae and all bacteria are too small to see without using a microscope, yet they do mighty things for us because God created them in a wonderful way.

Story for Unit 8

My Wilderness

My father owns a beautiful wilderness. It is alive with strange creatures and unfamiliar plants. I've just discovered it and have already spent several fascinating hours exploring its beauty. Come with me, and let's go exploring together.

No, you won't need hiking boots or a compass. My wilderness is not nearly large enough for a hike. Actually, it is on my father's thirteen-acre hobby farm. A few sheep and beef cattle here and there and a dog or sometimes two are all the animals most people see. But they haven't seen my wilderness. Bring your magnifying glass to observe details, and let's go. You won't need anything besides your five senses and your magnifying glass.

No, we're not going on a long journey. We need only to go across the lawn to the watering trough in the sheep pasture. Now strap on your magnifying-glass helmets and diving suits, and let's all shrink really small and plunge into this miniature wilderness here at the sheep's watering place.

First, let's examine the forest of tangled algae strands here on this side of the trough. Algae? Yes, that's algae. Remember that you are seeing everything through your magnifying glass and that things look larger than they are. Notice the sections of this bright green strand of algae. Each section of stalk has a spiral-shaped thing inside, which makes it look lacy. Now stroke it with your hand. Feel the slimy surface? They are all like that. The slime helps hold the strands together in clumps and bunches.

Lesson 44

Algae

Vocabulary

algae (al′·jē), *singular* **alga** (al′·gə), simple plantlike organisms that contain chlorophyll and produce food and oxygen through photosynthesis.

cell, the smallest unit of living matter, of which all living things are made.

chlorophyll (klôr′·ə·fil), a green chemical needed by plants and algae in the process of photosynthesis.

chloroplast (klôr′·ə·plast), a tiny structure that contains chlorophyll, found inside the cells of algae and plants.

filament (fil′·ə·mənt), a fine, threadlike strand of algae cells, often resembling a green hair.

fission (fish′·ən), a method of reproduction in which a cell divides into two cells.

photosynthesis (fō′·tō·sin′·thi·sis), the process of changing carbon dioxide and water into sugar and oxygen by the aid of sunlight.

What Are Algae Like?

You may never have heard of ***algae,*** yet you have surely seen these simple plantlike organisms many times. Algae are among the most abundant living things on earth. Where have you seen them? The green, slimy scum in ponds and aquariums is algae. So too is the green coating on the moist sides of trees.

A magnified view of green algae filaments

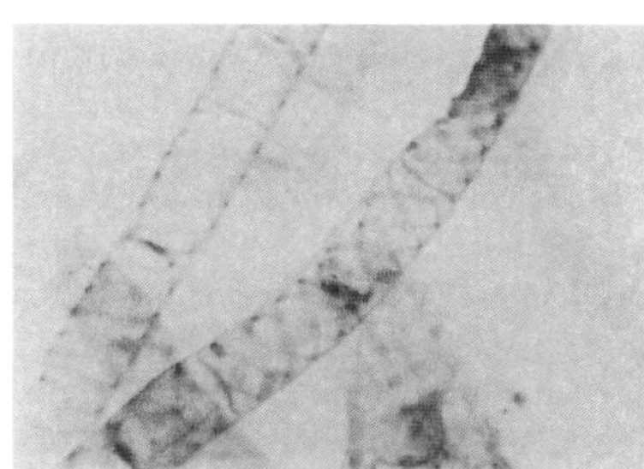

Green spirals of *Spirogyra*

Remember how scummy algae usually looks? This magnified view surely is better, isn't it. Now it looks more like something that people put in salads. Yes, miniature salad greens—that's what they are. And actually, all these little water bugs and creatures do eat the algae. That's all there is for them to eat in here, besides each other. It's no wonder the encyclopedia calls algae the pasture of the sea. Millions of things eat it.

We could study algae for hours without finding all the different kinds, but let's move on.

If I had a microscope, I could show you some exciting creatures in the rotting leaf that's floating here upside down. But they are too tiny for us to see with only magnifying glasses. The only way I can tell they're in there is by noticing how the leaf is rotting. Bacteria make that happen. Colorful little creatures some of them are.

Next we're in for a thrilling slide. This is where the water overflows. Algae grows thick here and makes a slick slide down to the moss below. Everyone coming? Here we go!

Welcome to moss country! Remember the familiar green moss carpet that grows down beside the trough? That's where we are, only we're looking through a magnifying glass now. The familiar moss carpet looks more like a pine forest, doesn't it! And each tall red stem coming from the top holds a spore case that reminds me of a tree house.

Over here under the trough is something strange. It is a musty-smelling mushroom. Look up under the spreading umbrella cap. See all the gills hanging there? Beautiful and delicate, aren't they. Here a sheep tramped beside it. That lets us see the main part of the plant. It grows under the surface of the soil. Right here—see this white rope? That's it. Without our magnifying glasses, it would look like a small thread.

Next to it is another interesting item. Someone help me flip this leaf. There, look at that! More little white strings! But these are smaller. They are making the leaf fall apart. What are they? You guessed it! These are mold fibers.

You found more algae? Sure enough. Here where the sheep tracks filled with water last night, algae are growing. Hundreds of tiny algae ponds. Fascinating, isn't it? I think so too. This is much more interesting than just sitting here twice a day doing nothing while I wait for the watering trough to fill!

Algae also grow well on swampy soil, forming a green film over its surface. Another kind of algae is the large seaweeds that you may have seen along the seashore.

Algae, like other living things, are made of cells. ***Cells*** are the building blocks that living things are made of. Most cells are much too small to see without a microscope.

Large algae, such as the giant seaweeds, are made of millions of cells. But the smallest algae have just one cell each and can be seen only with a microscope. Some one-celled algae float through the water by themselves. Others group together to form colonies of dozens or even millions of cells floating in small jellylike blobs in the water. Some algae cells link together end to end and form long strings called ***filaments.*** The filaments look like fine strands of green hair.

Corn plants reproduce by making corn seeds. Bean plants multiply by producing bean seeds. But algae are different from plants. Algae do not produce seeds. How then do they multiply? When a single-celled alga is large enough, it divides into two cells. This is called ***fission.*** Fission is one of the most common ways that algae multiply. As the algae cells in colonies multiply by fission, the colony grows larger. Some algae also multiply in other ways, but no algae produce seeds.

Many Kinds of Algae

There are thousands of species of algae—far too many to study in one lesson. But most, from the smallest to the largest, fit into one of five general groups. These five groups are easy to remember because they are named by their colors: green, golden, brown, red, and blue-green.

Green algae. Probably you are most familiar with the green algae that grow in pond water. Algae of this kind have fine, wispy filaments with a slippery coating that feels slimy. If possible, examine some green algae under a microscope. You will see how the cells join end to end to form a filament. Inside each cell, you will also see a green spiral, which explains why scientists call this alga *Spirogyra.* Green algae are very beautiful when viewed through a microscope.

Another familiar green alga grows on the moist sides of trees and fence posts. When it is moist, it becomes slippery. Both kinds of green algae may multiply by fission.

Golden algae. Most of the golden algae are diatoms (dī′·ə·tomz′). A diatom is a tiny single-celled alga with two glassy shells that fit neatly together like a miniature shoebox. They live in ponds, lakes, and oceans. There are thousands of different shapes of diatoms. Their glass shells are beautiful when enlarged by a microscope. Some diatoms look like lacy snowflakes.

Lesson 44

Lesson Concepts

1. Living things are made of tiny parts called cells.
2. Algae are the green, slimy material that grows in ponds and fish aquariums.
3. Many algae grow in long filaments that are only one cell thick.
4. One way algae reproduce is by cell division, called fission.
5. Algae are classified by color: green, golden, brown, red, and blue-green.
6. Seaweeds are several kinds of algae. Kelp is an example of brown algae.
7. Chlorophyll in the cells of algae enables them to make food in the presence of sunlight.

Teaching the Lesson

Algae, the pasture of the ocean. Algae, the scum of farm ponds. Algae, the habitat for myriads of minute creatures. Algae, the plague of urban water systems. Algae, an occasional problem but always the miraculous green-food-and-oxygen machine.

There are few places without naturally occurring algae. Moisture and sunlight are its two main requirements. Typically, where these two coexist in abundance, there you will find algae. Help the children find the green friends around them. Bring samples to class, and get a good, firsthand look at algae.

You may find in your studies that reference books disagree on the classification of different kinds of algae. For most of the twentieth century, biologists classified

When diatoms die, their glassy shells do not decay. In some places, deposits of diatom shells hundreds of feet thick have been discovered. These shells are used in polishes and filters.

Brown algae. The brown algae are seaweeds that live mostly in cool ocean waters. One kind of brown algae grows in tangled masses in the middle of the North Atlantic Ocean, scattered over an area of two million square miles. These seaweeds are not fastened to anything. They have many pea-sized pockets called air bladders, which make them float near the surface where sunlight can reach them.

The giant kelp is the largest algae. It produces a plantlike stalk that begins on the ocean floor, where it is fastened and grows upward toward the surface. Giant kelp grows as long as 200 feet (60 m). Kelp also has air bladders to keep it afloat.

Brown algae are a source of food, fertilizer, and vitamins.

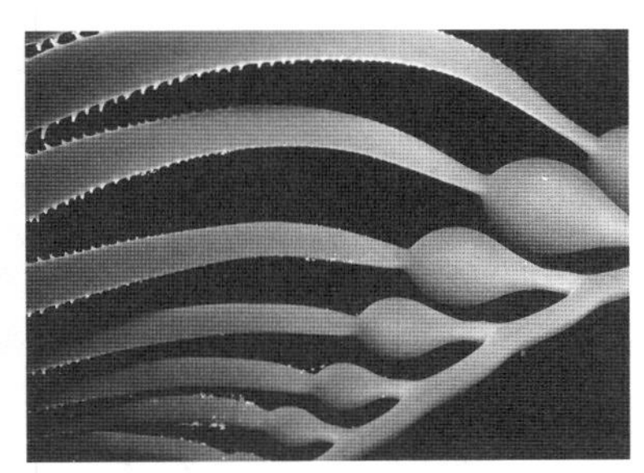

Kelp with air bladders

Red algae. The red algae are many different colors, pink to red and purple to black. These algae are also seaweeds, but most of them are smaller than the brown algae, and they prefer the warmer ocean waters.

Some common types of red algae grow on rocks in the shallow water along the coast. They are firmly anchored to rocks so that waves and tides do not wash them away. Some red algae are used in making certain foods.

Blue-green algae. Blue-green algae are usually helpful organisms, but sometimes they contaminate water supplies. They are single-celled and often form colonies of slimy blobs.

One thing special about some kinds of blue-green algae is their ability to take nitrogen from the air and fix it in compounds so that plants can use it. Water plants need that nitrogen to live.

Algae Produce Food and Oxygen

How can a seaweed produce food? When scientists say that algae produce food, they do not mean that algae produce a complete, prepared dinner that you could enjoy eating. Instead, they mean that algae can take water and carbon dioxide and a few nutrients from the soil or seawater; and with the aid of sunlight, they make molecules

algae as plants; but in more recent years, many have reclassified blue-green algae as bacteria, unicellular green algae and golden algae as plantlike protists, and the rest as plants. Others classify all algae (except blue-green algae) as plantlike protists. The distinction is not really important for fifth graders, but it does explain the use of the phrase "plantlike organisms."

Discussion Questions

1. Are algae a help or a hindrance to man?

   Algae are sometimes a help and sometimes a hindrance. On the positive side, they hold great importance in the aquatic food chain and in the production of oxygen. On the negative side is the contamination of drinking water, the unsightliness of stagnant, algae-ridden ponds, and the problem of algae in fish aquariums.

2. What part do algae play in providing food for man?

   Some algae are eaten directly by man. Other algae provide raw materials for many food products. By far the largest role in food production is the place of algae in the food chain. All seafood—including fish, clams, oysters, shrimp, and lobsters—either eat algae or feed on creatures whose ultimate food source is algae.

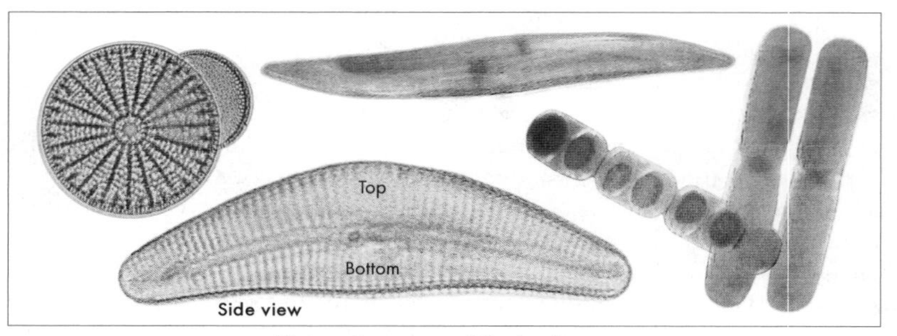

Diatoms have many different shapes.

of sugar. This remarkable process is called ***photosynthesis.***

Only algae and plants can produce food by photosynthesis. Animals cannot do that. For example, a cow cannot breathe air, drink water, and lick a salt block to grow and produce milk. She must also eat grass or other feed made from plants. Even scientists in their laboratories are unable to produce food from air and water and soil nutrients. Animals and humans all depend on plants and algae to produce food for them.

Another of the products of photosynthesis is oxygen. Animals and humans need oxygen to live. Thus, photosynthesis provides both the food we eat and the oxygen we breathe.

To carry on photosynthesis, plants and algae must have a green chemical called ***chlorophyll*** in their cells. The chlorophyll is contained in tiny structures called ***chloroplasts.*** The spirals inside the cells of green algae are chloroplasts.

Algae have been called the pasture of the sea. From the tiny diatoms to the giant kelps, algae use photosynthesis to produce food and oxygen for fish and other water creatures. Suppose it takes 100 pounds (45 kg) of algae to grow about 10 pounds (4.5 kg) of tiny sea creatures called krill. If a whale gains 1 pound by eating 10 pounds of krill, then 100 pounds of algae is needed for 1 pound of whale.

The largest of the whales is the blue whale, which lives mostly on plankton and krill. A large blue whale weighs as much as 200,000 pounds (90,900 kg). Can you imagine how much algae that whale must have eaten to grow so big? A large whale may eat over a ton of food per day. As you can see, millions of tons of algae are needed to feed all the ocean's creatures. They all depend on algae for food and oxygen. They eat either algae or other creatures that feed on algae.

Algae also help feed people. All the seafoods that we eat—including fish, crabs, and oysters—live on algae or on creatures that feed on algae. But some people eat the algae itself. For instance, the Japanese have eaten kelp for years. Many healthful, nourishing dishes can be prepared from these giant algae.

You too eat some algae products. Agar and carrageenan, which come from red algae, are added to many foods. They serve as thickeners for soups, puddings, and other foods. The ice cream you sometimes eat may contain algae to make it thick and creamy. God knew what He was doing when He made algae for us.

Algae produce food for many kinds of sea animals.

Lesson 44 Answers

Exercises

1. a. chlorophyll
   b. chloroplast
   c. photosynthesis
   d. filament
   e. cell
   f. food, oxygen
   g. (giant) kelp
   h. fission
2. green, golden, brown, red, blue-green
3. (Sample answers.) ponds, oceans, aquariums, sides of trees, top of soil
4. Algae provide food and oxygen. All sea creatures eat algae or feed on other creatures that eat algae. Man eats sea creatures fed by algae, and he also eats some kinds of algae.
5. They have chlorophyll that allows them to change carbon dioxide and water into sugar and oxygen by the aid of sunlight.

Review

6. a. 2 main body parts and 4 pairs of legs
   b. 3 main body parts and 3 pairs of legs
7. freeze it
8. falling buildings, fires, tsunamis
9. head (the distance that the water falls);
   flow (the amount of water running over the falls)
10. lubricant
11. Divide the length of the effort arm by the length of the resistance arm.

Activities

3. Diatomaceous earth can be purchased at a garden supply store, such as Agway.

Exercises

1. Write the correct words for these descriptions.
   a. A green chemical that algae use in the process of photosynthesis.
   b. A structure that contains chlorophyll within algae cells.
   c. The process that algae use to change carbon dioxide and water into sugar and oxygen.
   d. A strand formed by joining algae cells end to end.
   e. One of the building blocks that make up living things.
   f. Two important things that algae help to provide for other living things.
   g. The largest algae, with a length as great as 200 feet.
   h. The process of one alga cell dividing into two.
2. List the five general groups of algae.
3. What are four places where you can find algae?
4. In what ways are algae important to sea animals and man?
5. How do algae produce their own food?

Review

6. How many main body parts and pairs of legs do the following creatures have?
   a. arachnids
   b. insects
7. What is the best way to kill an insect for a collection?
8. What three things resulting from earthquakes are harmful to people?
9. What two things determine the amount of energy in a waterfall?
10. When there is too much friction between two surfaces, we use a ——— to reduce the friction.
11. Explain how to find the mechanical advantage of a lever.

Activities

1. Start a list of different places where you see algae. Look for the telltale scum everywhere there is sunlight and moisture. You will be surprised at what you find. Use a chart to record who saw it, when he saw it, and where he saw it. It would be interesting for you to bring samples of algae to school as you find them so that others can see them too.
2. Observe some algae close up as it floats in a pond. Do you see bubbles on it? What do you suppose the bubbles are? How many insects and other little creatures can you find in and around the algae? Do you see why algae are called the pasture of the sea?
3. Diatomaceous earth is a light-colored substance that contains diatom shells. Get some of this material, and use a microscope to see the diatoms in it.

Lesson 45

Mosses and Ferns

Vocabulary

alternation of generations, a cycle of reproduction in which one generation always differs from the next, yet every other generation is the same.

fiddlehead, a young fern leaf that is still partly coiled, so-called because it resembles the end of a fiddle (violin).

frond (frond), the compound leaf of a fern.

spore, a seedlike cell that can grow into a new plant or other organism.

zygote (zī′·gōt′), a seedlike cell produced by some plants that starts a new plant.

Remember the soft green carpet of moss that you found while hiking in the woods? It was growing in a damp, shady spot. Most likely, ferns were also growing nearby. In today's lesson we want to learn more about these charming plants: mosses and ferns.

Moss Plants

A green carpet of moss contains dozens, hundreds, or even thousands of moss plants, depending on how large the carpet is. Each tiny stalk of moss is a separate plant.

Moss plants are unlike most other plants. They do not have true stems and leaves as other plants do. The stalks do not have sap flowing through them as do the stems of other plants. Because of this, moss plants can never grow tall. Most plants have thick leaves with many special cells. But moss leaves have only the thickness of one cell. Moss plants also lack roots. Instead, they have thin filaments that anchor the plant and absorb water and minerals for the plant.

Like other plants, moss plants have chlorophyll. This allows them to make their own food from water and carbon dioxide. Moss plants carry on photosynthesis to produce their food.

Fern Plants

The main part of a fern plant is a stem that grows just under the surface of the soil and may become quite long. At various points along this stem, familiar clumps of ***fronds*** appear. Fronds begin as ***fiddleheads.*** The young fiddlehead is a tightly rolled stalk that slowly unrolls to become a

Lesson 45

Lesson Concepts

1. Mosses are common plants of damp, shady wooded areas.
2. The main part of a fern plant is a stem growing just under the ground, which sends leaves upward at various places.
3. Both mosses and ferns reproduce by spores.
4. The spores of mosses are produced on slender stalks.
5. The spores of ferns are produced in spore cases on the undersides of the fronds.
6. One kind of dead moss, called peat, is useful for packing and mulching plants because of its ability to absorb water.
7. Mosses are soil builders.

Teaching the Lesson

We all know what mosses and ferns are. Moss is that well-known green carpet of the moist, cool woods; and ferns are the feathery, fronded plants that populate the area in and around the green moss carpets. But have you ever examined moss closely enough to see the tiny, individual plants? Or have you ever noticed a fern's mysterious fiddleheads or its small dark spore cases? Have your students noticed this exquisite beauty? Show it to them, and point out the details visible through a large magnifying glass. To do this means either a class excursion to a wooded lot, or plenty of fresh samples brought to class.

Fern fronds and a fiddlehead

leafy frond. Fern fronds are green because they contain chlorophyll. Ferns also use photosynthesis to produce their food.

Fern plants have roots that extend downward from their underground stems. These roots absorb the water and minerals that ferns need for life.

New Moss and Fern Plants

Ferns have an unusual way of multiplying. One generation of ferns

Roots and underground stem of a fern

produces ***spores,*** and the next generation produces ***zygotes.*** Scientists call this process ***alternation of generations.***

Spore-producing generation of the fern. The best-known generation of ferns has the lacy fronds that are familiar to us. On the undersides of mature fronds are small dark spots that produce spores. When the spores land in a moist, suitable place, they develop into the next generation of ferns.

A frond with dark spots producing spores

Zygote-producing generation of the fern. The plant that grows from the spore does not look like a fern at all. Its tiny, heart-shaped leaf measures only about ¼ inch (6 mm) across and is easily missed. This strange-looking plant lives only a few weeks, but that is long enough for it to produce a zygote. The zygote is the beginning of a new fern plant with its graceful fronds.

As you can see, alternation of generations means that each generation

Discussion Questions

1. How is moss like other plants?
   Moss has leaves and stems.
   It takes in water and minerals from the soil.
   It is green.
   It has chlorophyll.
   It makes its own food by the aid of chlorophyll and sunlight.
   It produces oxygen.
2. What is the purpose of fiddleheads?
   A fiddlehead looks like a small button made of a rolled-up stem. Slowly the button unrolls, and the uncurled "stem" becomes a leafy frond.

is different from its parents but just like its grandparents.

Alternation of generations in mosses. Moss plants also reproduce by alternation of generations. But in mosses, the familiar moss plant produces a zygote. When the zygote matures, the plant of the next generation grows from the top of the original moss plant. This second generation looks as if it were a part of the original plant. It is a tall, narrow stalk with a tiny spore case on top of it. When these spores ripen and are blown to a suitable place, a new moss plant begins growing.

Value of Mosses

Moss plants are soil builders. In rocky, shady places, few things besides moss will grow. As the moss filaments grow downward, they help to dissolve rocks and break them apart. Then when the moss plants die, they help build a layer of soil for other plants to grow in.

A kind of moss known as peat grows in swamps and bogs. Over a number of years, extremely thick mats of peat moss may build up. Sometimes a thick, floating mat of mosses and plant roots is strong enough to support a person's weight. Walking on such a mat is dangerous, however, because a thin spot may allow you to break through.

Peat moss is very useful. Because it absorbs many times its weight in water, nurseries pack partly decayed moss called peat around the roots of seedlings before shipping them to customers. The peat provides moisture for the roots so the seedlings do not wither and die. Peat moss also makes excellent mulch because of its ability to hold water. Many people mulch their flower beds with peat moss to keep the weeds down and to keep the moisture in the soil.

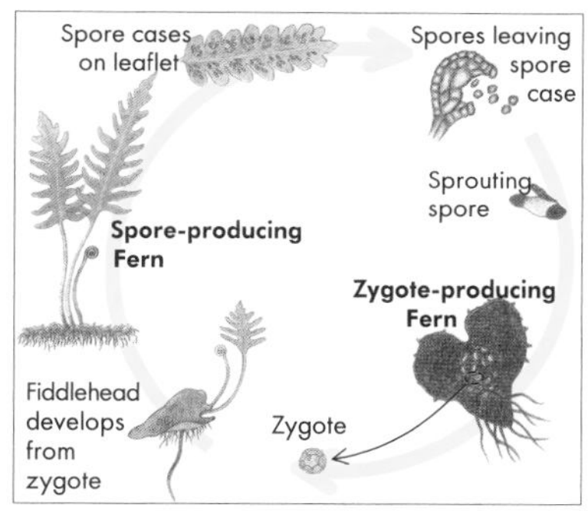

The life cycle of ferns

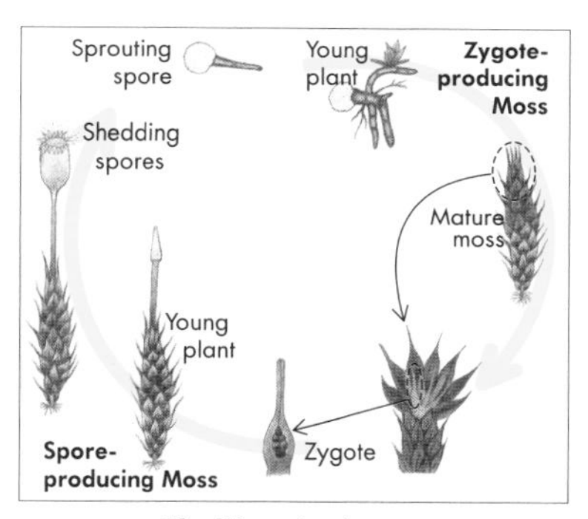

The life cycle of mosses

Lesson 45 Answers

Exercises

1. c
2. (Any two.)
   Moss plants do not have true stems and leaves as other plants do.
   The stalks do not have sap flowing through them as do the stems of other plants.
   Other plants have thick leaves, but moss leaves are only one cell thick.
   Moss plants do not have roots as other plants do.
3. a. frond
   b. fiddlehead
4. fern plant—in small dark spots on the undersides of mature fronds;
   moss plant—in a tiny spore case on top of a slender stalk that grows from the top of the original moss plant
5. When the spores settle in a moist place, they grow into tiny plants that live only a few weeks. The tiny plants produce zygotes, which develop into the familiar fern plants.
6. Many people put peat on their flower beds to keep the weeds down and to help keep moisture in the soil.

Review

7. element
8. compound
9. chemical
10. physical
11. fuel, oxygen, and heat
12. Acids, Bases

Activities

1. The fiddlehead is considered a delicacy.
2. Peat is a moss product that the Irish have used as fuel for hundreds of years. They use it for cooking, for heating their houses, and even for generating electricity. In the early 1980s, they harvested 4,000,000 tons of peat annually.
3. Supervise the building of the pupils' terrarium. Give any help needed for understanding the directions and for gathering specimens to include in the miniature landscape.

Exercises

1. A carpet of moss
   a. is one large, spreading moss plant.
   b. is the branches of an underground moss stem.
   c. is a whole forest of tiny moss plants.
2. List two ways moss plants are unlike most other plants.
3. What term is used for each of the following parts of a fern plant?
   a. leaf b. leaf bud
4. Where would you find spores on a fern plant? on a moss plant?
5. Explain how ferns multiply by alternation of generations.
6. How do many flower gardeners find moss useful?

Review

7. A substance whose molecules are made of only one kind of atom is called an ———.
8. A substance whose molecules are made of two or more different kinds of atoms is a ———.
9. Burning is an example of a (physical, chemical) change.
10. Cutting and melting are examples of (physical, chemical) changes.
11. What three things are always needed for a fire to burn?
12. (Acids, Bases) turn blue litmus paper red. (Acids, Bases) turn red litmus paper blue.

Activities

1. Find out what part of a fern plant is edible and considered a delicacy.
2. A certain moss product is very important to the Irish. See if you can find out what it is and why it is important.
3. Make a moss garden that looks like a small landscape. Begin by putting some soil and rocks in the bottom of an empty glass aquarium. Moisten the soil well, and shape it into slightly rolling hills. Place a small mirror in a low spot to look like a pond. After you have the mirror in place and the hills all formed, cover the hills with moist moss. Try to get large carpetlike pieces from a shady creek bank. Use different kinds of moss for variety, and fit them all together snugly.

   When the hills are all covered with moss, sprinkle your moss garden lightly with water. You will need to sprinkle it again from time to time so that the moss stays green and healthy. Did you remember to include ants and other small insects?

Lesson 46

Bacteria

Vocabulary

bacteria (bak·tir′·ē·ə), *singular* **bacterium,** single-celled organisms so small they can be seen only with the aid of a microscope.

microbe (mī′·krōb′), an organism that causes disease; a germ.

rot, to decay; to break down into simpler substances, as when leaves break down into soil nutrients and humus.

sterilize, to kill or remove all the bacteria from something.

What Are Bacteria?

Bacteria are strange creatures. They live around you, upon you, and even inside you, but you have never seen them. They are creatures that help to feed you, yet some take food away from you.

Bacteria are one-celled organisms that are found almost everywhere on earth. They are so tiny that we cannot see them without a powerful microscope to enlarge them hundreds of times. They float in the air we breathe, they rest on the food we eat, and they live on our bodies and on nearly everything we touch. Some bacteria cause diseases. We call them germs, or ***microbes.*** Other bacteria cause food to spoil, but still others help to produce food for us.

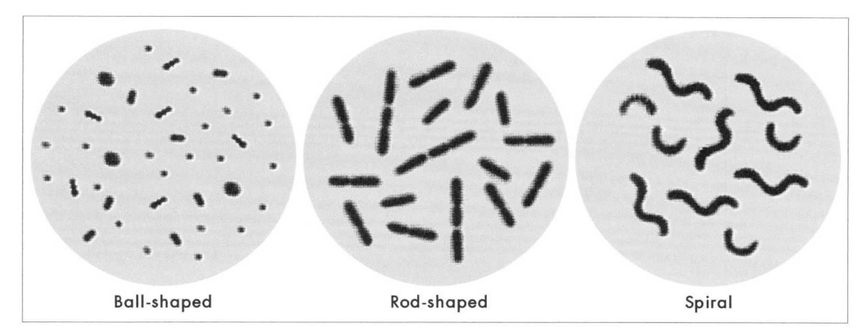

Three main groups of bacteria

Lesson 46

Lesson Concepts

1. Bacteria need food, warmth, moisture, and darkness to grow rapidly.
2. Since bacteria do not contain the green material to allow them to make food by sunlight, they must get their food from dead or living organisms.
3. Some bacteria cause disease and spoilage, some are harmless, and some are beneficial in making cheese, silage, and adding to soil fertility.
4. A material is sterilized when all bacteria are killed by heat, strong chemicals, or light.
5. Rotting is the breakdown of dead plants and animals and is partly the work of bacteria.

Teaching the Lesson

Because we cannot see them without magnifying them, sometimes we forget that bacteria exist. But evidence of their presence surrounds us. Prepared food spoils. Fresh fruit and vegetables become overripe and then turn brown and mushy. The contents of the compost bin get hot and break down into dark, soil-enriching compost. At times, the mysterious activities of bacteria hinder the efforts of man; and at other times, they are a very important asset to the efforts of man.

Since you may not be able to show your students live bacteria, make the lesson come alive by studying the needs and effects of growing bacteria. You can aptly demonstrate the presence and activities of bacteria by observing most anything that rots with exposure to room temperature air. Study some of these in class.

Four Things Bacteria Need to Live

1. Food. Algae, moss, and ferns can make their own food, but bacteria cannot. Like animals and people, they lack the chlorophyll that enables algae and plants to produce food from water and air. So they must find food to eat.

Some bacteria are parasites. They live in and feed on other living things, such as your body. Many bacteria feed on plant or animal matter that is no longer alive. Since we also eat plant and animal matter, we compete with bacteria for our food supply. If bacteria win the race, they spoil our food before we can eat it.

Bacteria are abundant in almost any plant or animal matter. We do not like when bacteria spoil our food, yet we are glad they live in the soil and eat the dead animal and plant matter there.

2. Warmth. Since your body is warm, it provides an excellent place for bacteria to grow. A compost pile of leaves and grass has bacteria that produce their own heat. In this warm place, the bacteria grow rapidly and consume great quantities of food. Many bacteria live in the soil. The sun shines on the soil and provides the warmth that soil bacteria need to live and grow.

3. Moisture. Bacteria cannot grow in a dry place. Since your body is moist inside, bacteria grow very well there. A compost pile of leaves and grass is also moist. Rainwater keeps the soil moist. And of course, the warm, murky water of many ponds encourages bacteria to grow.

4. Darkness. Bacteria find darkness inside your body. They find it in compost piles as well. Bacteria grow best in darkness.

Harmful Bacteria

Since bacteria cannot make their own food, they often eat food that is meant for people. This causes the food to ***rot.*** Rotting makes food unsafe to eat. Sour milk results from bacteria feeding on the milk and making it spoil. Bacteria also feed on meat, both fresh and cooked. They produce poisons in the meat and give it a bad flavor.

Meat that has been spoiled by bacteria is not safe to eat. Every year many people in the United States suffer from a disease called salmonella (sal′·mə·nel′·ə) poisoning. This disease

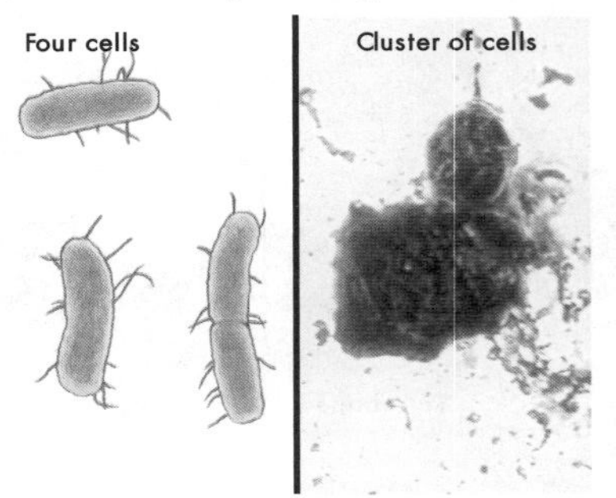

Salmonella bacteria

Discussion Questions

1. What are some ways that bacteria are different from mosses and ferns?
   Bacteria is not green, contains no chlorophyll, and therefore cannot make its own food by photosynthesis.
   Bacteria are also much smaller than mosses and ferns. You cannot see them without a microscope.
   Bacteria have only one cell each. Each moss or fern plant is made of many, many cells.
2. Is rotting harmful or helpful?
   Rotting can be either harmful or helpful. When food spoils, we consider rotting to be harmful. When dead plants and animals rot and produce compost, we consider rotting to be helpful and necessary.
3. Is food that is prepared by bacteria safe to eat?
   Of course! Cheese, yogurt, vinegar, and sauerkraut are all products of bacterial activity. Silage provides good cattle feed. It is prepared by bacteria.

is caused by the salmonella bacteria in improperly cooked chicken, eggs, and meat.

Many sicknesses are caused by bacteria. Your body is an excellent place for bacteria to live and grow. It provides bacteria with food, warmth, moisture, and darkness. When bacteria multiply too rapidly in your body, you become sick. For example, one kind of bacteria causes illnesses such as strep throat and scarlet fever. Sometimes it causes fatal blood poisoning.

God has created your body with the ability to fight harmful bacteria. Skin, mucus, white blood cells, and numerous other defenses protect your body by killing microbes or keeping them from entering your body.

To protect ourselves from harmful bacteria, we can kill them by ***sterilizing*** things. Sterilizing is especially important in hospitals so that diseases are kept from spreading. To avoid introducing harmful bacteria into a patient's body, surgeons sterilize the instruments they use during operations. The bandages used to protect wounds must be sterilized so that the wound is not infected with microbes.

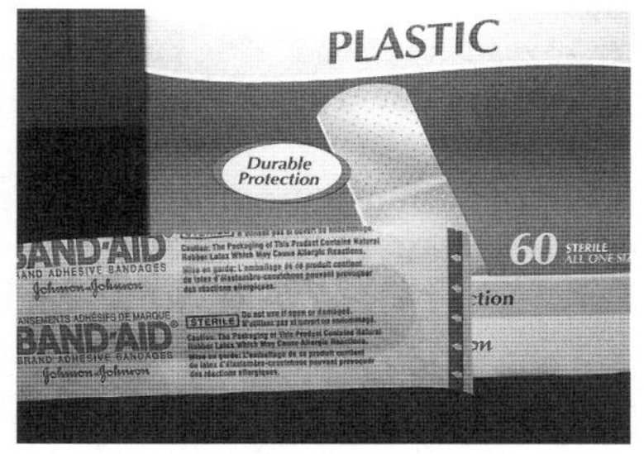

Band-Aids are sealed in paper strips to keep them sterile.

One of the best ways to kill harmful bacteria is with heat. Mothers sterilize baby bottles with boiling water to destroy the microbes that may be in them. Cooking and baking kill the bacteria in food. When your mother cans fruit, she covers the jars tightly and then cooks them in boiling water to kill the bacteria on the fruit. If outside air leaks into a sealed jar, bacteria enter and cause the fruit to spoil.

When heat cannot be used to kill bacteria, people may use chemicals instead. Nurses use alcohol to kill bacteria before giving someone an injection. Farmers use chemicals to control bacteria that cause plant diseases. Strong light may also kill bacteria. Sunlight helps to kill bacteria in the water at sewage treatment plants. We can sterilize things by using heat, chemicals, and strong light to kill the harmful bacteria in them.

Cooling and freezing do not kill bacteria, but low temperatures keep the bacteria from growing rapidly. This is why refrigerators and freezers are useful for preserving food.

Helpful Bacteria

You might wonder why God created bacteria. But not all bacteria are harmful. In fact, bacteria are very necessary for life on the earth. One of the most important ways that bacteria help us is by making things rot.

Is rotting helpful? Imagine what the earth would be like if nothing ever rotted. Great piles of dead plants and animals would litter the ground. Our world would be a miserable place to live. But bacteria help to prevent this, especially by causing plant matter such as leaves in forests to decay. This builds up humus and returns valuable nutrients to the soil. Humus helps the soil to hold water.

Some bacteria make soil fertile by adding nitrogen to it. The air contains much nitrogen, but plants cannot use it in that form. They need nitrogen that is joined with other elements in compounds. Nitrogen-fixing bacteria take nitrogen from the air and put it into a form that plants can use. These helpful bacteria grow in lumps on the roots of plants such as peas and alfalfa.

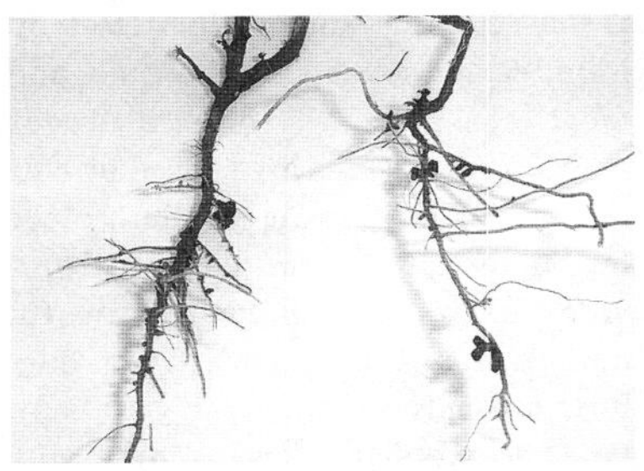

Lumps of bacteria on pea roots add nitrogen to the soil.

Have you ever eaten food prepared by bacteria? It is almost certain that you have. Bacteria change milk to cheese and yogurt. At least part of the flavor in these foods comes from the work of bacteria. Bacteria change apple juice to vinegar. Sometimes baking recipes call for sour milk. Milk is soured by bacteria.

Bacteria even help to prepare feed for cattle. When farmers chop corn or alfalfa and blow it into a silo, bacteria changes it to silage. This makes the fodder easier for cattle to digest.

Exercises

1. In what important way are bacteria different from algae and plants?
2. a. List the four things bacteria need to live.
   b. List four places that supply these needs and allow bacteria to grow well.
3. Give two ways in which bacteria are harmful.
4. List the three ways of killing harmful bacteria, and give an example of each. See if you can think of examples not mentioned in the lesson.
5. How is a sterilized bandage different from one that is not sterilized?
6. What is the purpose of freezing food?

Lesson 46 Answers

Exercises

1. Bacteria lack chlorophyll and cannot make their own food.
2. a. food, warmth, moisture, darkness
   b. (Sample answers.) human body, soil, compost pile, pond water
3. They cause food to spoil.
   They cause diseases.
4. (Examples will vary.)
   heat—by sterilizing, cooking, baking, canning;
   chemicals—with alcohol, sprays for bacterial plant diseases, antiseptics;
   strong light—in sewage treatment
5. If a bandage is sterilized, the bacteria on it have been killed.
   If a bandage is not sterilized, the bacteria have not been killed.
6. Freezing keeps the bacteria in the food from growing rapidly and causing the food to spoil.

7. How does bacteria help a crop farmer by building up the soil? Describe two ways.
8. List two foods you have eaten that bacteria helped to prepare.

Review

9. For what two purposes has God given us bones?
10. Since bones are stiff and hard, what has God put in our bodies to let us move?
11. Why are muscles important to our bodies?
12. Explain the work of each body part named below.
    a. the heart
    b. the digestive tract
    c. the diaphragm
13. What is the work of the nervous system?
14. Explain what first aid should be given in the following cases.
    a. severe bleeding
    b. a broken leg

Activities

1. Use an encyclopedia to find which of the three methods of killing bacteria are used to sterilize a surgeon's instruments before he performs an operation.
2. Get a small jar with a snug-fitting lid, some water, some dried beans, and a medicine dropper. Fill the jar halfway with tap water. Drop a dozen beans into the water to soak. Let them soak several days without a lid. Then put the lid on the jar, and keep it in a warm place. After a few days, open the jar, and use a medicine dropper to remove a few drops of water and scum from the surface of the liquid. Place several drops on a microscope slide, and observe it under both low and high power.

   What do you see? Some of those things are bacteria! The nutrients in the beans helped the bacteria to grow rapidly. In a week's time, they multiplied until there were thousands of them in the jar. These bacteria are the same harmful bacteria that cause food to spoil. That is just what happened. You left the beans (food) exposed to bacteria (in the air and water). Then you kept them in a warm place until the bacteria caused the beans to begin rotting.

7. Bacteria cause plant and animal matter to rot and add humus to the soil.
   Bacteria take nitrogen from the air and fix it in compounds so that plants can use it.
8. (Sample answers.) cheese, yogurt, sauerkraut, vinegar, sour cream, sour milk

Review

9. for support and protection
10. joints
11. Muscles move the bones. Muscles enable organs like the heart and stomach to do their work.
12. a. It pumps blood through the body.
    b. It digests the food.
    c. It increases the size of the chest, which causes the lungs to draw in air.
13. The nervous system carries messages from the brain to all parts of the body.
14. a. Stop the bleeding by using a clean cloth to put pressure on the wound.
    b. Use a splint to keep the broken bone from moving and causing further damage.

Activities

1. High-pressure steam (heat) is used to sterilize some surgical instruments. Poisonous gases may be used for items that would be harmed by the heat or moisture of steam.
2. As always with microscope activities, you as the teacher must be in charge to prevent damage to the microscope.

Lesson 47

Fungi—Molds, Mushrooms, and Yeasts

Vocabulary

ferment (fər·ment′), to go through a chemical change in which sugar changes to carbon dioxide and alcohol.

fungus (fung′·gəs), *plural* **fungi** (fun′·jī), a simple, plantlike organism that lacks chlorophyll and cannot make its own food.

mold, a fungus that usually grows as a fuzzy or hairy covering over its food source.

mushroom, a fungus that sends fleshy, umbrella-shaped growths above ground to produce spores.

yeast, one-celled fungi that are able to change sugar and other carbohydrates into carbon dioxide and alcohol.

Kinds of Fungi

A ***fungus*** is a plantlike organism. Unlike plants, though, fungi do not contain chlorophyll and are not able to make their own food. They get their nutrients from the plants and animals on which they grow, as bacteria do. Like bacteria, fungi cause their food source to rot. Fungi grow almost everywhere. Almost everyone has seen them and has been helped or harmed by them.

One of the most common fungi is ***mold.*** Perhaps you would like to grow some bread mold. Lay a piece of soft white bread on a table in your classroom. If it becomes dry and hard, water it lightly. In a few days, you will see mold with fine strands like spider webs growing on the surface of the bread. The mold grows not only on the surface but also deeper into the bread. Molds grow on bread, cheese, and other kinds of food.

Where did the mold come from? Like most other fungi, molds multiply by producing millions of spores. These spores are too small to see unless you have a microscope. They are so tiny that winds easily blow them around. Mold spores were floating in the air of your classroom, and some of them landed on the bread and started growing.

If the mold on your bread continues to grow, it will form tiny spore cases throughout the tangle of mold fibers. They will look like black dots. These spore cases produce microscopic spores to be released into the air and start new molds.

Lesson 47

Lesson Concepts

1. Since fungi do not contain chlorophyll for making food with sunlight, all fungi must get their food from dead or living organisms.
2. The main part of a mushroom is an underground mass of branching fibers, which remain alive in the soil or wood even when no spore body is present. The part above the ground is only the spore-producing body.
3. Molds and mushrooms reproduce by spores.
4. Yeasts reproduce by budding.
5. Many plant diseases are caused by fungi.
6. Rotting is also partly the work of fungi.
7. Yeast is used to make tiny bubbles in bread during rising.
8. Yeast causes fermentation, which produces alcohol. This alcohol has good uses, but sinful man has wrongly used it to make strong drink.

Teaching the Lesson

Many kinds of fungi surround us. Mildew plagues our basements and shower curtains. Molds of various kinds may ruin our fresh fruit supplies, but they also provide the delicate flavor in our cheese. And mushrooms—those charming, edible delicacies—find a warm spot in most people's hearts. We all enjoy bread that has been raised and made fluffy by yeast. We would not want to be without it. This is the world of fungi—some good and some bad, but all important to us.

Do not overemphasize the negative qualities of fungi. Some people seem to have a bias against them. Passing on such a bias to your students will be sure to keep them from seeing the beauty and significance of this part of God's creation. As much as practical, bring specimens to class for the students to observe. They will enjoy contributing to the collection also.

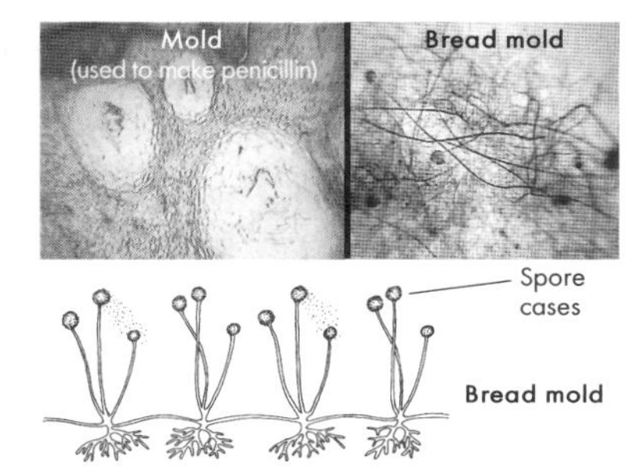

Another common fungus is the ***mushroom.*** Probably you think a mushroom is the familiar umbrella-shaped growth that you see in dark, damp places. You may be surprised to learn that the main part of a mushroom is not the part above the ground but the mass of fine threads that grow underground. These threads usually live for many years and occasionally send an umbrella-shaped structure above the ground.

The familiar umbrella-shaped structure that we call a mushroom or toadstool is the fruiting body of the mushroom. It lives only long enough to produce spores; then it decays. Under its cap, it has gills or tubes that produce spores. Often within hours after the fruiting body has finished growing, its spores mature and drop from the gills or tubes under the cap.

One common mushroom, the puffball, produces its spores inside a ball instead of a toadstool. When the ball ripens, the inside of the mushroom fills with powdery brown spores. If you step on a puffball, millions of spores will puff out of it. The spores rise in a little brown cloud to land elsewhere and start new puffball fungi.

Yeasts are tiny single-celled fungi with an oval shape. You need a microscope to see them. Instead of producing spores to multiply, yeast cells form buds. When a cell becomes large enough, a bulge begins to form on its side. This bulge grows until it forms a bud, which divides from the main cell and becomes a separate cell. The new cell also grows and forms buds.

The fruiting body of a mushroom

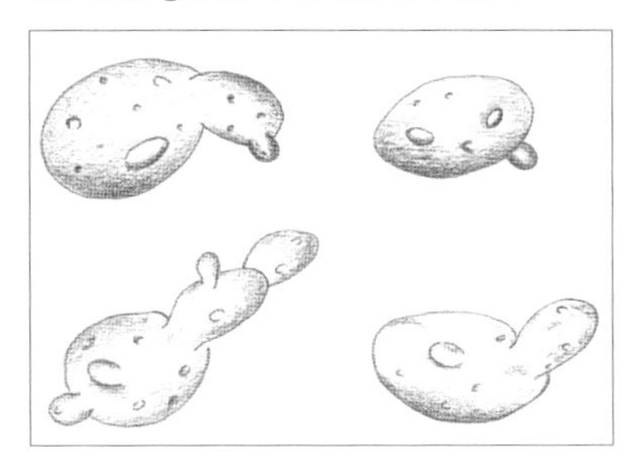

Yeast cells showing budding

Discussion Questions

1. What is the main part of a mushroom?
   The main part of a mushroom is an underground mass of branching fibers. The mushroom is only the fruiting body, the spore-bearing structure.
2. How do yeast cells make bread rise?
   The yeast cells actually use a small bit of the bread dough for food. As they do this, they produce bubbles of carbon dioxide in the bread dough, thus causing the dough to become spongy and to rise.
3. How can you tell whether or not a mushroom is safe to eat?
   Be sure to emphasize that there is no simple way to distinguish between poisonous and nonpoisonous mushrooms. You should never eat a wild mushroom unless an experienced person tells you it is all right. "If in doubt, don't."

Bud after bud, cell after cell, the yeast cells grow and spread throughout the food source. This is what the Bible means when it says, "A little leaven leaveneth the whole lump" (Galatians 5:9). The word *leaven* in this verse means "yeast."

Harmful Fungi

Many plant diseases are caused by fungi. Rust, smut, and mildew are all fungi that damage plants. Farmers work hard to keep these fungi from harming their crops. In the 1840s, a fungus blight killed most of the potato plants in Ireland. About 750,000 people starved, and many others left the country.

Fungi also cause human ailments. Ringworm, an itchy round patch on the skin, is a fungus. Often the patch is red and flaky. Another common fungus causes athlete's foot. It grows on people's feet, especially between the toes, and causes itching and flaking. Fungus diseases often cause the skin to crack, allowing the flesh below to become infected with bacteria.

Some fungi, especially molds and mildew, cause much harm by making things rot. Books, paper, wood, leather, and cloth are weakened or destroyed when mold and mildew feed on them. Keeping things dry is usually the best way to prevent this kind of damage.

Useful Fungi

Fungi are useful as well as destructive. They help us by causing leaves and other plant matter to rot. The decaying plant matter builds up the soil and makes it rich and well suited for growing plants.

Some fungi are used to make or flavor food. Many varieties of cheese are flavored by special molds that grow on and through them. The strength of the flavor depends on how long the cheese makers let the mold grow on the cheese.

Many people eat mushrooms. Some gather them in the wild, and others eat mushrooms grown at special mushroom farms. Since many wild mushrooms are poisonous, you need to be *extremely* careful when gathering them to eat. Many people prefer mushrooms from mushroom farms because they know these mushrooms are safe to eat. Others prefer wild ones because they have a better flavor. Mushroom lovers use words like *nutty* and *oyster-flavored* to describe the taste of various mushrooms.

One tasty mushroom found in some parts of the country is the common morel. This spongelike mushroom usually grows from two to five inches tall. The morel is considered one of the most delicious mushrooms, and it may be the most commonly gathered wild mushroom in the United States.

Another common edible mushroom that grows wild is the puffball. The size of this ball-shaped mushroom ranges from a golf ball to a basketball. When the spores mature, puffballs are powdery inside and are not good to eat. They must be gathered and eaten while they are still young and white inside.

Yeast is important in preparing many foods. Probably the most common use of yeast is in baking bread. As the yeast grows and spreads through the bread dough, it changes starch and sugar in the dough to alcohol. It also produces carbon dioxide gas at the same time. The carbon dioxide forms bubbles in the dough, causing it to rise into soft, light loaves. When the dough is baked, the heat kills the yeast and removes the alcohol. The carbon dioxide is harmless, so all you have left is delicious, fluffy bread.

Yeast is also used to produce alcohol. The alcohol forms as yeast feeds on starches and sugars in fruit juices or grain products. This process is called ***fermenting.*** Alcohol is valuable in many ways. It is used in fuel, in medicine, in paint thinner, and even in making plastic.

But fermenting also produces wine and other alcoholic drinks. These dangerous drinks may destroy the body and ruin the soul. "Wine is a mocker, strong drink is raging: and whosoever is deceived thereby is not wise" (Proverbs 20:1). Alcoholic drinks are very powerful and damaging. "At the last [wine] biteth like a serpent, and stingeth like an adder" (Proverbs 23:32).

Morels

Puffballs

Exercises

1. What are two ways that fungi are like bacteria?
2. Nearly all fungi reproduce by forming tiny cells called ———.
3. The main part of a mushroom grows (above, under) the ground.
4. What is the purpose of the umbrella-shaped part of a mushroom?

Lesson 47 Answers

Exercises

1. They lack chlorophyll and cannot make their own food.
   They cause things to rot.
2. spores
3. under
4. It produces spores.

5. When a yeast cell is large enough, a bud grows from the side of the cell. This bud splits off to become a new yeast cell.
6. (Sample answers.)
   They make bread unfit to eat.
   They kill plants. A fungus blight killed the potato plants in Ireland.
   Ringworm is an itchy, flaky skin disease caused by fungi.
   Mildew makes curtains rot in a damp basement.
   Mold can ruin schoolbooks in summer storage.
7. d
8. The growing yeast produces bubbles of carbon dioxide that cause the dough to rise.

Review

9. Sound is produced when something vibrates and makes the air around it vibrate.
10. An echo is produced when sound travels to a distant object, bounces back, and is heard a short time after the original sound.
11. a. frequency of vibration
    b. loudness or softness
12. hammer, anvil, stirrup
13. food (sugar), oxygen
14. green, golden, brown, red, blue-green
15. food, warmth, moisture, darkness
16. Bacteria cause dead animals and plants to rot.
    They help enrich the soil by adding nitrogen and making humus.
    Bacteria help to prepare certain foods for man and animals.

Activities

1. Help the students identify the mushrooms and fungi that they find. Teach them how to use a field guide. Caution them about the danger of eating unidentified or poorly identified mushrooms.
2. White paper works well for mushrooms with dark-colored spores. Provide a suitable place for the pupils to set their mushroom caps for gill prints. It should be a dark place without disturbances of any kind, including breezes.

5. Yeast are different from other fungi in the way they reproduce. Explain in your own words how yeast reproduce.
6. Name three specific ways in which fungi are harmful.
7. Fungi are good in all the following ways **except**
   a. they cause things to rot.
   b. they make and flavor food for us.
   c. they make our bread fluffy.
   d. they change fruit juice into strong drink.
8. Explain how yeast causes bread dough to rise.

Review

9. How is sound produced?
10. Explain how an echo is produced.
11. Write one or more words to complete each sentence.
    a. The pitch of a sound is determined by its ———.
    b. The volume of a sound is its ———.
12. List in order the three tiny bones in the middle ear.
13. Algae use photosynthesis to produce ——— and ———.
14. What are the five groups of algae?
15. List the four things bacteria need to live.
16. Give three ways in which bacteria are helpful.

Activities

1. Start a mushroom and fungus identification chart. The chart should have four columns: one to name the mushroom or fungus, and the others to tell who saw it, where he saw it, and when he saw it. Use an encyclopedia or field guide to help you decide what kinds of mushrooms and fungi you find. **CAUTION: Never eat any mushroom or fungus unless an older, experienced person is along to make sure it is all right. Some mushrooms are very poisonous.**
2. Pick several fresh mushrooms with gills on the underside of their caps. Break off the caps so that they lie flat. Cut small squares of black construction paper a little larger than the mushroom caps. Place the caps on the paper, and set them inside a cabinet. Be sure that no breeze disturbs them. After two days, carefully remove them from the cabinet and lift the caps from the paper. Do you see light-colored spokes on the construction paper? They are made of mushroom spores that fell from the cap.

Lesson 48

Unit 8 Review

Review of Vocabulary

algae	fiddleheads	mushroom
alternation of generations	filaments	photosynthesis
bacteria	fission	rot
cell	fronds	spores
chlorophyll	fungi	sterilize
chloroplasts	microbes	yeast
ferment	mold	zygote

The green scum floating on the surface of ponds is an example of _1_. Because algae provide food for many ocean creatures, they have been called the pasture of the sea. Some algae consist of a single _2_. The tiny green structures within algae cells are called _3_. They contain _4_, the green chemical that helps algae and plants to make food from water and carbon dioxide by the aid of sunlight. This chemical process is called _5_.

Some algae are single-celled, some form glob-shaped colonies of cells, and others form long strings of cells called _6_. When algae cells split in a process called _7_, the colony multiplies and becomes larger.

Moss is a carpet of familiar plants with tiny leaflets. Ferns are lacy plants with leaves called _8_. The leaves are tiny rolled-up buttons known as _9_ when they begin to open. What we see above the ground is only part of the fern plant. The main stem of the fern grows just under the surface of the soil.

Ferns multiply in an interesting way. The undersides of the fronds have dark spots that produce _10_, which float away to start new ferns. But from the spore grows a strange-looking plant instead of the one familiar to us. This plant is very short-lived and produces a seedlike _11_ that then produces another common, lacy fern. This two-step reproduction is called _12_.

_13_ are tiny single-celled organisms that cannot make their own food. When bacteria eat something and cause it to break down, they cause it to _14_, or decompose. Bacteria are both helpful and harmful. Helpful bacteria change apple juice to vinegar, cabbage to sauerkraut, and milk to yogurt. Harmful bacteria, called

Lesson 48 Answers

Review of Vocabulary

1. algae
2. cell
3. chloroplasts
4. chlorophyll
5. photosynthesis
6. filaments
7. fission
8. fronds
9. fiddleheads
10. spores
11. zygote
12. alternation of generations
13. Bacteria
14. rot

15. microbes
16. sterilize
17. fungi
18. mold
19. mushroom
20. Yeast
21. ferment

germs or _15_, cause disease, infection, and food spoilage. We _16_ an object when we kill all the bacteria on it.

Another group of living things that cannot make their own food is _17_. One example is the fuzzy mass of fibers called _18_, which grows on a moist food like bread. Another fungus, the _19_, has an underground mass of fibers and an umbrella-shaped fruiting body that produces spores.

_20_ is an interesting fungus that multiplies by budding. When it feeds on the starch and sugar in bread dough, it produces carbon dioxide bubbles that cause the bread to rise. When it feeds on the sugar in a fruit juice, it causes the juice to _21_ and change into strong drink.

Multiple Choice

1. d
2. c
3. b, c
4. a, c
5. b
6. a, c

Multiple Choice

1. All of the following are examples of algae **except**
   a. green pond scum.
   b. red seaweed.
   c. a green film on tree trunks.
   d. a green carpet of moss.
2. Algae are most useful to man because
   a. they provide a home for fish.
   b. they provide food for man.
   c. they provide food for edible water creatures.
   d. they provide carbon dioxide for man.
3. Moss plants are different from most other plants because (choose two)
   a. they are very small.
   b. they lack true roots.
   c. their leaves are only one cell thick.
   d. they have stems.
4. Mosses and ferns both reproduce by using (choose two)
   a. spores.
   b. fission.
   c. zygotes.
   d. budding.
5. The spores of ferns are produced
   a. in spore cases at the ends of tiny stalks.
   b. in dark spots on the undersides of the fronds.
   c. inside fern seeds.
   d. inside buds.
6. Many gardeners use peat moss for (choose two)
   a. mulching their flower beds.
   b. killing harmful insects.
   c. wrapping seedling roots to keep them moist.
   d. preventing plant diseases.

7. Bacteria need all of the following for life **except**
   a. warmth. c. moisture.
   b. food. d. sunlight.
8. All the following things help to provide food for us **except**
   a. yeast. c. salmonella.
   b. molds. d. mushrooms.
9. Bacteria are helpful by (choose two)
   a. making dead things rot and making food spoil.
   b. making dead things rot and changing milk to yogurt.
   c. building up the soil and causing diseases.
   d. building up the soil and changing corn fodder to silage.
10. All the following things can kill bacteria **except**
    a. heat. c. strong chemicals.
    b. freezing. d. sunlight.
11. The main part of a mushroom fungus is
    a. the umbrella-shaped part above the ground.
    b. an underground stem that sends up occasional fronds.
    c. the mass of underground fibers that live for many years.
12. Molds and mushrooms both reproduce by
    a. budding. c. zygotes.
    b. fission. d. spores.
13. Which of the following things is able to carry on photosynthesis?
    a. algae c. bacteria
    b. molds d. mushrooms
14. Yeast makes bread dough rise by
    a. producing alcohol, which makes the dough swell.
    b. budding and growing until the dough swells with yeast.
    c. producing carbon dioxide bubbles that make the dough spongy.
    d. causing the dough to ferment.

7. d
8. c
9. b, d
10. b
11. c
12. d
13. a
14. c

God's Wonderful World

Unit 1 Test **Score** ____________

Name ______________________________ **Date** ____________________

A. Fill in each blank on the left with the letter of the correct word from the list on the right. Not every word will be used. *(9 points)*

__f__ 1. An event that cannot be explained by natural laws.

__g__ 2. Something that God set up to control the universe and give it order.

__b__ 3. God's act of making everything out of nothing.

__e__ 4. A great, worldwide miracle that God used to punish sinful man.

__c__ 5. Information that is gathered and organized to help make a decision or discovery.

__h__ 6. Use of the five senses to study God's creation.

__d__ 7. A test made to discover some new truth about God's creation.

__j__ 8. A mistaken idea about God's world that is not according to natural laws or the Bible.

__i__ 9. The observation and study of God's creation.

a. chemistry
b. the Creation
c. data
d. experiment
e. the Flood
f. miracle
g. natural law
h. observation
i. science
j. superstition

B. Choose one of the following words to fill in each blank. *(4 points)*

astronomy biology botany chemistry physics

10. The branch of science dealing with the makeup of materials is ____chemistry____.
11. ____Biology____, another branch of science, is the study of living things.
12. The branch of science that deals with force and energy is ____physics____.
13. The study of stars and other heavenly bodies is called ____astronomy____.

C. Circle *T* or *F* to show whether each statement is true or false. *(6 points)*

(T) F 14. Only God can create things from nothing.

T (F) 15. Floodwaters make little change in the surface of the earth.

(T) F 16. God has promised that He will never again destroy the earth with a flood.

T (F) 17. We are sure that the Flood happened because scientists have discovered that the world was once covered by water.

(T) F 18. Science is divided into branches because science is such a large area of study.

T (F) 19. Only scientists make helpful observations.

D. Circle the letter of the correct choice. *(7 points)*

20. God created the entire universe in
    a. many years.
    b. six days. (circled)
    c. six weeks.
    d. one moment.

21. All the following things may have been formed by the Flood **except**
    a. mountains.
    b. canyons.
    c. plant life. (circled)
    d. soil layers.

22. An example of a natural law would be
    a. tulips are usually red.
    b. grass seed never produces bean plants. (circled)
    c. most people like grass in their lawns.
    d. some yards do not have grass.

23. God gave us five senses because He wants us to
    a. be able to do the things we enjoy.
    b. become great, important people.
    c. learn about His world so that we can serve Him better. (circled)
    d. learn about His world so that we can make life pleasant.

24. Which is most likely to lead to wrong ideas about the natural world?
    a. planning tests so you can observe more closely
    b. performing an experiment several times
    c. writing observations so you can remember them
    d. deciding what to think after seeing something happen one time (circled)

25. People form superstitions when
    a. they study the Bible.
    b. they study science too much.
    c. they test their ideas with experiments.
    d. they become careless in their observations. (circled)

26. Which one of these things would be **wrong** to test by experiments?
    a. A farmer wants to know which type of corn grows best.
    b. A carpenter needs to know which kind of wood is best suited for porch floors.
    c. You wonder if drinking and smoking are actually harmful to your body. (circled)
    d. You wonder if setting the mower a little higher will keep the grass from turning brown.

E. Write good, complete answers. *(7 points)*

27. Write a good definition of science.

    Science is the observation and study of God's wonderful world.

28. What two things should the study of science help us to do? (Any two.)

The study of science should lead us to praise God for His wonderful works.

It should help us to solve the everyday problems of life on earth.

It should help us to be good stewards of what God has entrusted to us.

29. List in order the four steps of an experiment.

(1) Write down the question you are trying to answer.

(2) Plan an experiment that will help to answer the question.

(3) Work the experiment and record your observations.

(4) Use the data from the experiment to answer your question.

(Total points: 33)

God's Wonderful World

Unit 2 Test **Score** ____________

Name ______________________________ **Date** ______________

A. Fill in each blank on the left with the letter of the correct word from the list on the right. Not every word will be used. *(10 points)*

__d__ 1. The stiff outer shell of an arthropod.
__l__ 2. The middle part of an animal with three main body parts.
__a__ 3. The back part of an animal with three body parts.
__g__ 4. The stages that an insect goes through in becoming an adult.
__i__ 5. To shed an exoskeleton.
__j__ 6. The young of some insects, which looks much like the adult but is smaller.
__f__ 7. The wormlike young of insects such as butterflies.
__k__ 8. The inactive stage of insects during which the last part of metamorphosis takes place.
__c__ 9. The hard case of the pupa stage of a butterfly.
__h__ 10. The great change that takes place in the lives of many insects.

a. abdomen
b. antennae
c. chrysalis
d. exoskeleton
e. gills
f. larva
g. life cycle
h. metamorphosis
i. molt
j. nymph
k. pupa
l. thorax

B. Choose one of the following words to fill in each blank. *(3 points)*

antennae gills lungs spinnerets

11. Crayfish use ____gills____ to get oxygen from the water.
12. Spiders use ____spinnerets____ on the tips of their abdomens to produce silk for building webs.
13. A pair of ____antennae____ attached to its head allows an insect to touch and smell things.

C. Circle *T* or *F* to show whether each statement is true or false. *(4 points)*

(T) F 14. Spiders are helpful creatures.
T (F) 15. Grasshoppers pass through four life stages.
T (F) 16. Insects lay eggs during the larva stage.
(T) F 17. Spider silk is stronger than cotton or wool.

D. Circle the letter of the correct choice. *(10 points)*

18. Crayfish, grasshoppers, and spiders are all
    a. insects.
    b. arthropods.
    c. crustaceans.
    d. arachnids.

19. Animals with exoskeletons
    a. are stiff and cannot move around very well.
    b. have a soft covering that allows them to move.
    c. have many joints that allow them to move.
    d. cannot move at all, but stay in one place all their lives.

20. Insects take in oxygen through
    a. lungs.
    b. gills.
    c. tiny holes.
    d. special pads.

21. Spiders have
    a. two body parts and eight legs.
    b. three body parts and eight legs.
    c. two body parts and six legs.
    d. three body parts and six legs.

22. All spiders
    a. build webs to catch their prey.
    b. can produce silk.
    c. can see well.
    d. have four spinnerets.

23. Insects have
    a. three body parts and eight legs.
    b. three body parts and six legs.
    c. two body parts and eight legs.
    d. two body parts and six legs.

24. Creatures with exoskeletons molt because
    a. their exoskeletons shrink.
    b. their exoskeletons get too stiff.
    c. their exoskeletons do not grow.
    d. they no longer need an exoskeleton.

25. The stages in a grasshopper's life are
    a. egg, larva, adult.
    b. egg, nymph, adult.
    c. egg, larva, nymph, adult.
    d. egg, larva, pupa, adult.

26. Which of the following is **not** an insect?
    a. a caterpillar
    b. a spider
    c. a grasshopper
    d. a butterfly

27. The best way to kill an insect for a collection is to
    a. cut its head off.
    b. poison it.
    c. freeze it.
    d. push a pin through it.

E. Write good, complete answers. *(7 points)*

28. What are the stages in the life of a butterfly? Name and describe each one.

Egg: The adult butterfly lays the egg on a suitable food plant.

Larva: A caterpillar hatches from the egg and grows rapidly, molting several times.

Pupa: The caterpillar attaches itself to a support, forms a chrysalis around itself, and enters a quiet stage during which it mysteriously changes into a butterfly.

Adult: The chrysalis splits open, and the adult butterfly emerges.

29. a. What is one way in which insects are harmful? (Answers may vary.)

Insects like houseflies and mosquitoes are harmful because they spread disease.

Some beetles injure crops.

Bees and wasps have painful stings.

b. What are two benefits that we receive from insects? (Any two.)

Some insects eat harmful insects.

Some produce useful things like honey, wax, shellac, and dyes.

Bees and some other insects pollinate garden plants and fruit trees.

Insects provide food for birds and other animals.

Some insects help to get rid of dead animals.

(Total points: 34)

God's Wonderful World

Unit 3 Test **Score** ________

Name ____________________ **Date** ____________

A. Fill in each blank on the left with the letter of the correct word from the list on the right. Not every word will be used. *(9 points)*

g	1. Molten rock that has reached the surface of the earth.	a. climate
b	2. A crack between two sections of the earth's crust.	b. fault
m	3. The condition of the atmosphere at a certain time.	c. flow
l	4. A powerful wave resulting from an earthquake that occurs under the ocean.	d. front
f	5. The height of a waterfall.	e. galaxy
k	6. An instrument that records tremors in the earth's crust.	f. head
d	7. The boundary between a warm air mass and a cold air mass.	g. lava
a	8. The average weather over a period of time.	h. magma
e	9. A large group of stars usually numbering in the billions.	i. magnitude
		j. ocean current
		k. seismograph
		l. tsunami
		m. weather

B. Write the missing words. *(4 points)*

10. When two sections of the earth's crust suddenly shift, an ___earthquake___ is the result.

11. The most common use for waterpower at dams and waterfalls is producing ___hydroelectricity___.

12. Since the stars are so far away, scientists measure their distances with the ___light-year___, a very large unit of measure.

13. A ___binary___ ___star___ is made of two stars that revolve around each other.

C. Circle *T* or *F* to show whether each statement is true or false. *(7 points)*

T (F) 14. Mountains affect the weather only a little.

(T) F 15. Two great dangers resulting from earthquakes are falling buildings and fire.

T (F) 16. Waterpower is seldom used today.

T (F) 17. Ocean waves affect climate by moving warm water northward to cold regions.

(T) F 18. Tornadoes develop from severe thunderstorms.

(T) F 19. High, wispy clouds are a sign that rain will come in a few days.

(T) F 20. We can see about 3,000 stars on a clear, dark night.

D. Circle the letter of the correct choice. *(8 points)*

21. The center of the earth is
    (a.) very hot.
    b. very cold.
    c. the same temperature as the crust.
    d. cold in some places and hot in others.

22. Choose the one that is **not** a sign of fair weather.
    a. fluffy, cottony clouds
    b. a rising barometer
    c. winds from west or north
    (d.) a ring around the moon

23. What determines how much energy a waterfall has?
    a. depth of water and width of waterfall
    (b.) height of waterfall and flow of waterfall
    c. depth of river and rate of flow
    d. width of river and height of waterfall

24. Oceans cover about ——— of the earth's surface.
    a. 50% (half)
    b. 80% (8 tenths)
    c. 30% (3 tenths)
    (d.) 70% (7 tenths)

25. Tornadoes and hurricanes are alike in that they both
    (a.) have swiftly rotating winds.
    b. form over land.
    c. are high-pressure areas.
    d. have a funnel-shaped cloud.

26. Weather forecasting is possible because
    (a.) weather follows the patterns that God established.
    b. scientists understand the weather completely.
    c. weather is very simple to understand.
    d. both *a* and *b*.

27. Magnitude is used to tell
    a. how far away a star is.
    b. how fast a star is moving.
    (c.) how bright a star appears.
    d. how large a star is.

28. Which is **not** connected with rainy weather?
    a. a thunderstorm
    b. a low-pressure area
    c. a falling barometer
    (d.) a high-pressure area

E. Write good, complete answers. *(3 points)*

29. How do mountain glaciers help to provide water for surrounding valleys?

    The glaciers slide slowly down into the valleys and melt as they reach lower altitudes.

30. In what two ways does a telescope help us see stars better?

    The lenses and mirrors in a telescope gather more light and make the stars appear brighter. They also make the stars appear larger and nearer.

(Total points: 31)

God's Wonderful World

Unit 4 Test **Score** ____________

Name ________________________ **Date** ____________

A. Fill in each blank on the left with the letter of the correct word from the list on the right. Not every word will be used. *(11 points)*

m 1. A tube that transfers a liquid over the edge of a container and down to a lower level.
n 2. Friction that keeps a tire from slipping as it pulls across a surface.
i 3. The amount of force applied to a liquid.
f 4. Something that reduces friction between two surfaces.
h 5. The number of times a machine multiplies an effort.
o 6. Something measured in foot-pounds.
k 7. Any force that hinders the motion of an object.
j 8. A device that forms part of a block and tackle.
e 9. The support on which a lever pivots.
b 10. The amount of force put into a machine.
a 11. A device that uses friction to slow or stop a vehicle.

a. brake
b. effort
c. force
d. friction
e. fulcrum
f. lubricant
g. machine
h. mechanical advantage
i. pressure
j. pulley
k. resistance
l. screw
m. siphon
n. traction
o. work

B. Write the missing words. *(11 points)*

12. A force is required to move something or to stop it.

13. The four kinds of resistance that hinder movement are friction, gravity, inertia, and molecular attraction.

14. The two things that affect the amount of friction are pressure and surface roughness (or smoothness).

15. Machines help us do work by changing the direction, amount, or speed (and distance) of a force.

16. A centrifugal pump has an impeller that uses inertia to build up pressure in a liquid.

C. Circle *T* or *F* to show whether each statement is true or false. *(5 points)*

T (F) 17. Some objects can move without a force.

(T) F 18. Friction is the greatest resistance to be overcome in pulling a nail.

T (F) 19. The mechanical advantage of a lever is found by dividing the resistance-arm length by the effort-arm length.

(T) F 20. A pump can usually move water faster than a siphon can.

(T) F 21. If an effort of 45 pounds makes a 400-pound trailer move 2 feet, 90 foot-pounds of work has been done.

D. Circle the letter of the correct choice. *(7 points)*

22. Wheels help to reduce
    a. gravity.
    (b.) friction.
    c. inertia.
    d. molecular attraction.

23. Lightening a load helps to reduce
    a. force and inertia.
    (b.) gravity and inertia.
    c. gravity and force.
    d. molecular attraction and friction.

24. We can reduce friction in all the following ways **except**
    a. by using lubricants.
    b. by smoothing the surfaces.
    c. by using wheels.
    (d.) by adding pressure.

25. Friction is important because it
    a. is a good source of heat for homes.
    (b.) provides traction for fan belts and car tires.
    c. can be used to cut things.
    d. is used to reduce resistance.

26. A bearing reduces friction by
    a. smoothing the surface of the shaft.
    b. lubricating the shaft.
    c. keeping an air space around the shaft.
    (d.) providing balls or rollers for the shaft to spin on.

27. Which of the following actions will increase the mechanical advantage?
    a. Using a smaller effort wheel in a wheel and axle.
    (b.) Making the slope of an inclined plane more gradual.
    c. Making a wedge shorter and thicker.
    d. Shortening the effort arm of a lever.

28. A gear pump is well suited for
    a. pumping at high pressures and speeds.
    b. pumping large amounts very rapidly.
    c. pumping gritty liquids like mud and sludge.
    (d.) pumping thick, sticky liquids with a strong, even flow.

E. Write good, complete answers. *(5 points)*

29. What are three of the simple machines you have studied? Give an example of each.

(Any three; examples will vary.)

lever, inclined plane, screw, wedge, wheel and axle, pulley

30. What is the mechanical advantage of each machine?

a. lever: effort arm—9 ft.; resistance arm—3 ft. 3

b. a single movable pulley 2

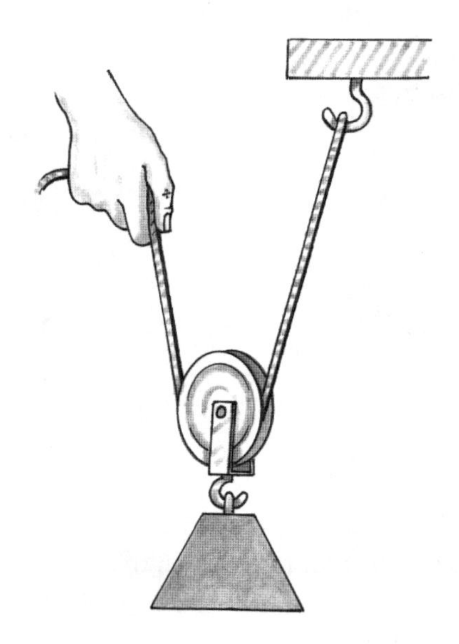

(Total points: 39)

God's Wonderful World

Unit 5 Test **Score** ____________

Name ________________________________ **Date** ____________________

A. Fill in each blank on the left with the letter of the correct word from the list on the right. Not every word will be used. *(10 points)*

g	1. A material made of only one kind of atom.	a. acid
i	2. The smallest possible particle of a compound.	b. base
l	3. A source of chemicals as it is found in nature.	c. chemical change
e	4. One or two letters that stand for an element.	d. chemical formula
d	5. A group of letters that stand for a compound.	e. chemical symbol
a	6. A sour chemical that turns blue litmus paper red.	f. compound
k	7. A characteristic such as shininess.	g. element
j	8. A change caused by freezing, melting, or crushing.	h. fuel
m	9. Poisonous.	i. molecule
h	10. One of the three things needed for a fire to burn.	j. physical change
		k. property
		l. raw material
		m. toxic

B. Write the missing words. *(4 points)*

11. All substances are made from about one hundred different materials called ____elements____.

12. If a substance has two or more kinds of atoms chemically joined as molecules, it is called a ____compound____.

13. In order to burn, fuel must be raised to its ____kindling____ ____temperature____ in the presence of oxygen.

14. Taking the ____antidote____ for a poison will help to counteract the poison.

C. Circle *T* or *F* to show whether each statement is true or false. *(6 points)*

(T) F 15. Some chemical changes are necessary for life to continue.

T (F) 16. Rusting and burning are both physical changes.

(T) F 17. Water is one common product of burning.

T (F) 18. We seldom use chemical changes in our homes.

(T) F 19. Litmus paper shows whether a chemical is an acid or a base.

(T) F 20. Pollution causes harm to living things.

D. Circle the letter of the correct choice. *(9 points)*

21. What is the chemical symbol for copper?
    a. C
    b. Co
    c. Cp
    (d.) Cu

22. The three elements in $FeCO_3$ are
    a. sodium, carbon, and oxygen.
    b. nitrogen, calcium, and oxygen.
    (c.) iron, carbon, and oxygen.
    d. hydrogen, calcium, and nitrogen.

23. Three examples of chemical change are
    a. fading, cutting, and melting.
    (b.) rusting, burning, and fading.
    c. burning, splitting, and freezing.
    d. cutting, melting, and freezing.

24. A fire can be put out in all the following ways **except**
    (a.) being careful with matches.
    b. cooling the fuel below its kindling temperature.
    c. removing the fuel from the fire.
    d. removing the oxygen from the fire.

25. Which one of these is **not** a true statement about chemical changes?
    a. Some chemical changes work very slowly.
    b. Light enables plants to change water and carbon dioxide to sugar and oxygen.
    (c.) Combining any two different materials will cause a chemical change.
    d. Electricity can break apart water molecules.

26. Chemical changes take place in all the following things **except**
    a. baking a cake.
    (b.) melting ice.
    c. curing concrete.
    d. mixing vinegar and baking soda.

27. If a liquid turns red litmus paper blue,
    a. it is an acid.
    b. it is a neutral liquid.
    (c.) it is a base.
    d. it is a metal.

28. If soil in a field has too much acid, the problem may be corrected by
    a. plowing or disking.
    b. spreading fertilizer.
    c. planting corn.
    (d.) spreading lime.

29. Which one of these is **not** a good rule for using strong chemicals?
    a. Keep the chemicals away from children.
    b. Use the chemicals in a well-ventilated place.
    (c.) Pour the leftover chemicals into the creek.
    d. Keep the chemicals from touching your skin.

E. Write good, complete answers. *(4 points)*

30. Write the chemical formula for each compound named below, and list the elements in that compound.

    a. water

    H_2O—hydrogen and oxygen

    b. carbon dioxide

    CO_2—carbon and oxygen

31. If your clothes catch fire, what is the best thing to do?

    Drop to the ground and roll.

32. Why should we avoid polluting the earth?

    We need to take good care of the earth that God created.

(Total points: 33)

God's Wonderful World

Unit 6 Test **Score** ____________

Name ______________________________ **Date** ____________

A. Fill in each blank on the left with the letter of the correct word from the list on the right. Not every word will be used. *(10 points)*

__n__ 1. The bones that form the spine or backbone.
__b__ 2. The padding in a joint.
__i__ 3. A tough band that holds the bones together in a joint.
__h__ 4. The place where two bones meet.
__c__ 5. To become shorter.
__m__ 6. A tough cord that connects a muscle to a bone.
__d__ 7. The passage in the body where food is digested.
__j__ 8. A bundle of fibers that carry messages between the body and the brain.
__f__ 9. Simple care given to an injured person before a doctor or nurse can help him.
__g__ 10. A crack or break in a bone.

a. antiseptic
b. cartilage
c. contract
d. digestive tract
e. esophagus
f. first aid
g. fracture
h. joint
i. ligament
j. nerve
k. skeletal
l. sternum
m. tendon
n. vertebrae

B. Write the missing words. *(4 points)*

11. The ____arteries____ help the heart to pump blood to the rest of the body.

12. When we swallow, the muscles of the esophagus work with a wavelike motion called ____peristalsis____.

13. The skeletal muscles are called ____voluntary____ muscles because we control them consciously, but the heart and stomach are ____involuntary____ muscles that work automatically.

14. Healthy muscles have good ____muscle____ ____tone____; that is, they contract slightly even when relaxed.

C. Circle *T* or *F* to show whether each statement is true or false. *(4 points)*

(T) F 15. Preventing accidents is better than knowing how to give first aid.
T (F) 16. The shoulder and knee are examples of ball-and-socket joints.

(T) F 17. God created man to be most healthy in the upright position.

(T) F 18. If a person has back or neck injuries, you should avoid moving him before an ambulance arrives.

D. Circle the letter of the correct choice. *(9 points)*

19. The lungs are protected by
    a. the vertebrae.
    b. the sternum.
    c. the skull.
    (d.) the ribs.

20. Our bones are stiff and hard, but ——— allow body parts to bend.
    (a.) joints
    b. cartilage
    c. muscles
    d. tendons

21. The muscles that move each part of a limb are located
    a. in that part itself.
    b. in the main part of the body.
    (c.) in the next part closer to the body.
    d. in the next part away from the body.

22. The heart moves blood through the body
    a. by means of peristalsis.
    (b.) by contracting and relaxing.
    c. with the help of the diaphragm.
    d. by both *b* and *c*.

23. The diaphragm is a large muscle that
    (a.) enlarges the chest to draw air into the lungs.
    b. controls the amount of digested food leaving the stomach.
    c. helps the heart to pump blood.
    d. mixes the digesting food in the stomach.

24. Which of these is part of poor walking posture?
    a. arms swinging briskly in time with the legs
    b. head held upright
    (c.) shoulders pulled forward and down
    d. all the things above

25. When lifting a heavy object, you should
    a. keep your back straight as you bend over to pick up the object.
    (b.) bend your knees, squat to grasp the object, and then lift with your legs.
    c. keep your legs straight, bend at your waist, and lift the object by straightening your back.
    d. bend over the object and lift it with your arms.

26. If a person has several injuries, which should be done first?
    a. splinting a fracture
    b. applying a bandage
    (c.) stopping heavy bleeding from a deep cut
    d. washing small scratches and cuts

27. First aid for a small cut or scratch means
    (a.) washing it, applying an antiseptic, and covering it with a clean bandage.
    b. applying direct pressure to stop the bleeding and then covering it with a clean bandage.
    c. applying an antiseptic, washing the cut, and then stopping the bleeding.
    d. avoiding things that might cause a cut or scratch.

E. Write good, complete answers. *(6 points)*

28. Muscles are arranged in pairs. Give an example of this, and tell why it is necessary.

    (Sample answer.)

    The biceps and triceps of the upper arm are one example. Both muscles are needed so that we can bend and unbend the elbow.

29. Describe good standing posture. Be sure to tell the position of the head, the stomach, the shoulders, the legs, and the back.

    head upright and balanced;

    stomach flat;

    shoulders back (square);

    legs and back straight

30. Harold tripped and fell down the stairs at school. He received a deep cut on his cheek, which is bleeding severely. What should be done?

    Press firmly on the cut with a clean cloth (or your hand if no cloth is available) to stop the bleeding. Get the help of a doctor; the cut probably needs stitches.

(Total points: 33)

God's Wonderful World

Unit 7 Test **Score** ____________

Name ______________________________ **Date** ____________________

A. Fill in each blank on the left with the letter of the correct word from the list on the right. Not every word will be used. *(10 points)*

_k_ 1. A fast back-and-forth motion.
_h_ 2. The voice box, where the vocal cords are.
_e_ 3. The reflection of sound from a distant object.
_i_ 4. The highness or lowness of a sound.
_l_ 5. The loudness or softness of a sound.
_d_ 6. The thin membrane at the inner end of the ear canal.
_g_ 7. The tiny bone in the middle ear that is connected to the eardrum.
_j_ 8. The tiny bone in the middle ear that vibrates in the oval window.
_c_ 9. The organ in the inner ear that changes sound waves to nerve signals for the brain.
_b_ 10. Something that carries nerve signals from the inner ear to the brain.

a. anvil
b. auditory nerve
c. cochlea
d. eardrum
e. echo
f. frequency
g. hammer
h. larynx
i. pitch
j. stirrup
k. vibration
l. volume
m. wave

B. Match the letters on the diagram of the ear to the terms listed here. *(8 points)*

_d_ 11. anvil
_f_ 12. auditory nerve
_e_ 13. cochlea
_a_ 14. ear canal
_b_ 15. eardrum
_c_ 16. hammer
_h_ 17. oval window
_g_ 18. stirrup

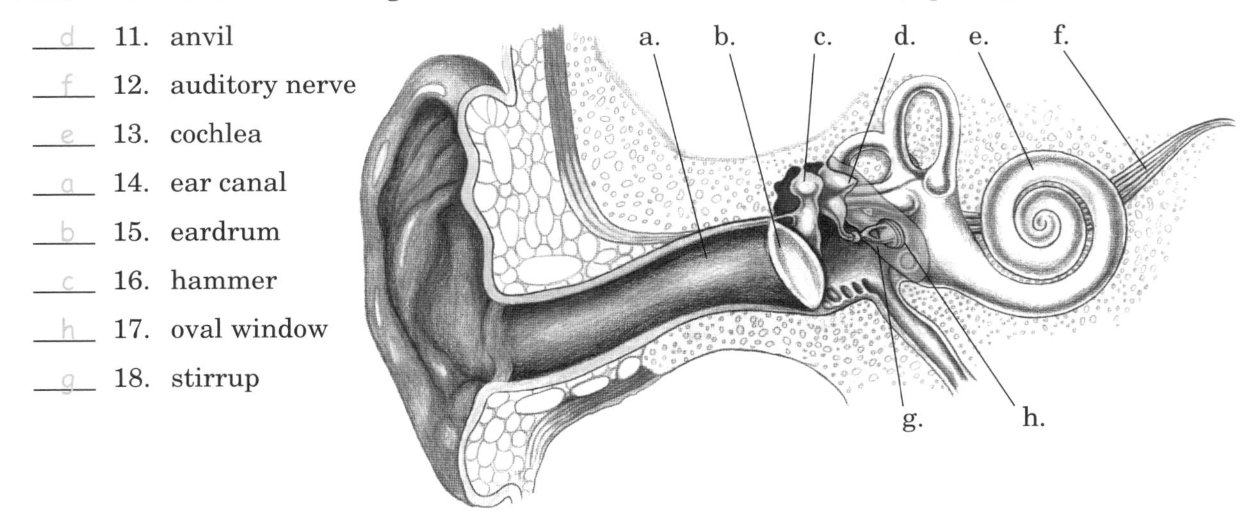

C. Write the missing words. *(4 points)*

19. The tube called the __ear__ __canal__ is part of the outer ear.

20. Your nose and mouth are __resonators__ that change and strengthen the sound of your voice when you speak.

21. The __frequency__ of sound is measured in vibrations per second.

22. Your ear has three main sections: the __outer__ __ear__, the __middle__ __ear__, and the __inner__ __ear__.

D. Circle *T* or *F* to show whether each statement is true or false. *(4 points)*

(T) F 23. Sound is a form of energy.

(T) F 24. Sound is vibration.

T (F) 25. Sound travels with a wavelike motion through a vacuum.

T (F) 26. Different combinations of pitch and volume help us to tell what direction a sound is coming from.

E. Circle the letter of the correct choice. *(7 points)*

27. Sound is produced
 a. by almost everything.
 b. when an object has energy.
 c. after objects stop vibrating.
 (d.) by vibrating objects.

28. When you speak,
 (a.) your vocal cords cause the air to vibrate and make sound.
 b. muscles in your larynx make the air vibrate.
 c. your throat vibrates and your larynx works as a resonator to change the sound.
 d. you use your lungs to make your breath vibrate and produce sound.

29. All the following things will reduce sound **except**
 a. tightening a loose guard on a piece of machinery.
 b. putting carpets and curtains in a living room.
 c. setting a rattly fan on a soft rug.
 (d.) covering the walls of a room with wood paneling.

30. Sound travels fastest through
 (a.) steel.
 b. water.
 c. the air.
 d. a vacuum.

31. When the frequency of a sound is higher,
 (a.) the pitch is higher.
 b. the volume is higher.
 c. the pitch is lower.
 d. the volume is lower.

32. Your voice becomes more high-pitched when
    a. your vocal cords are looser.
    (b.) your vocal cords are tighter.
    c. your vocal cords vibrate less.
    d. your vocal cords stop vibrating.

33. If our voices could not change in pitch or volume, we would not be able to
    a. work.
    b. play.
    (c.) sing.
    d. pray.

F. Write good, complete answers. *(3 points)*

34. How does an echo work?

    An echo is produced when sound is reflected from a surface some distance away, and we hear the reflected sound a little after the original sound.

35. What are two benefits of having two ears rather than just one?

    Two ears enable us to hear sounds coming from any direction.
    Two ears help us to tell what direction a sound is coming from.

(Total points: 36)

God's Wonderful World

Unit 8 Test **Score** ____________

Name ______________________________ **Date** ____________________

A. Fill in each blank on the left with the letter of the correct word from the list on the right. Not every word will be used. *(8 points)*

d 1. The tiny building blocks of which all living things are made.

g 2. A method of reproduction in which a cell divides into two.

i 3. The process used by algae and plants to make sugar and oxygen from water and carbon dioxide by the aid of sunlight.

e 4. The green chemical algae need in order to produce food.

a 5. The kind of reproduction that ferns have, in which new fern plants are different from their parents but like their grandparents.

j 6. To kill all the bacteria on an object.

h 7. A fungus with a mass of fibers underground and an umbrella-shaped growth above the ground.

k 8. One-celled fungi that are able to change sugar into alcohol and carbon dioxide.

a. alternation of generations
b. bacteria
c. budding
d. cells
e. chlorophyll
f. chloroplasts
g. fission
h. mushroom
i. photosynthesis
j. sterilize
k. yeast

B. Write the correct word in each blank. *(4 points)*

9. Sometimes many algae cells are joined end to end in a long threadlike filament.

10. Young fern leaves that are only partly open are called fiddleheads.

11. Fern leaves are called fronds.

12. When fruit juices ferment, there is a chemical change in which the sugar turns into carbon dioxide and alcohol.

C. Circle *T* or *F* to show whether each statement is true or false. *(4 points)*

(T) F 13. Giant kelp is an example of algae.

T (F) 14. Moss spores are formed in special structures on the undersides of the leaves.

T (F) 15. Bacteria are always harmful to the human body.

(T) F 16. The part of the mushroom fungus above the ground is only a spore-producing structure of the main fungus.

D. Circle the letters of the correct choices. *(11 points)*

17. Algae are grouped as
    a. kelp, seaweed, and scum.
    b. giant algae, small algae, and microscopic algae.
    c. green algae, seaweed, and kelp.
    (d.) green algae, golden algae, brown algae, red algae, and blue-green algae.

18. Algae are most important to mankind because
    a. man can eat algae.
    b. algae contain many vitamins.
    (c.) algae feed the sea creatures that man eats.
    d. algae provide a safe home for many fish.

19. Fern plants multiply by producing special cells called (choose two)
    (a.) spores.
    (b.) zygotes.
    c. buds.
    d. seedlings.

20. Various kinds of mosses are useful for (choose two)
    a. eating in salads.
    (b.) building up the soil.
    (c.) wrapping seedling roots to keep them moist.
    d. feeding fish.

21. ——— need moisture, warmth, food, and darkness in order to live.
    (a.) Bacteria
    b. Mold
    c. Algae
    d. Moss

22. Which of the following things does not kill bacteria, but slows its growth?
    a. sunlight
    (b.) freezing
    c. cooking
    d. strong chemicals

23. Which one of the following statements is **not** correct?
    a. Both fungi and bacteria make things rot.
    b. Both fungi and bacteria help to flavor some foods.
    (c.) Both fungi and bacteria are eaten as food.
    d. Both fungi and bacteria sometimes cause disease.

24. Molds and mushrooms both
    a. have large spore cases.
    b. have small spore cases.
    (c.) reproduce by spores.
    d. reproduce by zygotes.

25. Bread dough rises because
    a. bacteria cause the starch to ferment and make bubbles of alcohol in the dough.
    (b.) yeast cells in the dough produce carbon dioxide bubbles that make the dough spongy.
    c. yeast cells grow rapidly in the dough and produce alcohol.
    d. mold feeds on the dough and produces carbon dioxide bubbles that raise it.

E. Write good, complete answers. *(10 points)*

26. Describe the three steps in the life cycle of a fern. Use the correct vocabulary words in your description.

    Spores come from the dark spots on the undersides of fern leaves. The spores produce small plants that do not look like ferns at all. These plants produce zygotes, from which new ferns grow.

27. List the four things that bacteria need to live. For each item, name an especially good place for bacteria to find it. You should give a different place for each item.

    (Places will vary.)

    food—dead plants;

    warmth—human body;

    moisture—pond water;

    darkness—soil

28. What are two ways that fungi are helpful to man? What is one way that fungi are harmful?

    Helpful—causing plant matter to rot, flavoring food, providing food, causing bread dough to rise.

    Harmful—causing disease, causing useful things to rot.

(Total points: 37)

Making an Insect Net

Hoop and Handle

You will need about two feet of an old broom handle, 50 inches of #10 steel wire, two 1-inch hose clamps, and two pieces of limber fabric from a well-worn T-shirt, each measuring 19 inches by 24 inches. The fabric must be porous enough to let air flow freely through it.

First enlarge the pattern 300%. Then cut out two pieces of fabric according to the pattern. Sew all the vertical seams together. This produces a cone-shaped piece with the wider end being the hoop end. At the hoop end, fold the fabric back to make a sleeve ½ inch wide for the hoop wire. Sew around most of the circumference except for a 1½-inch space used to insert the hoop wire.

Grind the ends of the wire smooth. Bend the wire into a circle about 14 inches in diameter. Feed it through the sleeve. Have at least three inches extending beyond the sleeve on either side. Bend each three-inch section outward at a right angle. Straddle these sections on either side of the broom handle, and clamp the wire firmly to the handle with two hose clamps. Tighten the clamps until the wire sinks partly into the wood.

When you capture an insect, immediately give the handle a quick twist to close the opening of the net.

Depending on the students' skill, you may end up doing most of the project yourself. They will still enjoy helping you. (You may make photocopies of this activity for your students.)

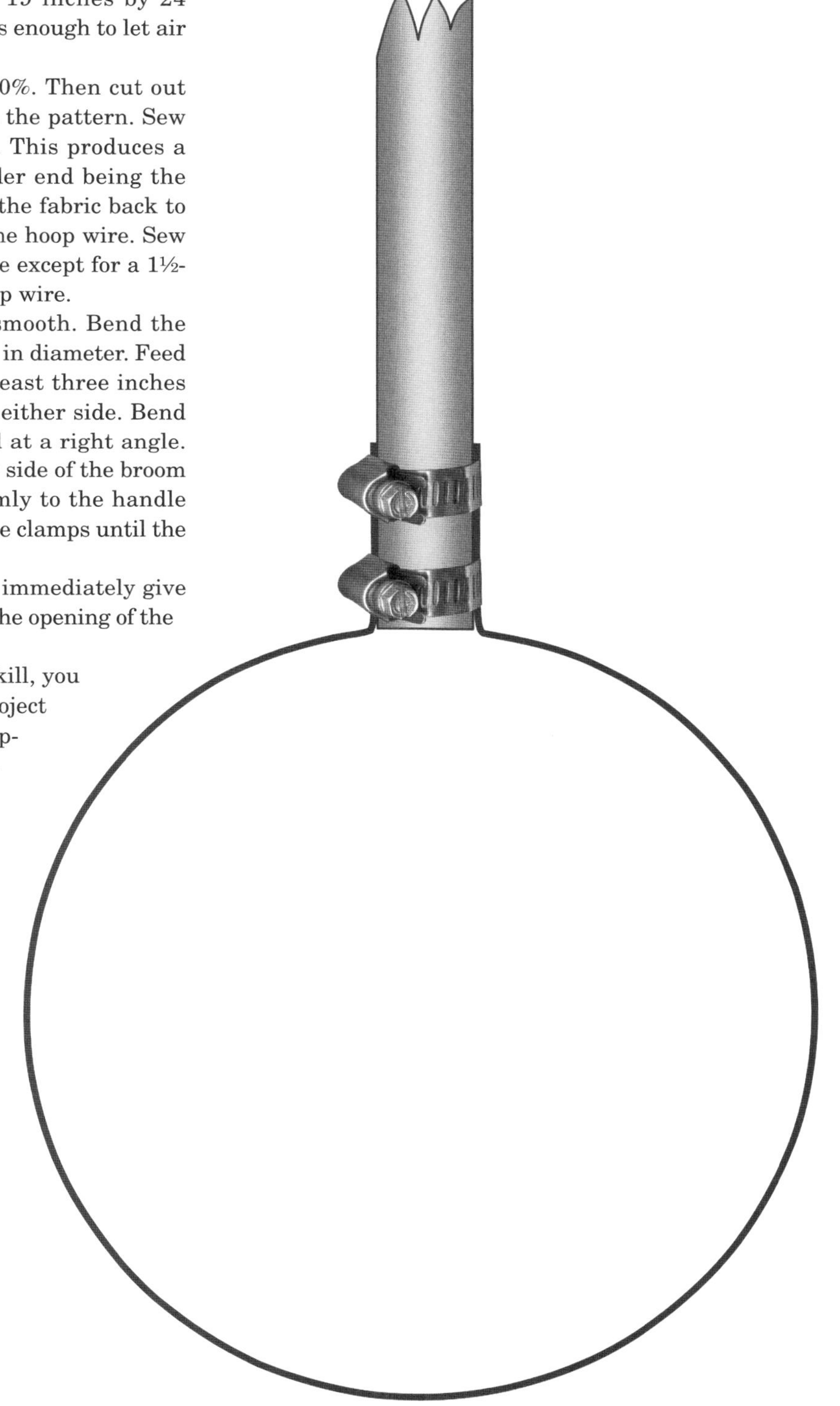

Fabric Pattern

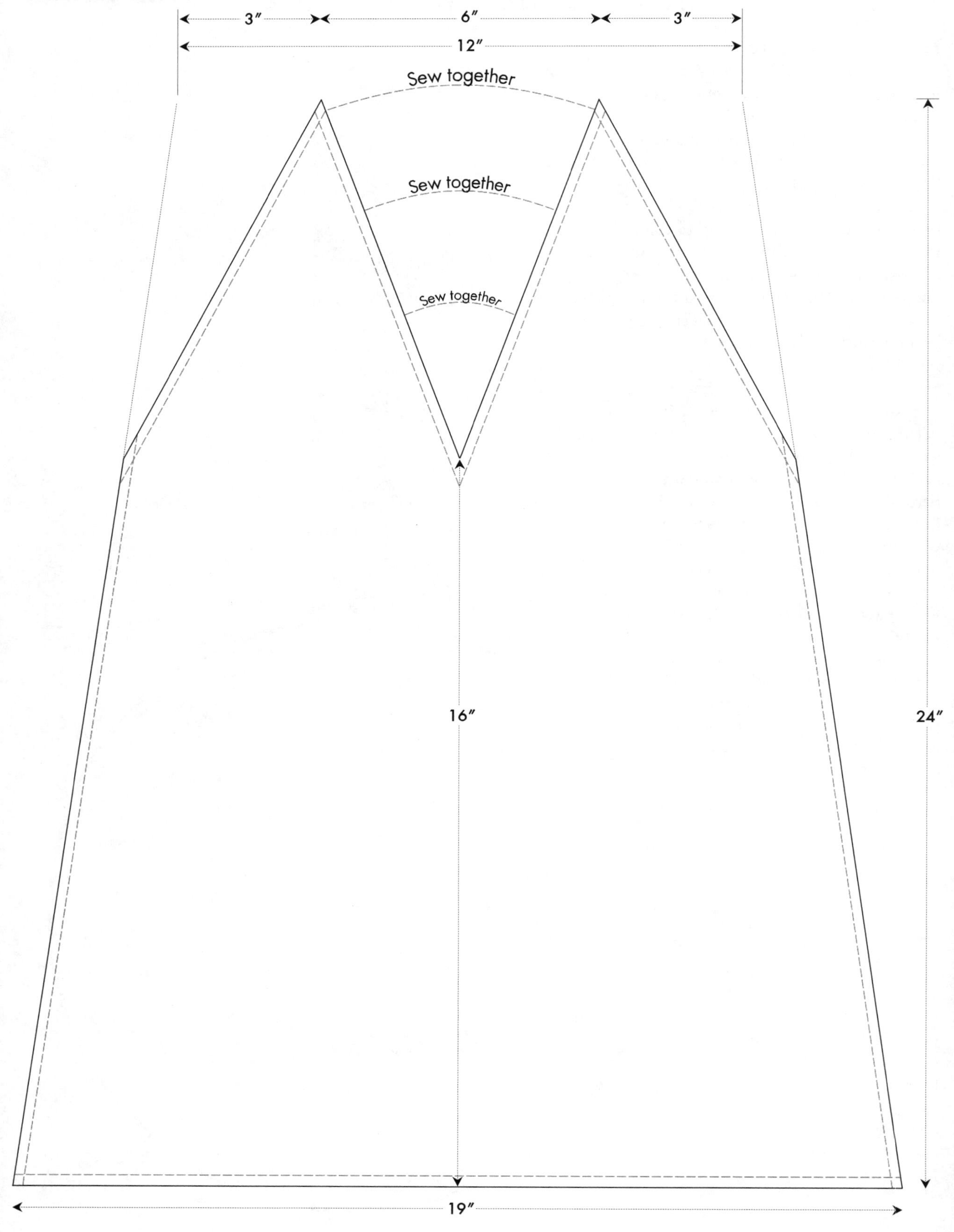

Index

Italics indicates vocabulary words and the pages on which the definitions are found. Some illustrations are also indexed.